Library of Congress Cataloging-in-Publication Data

Balaguru, Perumalsamy N.
    Fiber-reinforced cement composites / P. Balaguru and Surendra P.
Shah.
        p.   cm.
    Includes bibliographical references.
    ISBN 0-07-056400-0
    1. Fibrous composites.   2. Cement composites.   3. Porland cement-
-Additives.   I. Shah, S. P. (Surendra P.)   II. Title.
TA418.9.C6B354   1992
620.1'35—dc20   92-11927
                                                    CIP

1 2 3 4 5 6 7 8 9 0   DOC/DOC   9 8 7 6 5 4 3 2

ISBN 0-07-056400-0

*The sponsoring editor for this book was Larry Hager, and the
production supervisor was Donald Schmidt. This book was set in
Century Schoolbook by Publication Services, Inc.*

Printed and bound by R. R. Donnelley & Sons Company

Fiber-Reinforced Cement Composites

Perumalsamy N. Balaguru

Professor of Civil Engineering
Rutgers, The State University
New Brunswick, New Jersey

Surendra P. Shah

Walter P. Murphy Professor of Civil Engineering
Director, NSF Center for Advanced
Cement-Based Materials
Northwestern University
Evanston, Illinois

McGraw-Hill, Inc.

New York St. Louis San Francisco Auckland Bogotá
Caracas Lisbon London Madrid Mexico Milan
Montreal New Delhi Paris San Juan São Paulo
Singapore Sydney Tokyo Toronto

Dedicated to Our Parents
(Late) Perumalsamy Nadu
and Kengammal, and
(Late) Poonamchand Shah
and Maniben Shah

Contents

Preface

Reinforcing brittle matrices to improve their mechanical properties is an age-old concept. However, the modern development of fiber-reinforced cement composites dates back only to the 1960s. In the beginning, only straight steel fibers were used. The acceptance of fiber-reinforced concrete by the construction industry has led to a number of developments. Among these developments are new fiber types made of steel, stainless steel, polymeric and mineral materials, and naturally occurring materials. New manufacturing techniques and applications have also been developed. A large number of researchers around the world have investigated the various aspects of fiber-reinforced composites [FRC].

The primary purpose of this book is to introduce the reader to various portland cement–based fiber composites and to provide information on their constituent materials, fabrication, mechanical and long-term properties, applications, and field performance. The book is geared toward advanced undergraduate and graduate students, professional engineers, field engineers, fiber manufacturers, precast fiber-reinforced structural and nonstructural component manufacturers, and engineers involved with user agencies such as various departments of transportation. The book can be used as a reference text for fiber-reinforced composites.

The chapters in the book are conveniently arranged for readers with varied interests. For example, readers interested in glass fiber–reinforced composites can concentrate on the first few chapters, dealing with various mechanical properties, and on Chapter 13, dealing with the fabrication, properties, and applications of glass fiber–reinforced composites.

Chapter 1 provides a historical development of fiber-reinforced cement composites and the various types of composites that are currently used. This chapter also provides information on the various professional and research organizations that periodically update the state of the art.

Chapters 2, 3, and 4 cover the basic concepts and are geared toward graduate students. These chapters deal with the latest testing and modeling developments and with promising research directions. These chapters are also useful for design professionals who are interested in the basic concepts.

Chapters 5 through 11 deal with conventional fiber-reinforced concrete. The majority of applications involve the use of either steel or polymeric fibers. The chapters cover the designing of mixes and the properties of plastic (fresh) and hardened concrete. Matrix compositions and fiber contents normally used in the field are covered in these chapters. Typically, the matrix contains coarse aggregate and the fiber volume fraction is less than 2 percent. Although these chapters are written mainly for professionals involved in FRC use, students will greatly benefit by learning about real-life situations.

Chapter 12 deals with the shotcreting method of construction using FRC. A great deal of practical applications have been devised in this area for tunnel and canal linings and for the lining of waste dispoal sites. Both steel and polymeric fibers have been used. The use of the shotcreting technique, special requirements for mix proportions, additives such as silica fume and high-range water-reducing admixtures, and plastic and drying shrinkages are covered in this chapter.

Chapter 13 specifically deals with the use of glass fibers. This is a growing industry, with more than $100 million in sales per year in the United States alone. Constituent materials, construction methods, and problems with long-term durability that are unique to glass fiber–reinforced concrete (GFRC) are discussed in this chapter.

Chapter 14, which deals with other thin-sheet products, includes the composites developed primarily to replace asbestos fiber–reinforced sheets. This is also a growing field worldwide. Products included in this chapter are thin sheets reinforced with polymeric fabrics and meshes and with short fibers (pulps) including wood fiber–reinforced products. The recent developments in the area of polymeric pulp and the advances made to improve the performance of wooden fibers are also discussed in this chapter.

The chapter on slurry-infiltrated fiber concrete (SIFCON) deals with composites with high volume fractions of fibers. These composties have some unique properties and applications for blast-resisting structures.

Chapter 16, dealing with the use of FRC in structural components, provides details for designing beams, columns, and slabs. Fiber-reinforced concrete was found to provide notable improvements in the area of shear, ductility under cyclic loading, and impact and fatigue loading. It shows good potential for earthquake-resistant structures because of the ductility it provides compared to plain concrete. Re-inforcement congestion can also be reduced by using FRC and less continuous reinforcement in the junctions of beams and columns and other critical locations.

The chapter on field performance and case studies provides examples of real-life applications and the performance of FRCs in the field.

The authors would like to add the following note for the readers. Selecting a system of units of measure for the text, either the metric system or the U.S. and avoirdupois systems, was a problem. After considerable thought we decided to use the units that were used in the publications from which the information was taken. This decision led to the use of both systems. Conversions are provided so that the reader can have a feel for the dimensions. We would like to mention that the conversions are not as accurate (say, to three digits) and also not as complete as we would like them to be. We had to choose between clarity (readability) and accuracy and we choose the clarity, since the readers can always obtain an accurate conversion if they need one. A complete conversion table is provided at the end of the text.

We would also like to inform the readers that the tables and figures are not exactly the same as those presented in the sources cited. They were modified to improve the clarity. Some of the illustrations were taken from the original reports rather than the references mentioned. Since the reports are difficult to obtain, the published papers are used for references.

P. Balaguru
Surendra P. Shah

Acknowledgments

The authors gratefully acknowledge the contributions of Dr. Barzin Mobasher, Mr. Peter C. Tatnall, Dr. Regi John, Mr. Anil Khajuria, Mr. James I. Daniel, Dr. Parvis Soroushian, Dr. Nemkumar Banthia, and Dr. Gordon B. Batson, who reviewed various chapters. Special thanks are due to Dr. Barzine Mobasher, who made considerable contributions to Chapters 2 and 3; Mr. Anil Khajuria, who painstakingly reviewed the entire manuscript, helped in preparing the figures, and contributed to both editing and proofreading; and Mr. Peter C. Tatnall, who reviewed Chapters 5 through 10, dealing with fiber-reinforced composites containing coarse aggregates. The contributions of Miss Alison Chien, Dr. Chengsheng Ouyang, and Mr. Edward Wass at the final stage of proofreading are acknowledged with thanks. The authors thank Ms. Donna Foster, who typed and retyped the manuscript, and Ms. Sheryl Lilke, who was the production manager for the book.

The authors wish to acknowledge the contributions of their graduate students for their research efforts and the various publications from which some of the figures and tables were adapted.

The first author gratefully acknowledges the support and contributions of the National Science Foundation Center for Advanced Cement-Based Materials and the Department of Civil Engineering, Northwestern University, where he spent a semester and a summer, during which time most of the first draft was completed. He expresses his gratitude to Professor Yong S. Chae, Chairman of the Department of Civil and Environmental Engineering, and Professor Ellis H. Dill, Dean of the College of Engineering at Rutgers-The State University, for their encouragement and support. He also acknowledges with grateful thanks Professors V. Ramakrishnan, Antoine E. Naaman, and Edward G. Nawy for both their professional and personal support.

The authors acknowledge the dedication and support of their families. The book is as much their accomplishment as ours.

Introduction

The use of randomly oriented, short fibers to improve the physical properties of a matrix is an age-old concept. For example, fibers made of straw or horsehair have been used to improve the properties of bricks for thousands of years. In modern times, fiber-reinforced composites are being used for a large variety of applications. The composite could be a clay brick reinforced with natural fibers or a high-strength, fiber-reinforced ceramic component used in space shuttles. This book deals with the fiber-reinforced composites made with primarily portland cement–based matrices. These matrices can consist of any of the following:

1. Plain portland cement

2. Cement with additives such as fly ash or condensed silica fume

3. Cement mortar containing cement and fine aggregate

4. Concrete containing cement, fine and coarse aggregates

In certain applications, the matrix may also contain admixtures and polymers. Composites containing non-portland cement-based matrices, which are primarily used for rapid repairs, are also briefly discussed.
The fibers can be broadly classified as

1. Metallic fibers

2. Polymeric fibers

3. Mineral fibers

4. Naturally occurring fibers

Metallic fibers are made of either steel or stainless steel. The polymeric fibers in use include acrylic, aramid, carbon, nylon, polyester, polyethylene, and polypropylene fibers. Glass fiber is the predominantly used mineral fiber. Various types of organic and inorganic naturally occurring fibers such as cellulose are also being used to reinforce the cement matrix.

The subject areas discussed in this book include the basic mechanism of fiber contribution, the physical properties of plastic and hardened cement composites made with different types of matrices and fibers, the applications for these composites, and the experience gained in using these composites in the field. A brief review of the material covered is presented in Section 1.2. Throughout the book, the abbreviation FRC is used for all types of cement composites irrespective of matrix composition.

1.1 Historical Development

Even though reinforcing a brittle matrix with discrete fibers is an age-old concept, modern-day use of fibers in concrete started in the early 1960s [1.1]. In the beginning, only straight steel fibers were used. The major improvement occurred in the areas of ductility and fracture toughness, even though flexural-strength increases were also reported. The law of mixtures was applied to analyze the fiber contributions [1.2]. Shah and Rangan established that fiber-reinforced concrete can be designed to obtain a specific ductility or energy absorption [1.2, 1.3].

For straight steel fibers, the primary factors that controlled the properties of the composite were fiber volume fraction and length/diameter, or aspect, ratio of the fibers. The amount of fiber used ranged from 150 to 200 lb/yd^3 (90 to 120 kg/m^3) of concrete. The aspect ratios were in the range of 60 to 100. The major problems encountered in the early stages were difficulty in mixing and workability. At higher volume fractions, fibers were found to ball up during the mixing process. This process, called *balling,* was found to occur frequently for longer fibers. The size of the coarse aggregate was normally restricted to facilitate the use of shorter fibers and to avoid balling of fibers. Additionally, the mortar fraction of concrete was increased to combat the balling problem. There was always a reduction in workability with the addition of fibers. This tends to affect the quality of concrete in place, especially for higher fiber volume fractions.

The advent of deformed fibers and high-range water-reducing admixtures provided a big boost to the fiber-reinforced concrete use in the field. Ramakrishnan and his colleagues established that fibers with hooked ends can be used at much lower volume fractions than straight steel fibers, producing the same results in the area of ductility and toughness [1.4, 1.5]. These fibers were also glued together at the edges with water-soluble glue. When added to the concrete, the fibers had a much lower (apparent) aspect ratio. During mixing, the fibers got separated and dispersed as individual fibers. The gluing and dispersal, in combination with a lower volume fraction of fibers, resulted in virtual elimination of balling [1.4, 1.5]. The use of high-range water-reducing admixtures eliminated the problem associated with reduction in workability. Since these admixtures were so effective at very low dosages, controlling the workability of fiber-reinforced concrete was no longer a problem. A number of other fiber shapes such as crimped, paddled, and enlarged ends were also developed.

The development of high-range water-reducing admixtures also aided the use of fibers in shotcrete [1.6, 1.7]. This admixture made it possible to proportion flowable mixes with low water-cement ratio. In recent applications, microsilica (silica fume) has often been used in shotcrete. Use of silica fume was found to make the mix cohesive, allowing workers to build greater thickness in a single pass [1.7]. In addition, silica fume shotcrete has lower permeability and higher strength.

In the case of polymeric fibers, the concept was tried in 1965 [1.8]. However, large-scale use of polymeric fibers in concrete did not happen until the later 1970s. Currently, polymeric fibers are used in very low volume fraction (about 0.1 percent by volume), primarily to control cracking in the early stages of setting, typically less than three hours after casting. This application was developed using mainly polypropylene fibers [1.9]; currently, other polymeric fibers made of nylon, polyester, polyethylene; and cellulose are also being used. For the improvement in properties of hardened concrete such as resistance to cracking caused by drying shrinkage, a volume fraction substantially higher than 0.1 percent is necessary.

Use of carbon fibers in concrete attracted attention when these fibers could be manufactured at a lower cost using petroleum pitch, compared to the high cost of conventional carbon fibers. These carbon fibers have higher strength and a higher modulus compared to other polymeric fibers. They have been successfully used to fabricate exterior wall panels. Use of these fibers is still in its early developmental stage even though a number of buildings in Japan have already used carbon fiber–reinforced cement sheets as cladding.

Use of asbestos fibers is being severely discouraged because of health risks. A number of industrialized countries have already banned the use of these fibers for general applications. The absence of asbestos fibers has created a need for fibers that can be used in the Hatschek process. In this process a cement slurry is reinforced with a high volume fraction of thin, short fibers. Researchers have developed various schemes to replace asbestos-cement sheets by using sheets made with a large volume fraction (5 to 15 percent) of polymeric fibers, wood fibers, alkali-resistant glass fibers, continuous polymeric fibers, and fiber meshes. The search for a perfect replacement for asbestos fibers is continuing.

A number of naturally occurring fibers are being investigated for manufacturing reinforced-cement sheets. Cellulose fibers seem to show promise for large-scale use. Other types include sisal, coconut, jute, and bamboo fibers. A possible problem with some of the naturally occurring fibers is their lack of durability in the alkaline environment of concrete, unless modifications are done either to the fiber surfaces or to the matrix composition.

In the area of thin-sheet product development, glass fibers are being used extensively. The development is continuing even though the fibers have been in use for more than a decade. Again the major problem is lack of fiber durability in the alkaline environment. Alkali-resistant fibers and additives for matrix modification are being developed.

In the beginning, fiber-reinforced concrete was primarily used for pavements and industrial floors. But currently, the fiber-reinforced cement composite is being used for a wide variety of applications including bridges, tunnel and canal linings, hydraulic structures, pipes, explosion-resistant structures, safety vaults, cladding, and roller-compacted concrete. The various applications are discussed in Chapters 12 through 17.

Essentially, fiber composites can be grouped into three categories. The low (fiber) volume composite, containing less than 1 percent fiber, is typically used for bulk field applications involving large volumes of concrete. The matrix in this case is usually concrete containing coarse aggregate. The higher (fiber) volume composite is mainly used for thin sheets with either cement or cement-mortar matrix. The fiber volume fraction ranges from 5 to 15 percent. The third category, with moderate volume content of 1–5 percent, is used only for very special applications such as safety vaults.

Research is being conducted for using fiber-reinforced concrete in structural applications such as beams, columns, connections, plates, and prestressed concrete structures [1.10–1.12]. High-performance composites containing large volume fractions of steel fibers are also being developed for special applications. One such high-volume fiber composite, known as SIFCON (Chapter 15), can withstand flexural stresses of almost an order of magnitude higher [1.13] and can absorb large amounts of energy before failure.

Some of the new approaches being investigated may alter the way we design and construct concrete structures [1.14–1.17]. With the advent of the high-range water-reducing admixtures, it is now possible to incorporate a fiber volume of 15 percent in cement matrices. Fibers in such large quantities seem to fundamentally alter the nature of cementitious matrices. Typical tensile stress-strain responses obtained for various types of fibers are shown in Figure 1-1. The responses clearly indicate the enormous increase in both strength and ductility. In addition, tensile properties of the matrix itself are enhanced in the presence of fibers. This can be seen in Figure 1-2, which shows the contribution of the matrix obtained by subtracting the contribution of the fibers from the response of the composite.

The high-performance fiber composite also improves both the strength and ductility of reinforced-concrete members tremendously. Figure 1-3 shows two curves, one obtained for conventionally reinforced

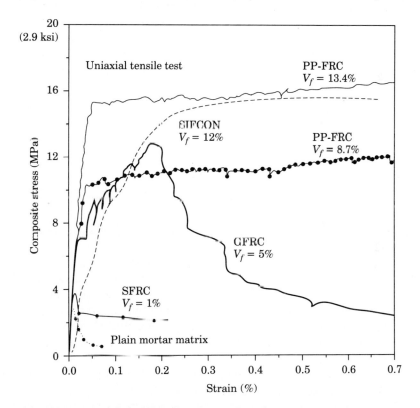

V_f = volume fraction of fibers in cement composite

PP-FRC = polypropylene fiber–reinforced concrete

SIFCON = slurry-infiltrated fiber concrete

GFRC = glass fiber–reinforced concrete

SFRC = steel fiber–reinforced concrete

Figure 1-1   Uniaxial tensile stress-strain curves for various cement-based composites [1.17].

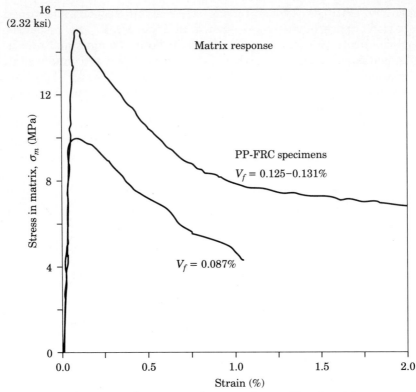

Figure 1-2 Enhanced tensile stress-strain curve of cement-based matrix as a result of fiber reinforcement [1.17].

concrete beams and the other—labeled Compact reinforced composite—for a beam made using a large volume of conventional bars and a high volume of steel fibers [1.17]. It can be seen that the properties of the high-performance reinforced-concrete beam approaches that of structural steel.

In the area of theoretical analysis, much progress is still needed. Most designs are based on experimental results and empirical relations. The law of mixtures was used in earlier studies to model the behavior of the composite. Because of the random distribution of fibers, variability in fiber types, geometry, and matrix composition, generalized relationships could not be obtained based on the law of mixture. The law does not apply to large-volume fiber matrices because the fibers affect the matrix properties as discussed above.

The analytical models available for FRC can be categorized as models based on the theory of multiple fracture, composite models, strain-relief models, fracture mechanics models, interface mechanics

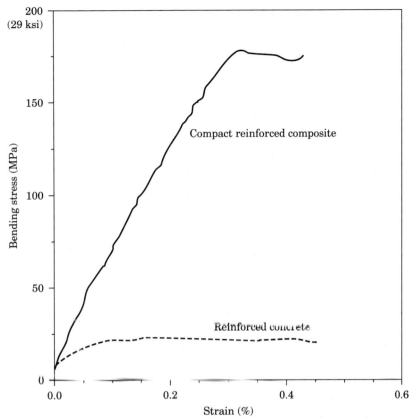

Figure 1-3 Comparison of the flexural responses of a conventionally reinforced concrete beam and a high-performance reinforced-concrete beam [1.17].

models, and micromechanics models [1.18]. Most of these models are still being refined. Models based on fracture mechanics and interface mechanics seem to have a good potential for dealing with the inelastic process that occurs in FRC. The development of these models is briefly reviewed in the following paragraphs.

In its simplest form, linear elastic fracture mechanics was used to solve the problem of crack initiation, growth, arrest, and the stability in the presence of fibers through appropriate changes in the stress-intensity factor. In earlier models, perfect bond was assumed to exist between the fiber and the matrix. In the development that followed, the inelastic interface response during the crack growth was incorporated using a nonlinear stress-displacement relationship for the fiber-bridging zone [1.19]. This approach has come to be known as the *cohesive crack* model. This basic approach was used by several researchers for developing prediction procedures [1.19–1.22]. The researchers used different assumptions for crack initiation, singularity at the crack tip, and crack growth and stability.

A two-parameter model was proposed by Jenq and Shah to improve the accuracy and capability of the fracture mechanics model [1.23]. In this model, the following two material parameters are used: (1) a modified critical stress-intensity factor based on linear elastic fracture mechanics and the effective crack length and (2) the critical crack-tip opening displacement. This model is being adopted as a RILEM (International Union of Testing and Research Laboratories for Materials and Structures) recommendation [1.24].

All the cohesive crack models rely on the stress–crack width relations obtained experimentally. The basic material parameters needed for the model are obtained using the behavior of the plain matrix and the fiber pull-out resistance for a given fiber and matrix composition.

Procedures have also been formulated to predict macroscopic stress–crack width relationships using the behavior of fiber-matrix interface [1.25–1.30]. These can be grouped as models based on shear-lag theory [1.25–1.27, 1.30] and fracture mechanics [1.28, 1.29]. These models have been successful to varying degrees in predicting the maximum pull-out load and load-slip response for aligned fibers. Significant research efforts are still needed to account for variabilities such as fiber orientation.

1.2   Synopsis of the Topics Discussed in the Book

This section provides a brief outline of the various topics covered in Chapters 2 through 17. Since the book covers topics for readers interested in a variety of cement composites, the chapters are written without using too much cross-reference. As a consequence, the reader who goes through the entire book might encounter minor repetitions, especially in Chapters 12 to 15.

Chapter 2 deals with the mechanism of fiber-matrix interaction. The various models that are used to identify and compute the fiber-matrix bond are reviewed. The bond between the fiber and the matrix plays a major role in the composite behavior. Hence characterization of interfacial bond is also discussed in detail. Chapters 3 and 4 provide basic concepts used in analyzing the fiber composite subjected to tension and bending. Test methods, interpretation of experimental results, and theoretical prediction models are presented. Together, Chapters 2 through 4 are intended to introduce the basic concepts of fiber contribution to the brittle or quasi-brittle matrix.

Chapters 5 through 11 deal with the current practice of fiber-reinforced concrete. The majority of applications involve the use of either steel or polymeric fibers. The material covered deals with designing the mixes and the properties of plastic (fresh) and hardened concrete. Matrix composition and fiber volume fraction (content) normally used in the field are covered in these chapters.

Chapter 12 deals with the shotcreting method of construction using FRC. A great deal of practical applications have occurred in this area for tunnel and canal linings as well as for linings and intermittent covers for waste-disposal sites. Both steel and polymeric fibers have been used in these applications. Shotcreting techniques, special requirements for mixture proportions, additives such as silica fume and high-range water-reducing admixtures, plastic and drying shrinkages, and precautions to be used during construction are covered in this chapter.

Chapter 13, on glass fiber–reinforced concrete (GFRC) deals with the use of glass fibers for manufacturing thin sheets. This is a growing industry with more than $100 million in sales per year in the United States alone. Constituent materials, construction methods, and problems with long-term durability that are unique to GFRC are discussed in this chapter.

Chapter 14, on other thin-sheet products, deals with the composite developed primarily to replace asbestos-cement sheets. This is also a growing field worldwide. Products included in this chapter are thin sheets reinforced with polymeric fabrics, meshes, and short fibers (pulps), including wood fiber–reinforced products. Recent developments in the area of polymeric pulp and advances made to improve the performance of wooden fibers are discussed in this chapter.

The chapter on SIFCON, Chapter 15, covers the composite with a high volume fraction of steel fibers. This composite has some unique properties and useful applications for explosion-resisting structures.

Chapter 16, dealing with the use of FRC in structural components, provides a review of the behavior of beams, columns, and connections subjected to monotonic and cyclic loads. Fiber-reinforced composite was found to provide notable improvements in the areas of shear, impact, ductility under cyclic loading, and fatigue loading. It shows good potential for earthquake-resistant structures because of the ductility it provides compared to plain concrete. Reinforcement congestion could also be reduced by using FRC and less-continuous reinforcement in junctions of beams and columns and other critical locations.

Chapter 17 provides examples of real-life applications and the performance of FRC in the field.

1.3   Relevant Specifications, Journals, and Special Publications

Fiber cement composite is still in the active research-and-development stage. A number of researchers are working on various aspects of this composite. The following list of references is provided so that the reader can refer to the latest information available on a particular subject.

1.3.1   Specifications and Recommended Procedures

Recommendations and guidelines regarding the various aspects of FRC can be found in the following American Concrete Institute documents, developed by ACI Committee 544 on Fiber Reinforced Concrete.

544.1R-82	State-of-the-Art Report on Fiber Reinforced Concrete (Reapproved 1986)
544.2R-89	Measurement of Properties of Fiber Reinforced Concrete
544.3R-90	Guide for Specifying, Mixing, Placing, and Finishing Steel Fiber Reinforced Concrete
544.4R-88	Design Considerations for Steel Fiber Reinforced Concrete

In addition, the following two reports published by the Committee on Shotcrete are also applicable to FRC.

506.1R-84	State-of-the-Art Report on Fiber Reinforced Shotcrete
506.2-77	Specification for Materials, Proportioning, and Applications of Shotcrete (Revised 1983)

These publications can be obtained from American Concrete Institute, P.O. Box 19150, Detroit MI 48219–0150, USA.

The following ASTM (American Society for Testing and Materials) specifications are applicable to FRC.

A 820-85	Standard Specification for Steel Fibers for Fiber Reinforced Concrete
C 31-90	Standard Practice for Making and Curing Concrete Test Specimens in the Field
C 39-86	Standard Test Method for Compressive Strength of Cylindrical Concrete Specimens
C 42-87	Standard Method of Obtaining and Testing Drilled Cores and Sawed Beams of Concrete
C 78-84	Standard Test Method for Flexural Strength of Concrete (Using Simple Beam with Third-Point Loading)
C 138-81	Standard Test Method for Unit Weight, Yield, and Air Content (Gravimetric) of Concrete

C 157-89   Standard Test Method for Length Change of Hardened Hydraulic-Cement Mortar and Cement

C 173-78   Standard Test Method for Air Content of Freshly Mixed Concrete by the Volumetric Method

C 192-90a   Standard Method of Making and Curing Concrete Test Specimens in the Laboratory

C 231-89a   Standard Method for Air Content of Freshly Mixed Concrete by the Pressure Method

C 293-79   Standard Test Method for Flexural Strength of Concrete (Using Simple Beam with Center-Point Loading)

C 341-84   Standard Test Method for Length Change of Drilled or Sawed Specimens of Cement Mortar and Concrete

C 418-89   Standard Test Method of Abrasion Resistance of Concrete by Sandblasting

C 469-87a   Standard Test Method for Static Modulus of Elasticity and Poisson's Ratio of Concrete in Compression

C 470-87   Standard Specification for Molds for Forming Concrete Test Cylinders Vertically

C 496-90   Standard Test Method for Splitting Tensile Strength of Cylindrical Concrete Specimens

C 512-87   Standard Test Method for Creep of Concrete in Compression

C 666-84   Standard Test Method for Resistance of Concrete to Rapid Freezing and Thawing

C 779-89a   Standard Test Method for Abrasion Resistance of Horizontal Concrete Surfaces

C 827-87   Standard Test Method for Change in Height at Early Ages of Cylindrical Specimens from Cementitious Mixtures

C 995-86   Standard Test Method for Time of Flow of Fiber Reinforced Concrete Through Inverted Slump Cone

C 1018-89   Standard Test Method for Flexural Toughness and First-Crack Strength of Fiber Reinforced Concrete (Using Beam with Third-Point Loading)

ASTM publications can be obtained from ASTM, 1916 Race St, Philadelphia PA 19183.

The British standard that is applicable is BS 181: Part 2, Methods of Testing Concrete. The address is British Standards Institution, Linford Wood, Milton Keynes MK14 6LE, England.

1.3.2 Journals and Special Publications in the Area of FRC

A number of journals contain papers dealing with fiber-reinforced cement composites. In addition, international conferences are held at frequent intervals. The proceedings of these conferences are good sources of information for the reader who wants in-depth discussions of a particular subject. The following list of journals and proceedings represent the major sources. The list does not include all the sources.

Proceedings:

- *Fiber Reinforced Concrete,* ACI-SP44, 1974.
- *RILEM Symposium on Fiber Reinforced Concrete,* London, 1975.
- *Testing and Test Methods of Fiber Cement Composites,* RILEM Symposium, R. N. Swamy (ed.), The Construction Press Ltd., U.K., 1978.
- *Fracture Mechanics of Concrete,* S. P. Shah and A. Carpinteri, (eds.), Chapman and Hall, 1991.
- *Fiber Reinforced Concrete International Symposium,* ACI-SP81, G. C. Hoff (ed.), 1984.
- *Steel Fiber Concrete,* S. P. Shah and A. Skarendahl (eds.), Elsevier, 1985.
- *RILEM Third International Symposium on Developments in Fiber Reinforced Cement Concrete,* R. N. Swamy (ed.), 1986.
- *Fiber Reinforced Concrete: Properties and Applications,* ACI-SP105, S. P. Shah and G. B. Baton (eds.), 1987.
- *Proceedings of the International Symposium on Fiber Reinforced Concrete,* Madras, India, V. Parameswaran (ed.), 1987.
- *Bonding in Cementitious Composites,* S. Mindess and S. P. Shah (eds.), Material Research Society, Boston, 1988.
- *Fiber Reinforced Cements and Concretes: Recent Development,* R. N. Swamy and B. Barr (eds.), Elsevier, 1989.
- *Thin-Section Fiber Reinforced Concrete and Ferrocement,* ACI SP-124, J. J. Daniel and S. P. Shah (eds.), 1990.

Journals:

American Concrete Institute:	*ACI Materials Journal*
	ACI Structural Journal
	Concrete International
American Society of Civil Engineers:	*Journal of Structural Engineering*
	Journal of Materials in Civil Engineering
	Journal of Engineering Mechanics
American Society for Testing and Materials:	*Cement, Concrete and Aggregates*

Concrete:	*Journal of the Concrete Society,* London
	Cement and Concrete Research, Pergamon Press
	Indian Concrete Journal, India
	Cement and Concrete, Composites, Elsevier Applied Science
	Magazine of Concrete Research, Cement and Concrete Asscociation (CCA), London
	RILEM Materials and Structures
	Transactions of the Japan Concrete Institute

1.4 Research Needs

The following research needs are intended as a spectrum of areas that need immediate attention. Researchers might already be working on some of these areas:

- Development of high-performance composites with novel processing techniques, superior matrices, better interfacial characteristics, and suitable fiber types
- Development of economical fabrication and rational design procedures for the use of fiber-reinforced concrete for structural applications such as beams, slabs, and connections
- Investigate ductility characteristics for potential application in seismic design and construction
- Properties of FRC with high-strength cement matrices
- Properties of FRC at high and low temperatures
- Methods to improve bond between fibers and cement-based matrices
- Methods to reduce corrosion of steel fibers and improve the fiber durability for naturally occurring and some polymeric and mineral fibers
- Develop theoretical models for stiffness degradation, stress analysis under static and dynamic loading, and interface mechanics for the pull-out of fibers.

1.5 References

1.1 *Fiber Reinforced Concrete,* SP-44, American Concrete Institute, Detroit, Michigan, 1974, 554 pp.
1.2 Shah, S. P.; and Rangan, B. V.; "Fiber Reinforced Concrete Properties," *ACI Journal,* Vol. 68, No. 2, 1971, pp. 126–135.
1.3 Shah, S. P.; and Rangan, B. V. "Ductility of Concrete Reinforced with Stirrups, Fibers and Compression Reinforcements," ASCE, *Journal of the Structural Division,* Vol. 96, No. 6, 1970, pp. 1167–1184.

1.4   Ramakrishnan, V.; Brandshaug, T.; Coyle, W. V.; and Schrader, E. K. "A Comparative Evaluation of Concrete Reinforced with Straight Steel Fibers and Deformed End Fibers Glued Together into Bundles," *ACI Journal,* Vol. 77, No. 3, May-June 1980, pp. 135–143.

1.5   Ramakrishnan, V.; Coyle, W. V.; Kulandaisamy, V.; and Schrader, E. K. "Performance Characteristics of Fiber Reinforced Concrete with Low Fiber Contents," *ACI Journal,* Vol. 78, No. 5, 1981, pp. 384–394.

1.6   Ramakrishnan, V.; Coyle, W. V.; Dahl, L. F.; and Schrader, E. K. "A Comparative Evaluation of Fiber Shotcrete," *Concrete International: Design and Construction,* Vol. 3, No. 1, January 1981, pp. 56–59.

1.7   Morgan, D. R.; et al. "Evaluation of Silica Fume Shotcrete," Proceedings, CAN-MET/CSCE International Workshop on Silica Fume in Concrete, Montreal, Quebec, Canada, May 4, 1987.

1.8   Goldfein, S. "Fibrous Reinforcement for Portland Cement," *Modern Plastics,* Vol. 42, No. 8, 1965, pp. 156–160.

1.9   Zollo, R. F. "Collated Fibrillated Polypropylene Fibers in FRC," Fiber Reinforced Concrete International Symposium, SP-81, American Concrete Institute, Detroit, Michigan, 1984, pp. 397–409.

1.10 Craig, R. "Flexural Behavior and Design of Reinforced Fiber Concrete Members," Fiber Reinforced Concrete Properties and Applications, SP-105, American Concrete Institute, Detroit, Michigan, 1987, pp. 517–563.

1.11 Swamy, R. N.; and Bahia, H. M. "The Effectiveness of Steel Fibers as Shear Reinforcement," *Concrete International: Design and Construction,* Vol. 7, No. 3, March 1985, pp. 35–40.

1.12 Balaguru, P.; and Ezeldin, A. S. "Behavior of Partially Prestressed Beams Made with High Strength Fiber Reinforcement Concrete," *Fiber Reinforced Concrete Properties and Applications,* SP-105, American Concrete Insitute, Detroit, Michigan, 1987, pp. 419–436.

1.13 Balaguru, P.; and Kendzulak, J. "Mechanical Properties of Slurry Infiltrated Fiber Concrete (SIFCON)," *Fiber Reinforced Concrete Properties and Applications,* SP-105, American Concrete Insitute, Detroit, Michigan, 1987, pp. 247–268.

1.14 Krenchel, H.; and Stang, H. "Stable Microcracking in Cementitious Materials," Proceedings, 2nd International Symposium on Brittle Matrix Composites, Cedzyna, September 1988.

1.15 Stang, H.; Mobasher, B.; and Shah, S. P. "Quantitative Damage Characterization in Polypropylene Fiber Reinforced Concrete," *Cement and Concrete Research,* Vol. 20, 1990, pp. 540–558.

1.16 "Testing Forum: Report Available on Compact Reinforced Composite," ASTM *Cement, Concrete, and Aggregate,* Vol. 9, 1987, pp. 160.

1.17 Shah, S. P. "Fiber Reinforced Concrete," *Concrete International,* Vol. 12, No. 3, 1990, pp. 81–82.

1.18 Gopalaratnam, V. S.; and Shah, S. P. "Failure Mechanisms and Fracture of Fiber Reinforced Concrete," *Fiber Reinforced Concrete Properties and Applications,* SP-105, American Concrete Insitute, Detroit, Michigan, 1987, pp. 1–25.

1.19 Hillerborg, A. "Analysis of Fracture by Means of the Fictitious Crack Model, Particularly for Fiber Reinforced Concrete," *International Journal of Cement Composites,* Vol. 2, No. 4, November 1980, pp. 177–185.

1.20 Petersson, P. E. "Fracture Mechanical Calculations and Tests for Fiber-Reinforced Concrete," Proceedings, Advances in Cement Matrix Composites, Materials Research Society Annual Meeting, Boston, MA, November 1980, pp. 95–106.

1.21 Wecharatana, M.; and Shah, S. P. "A Model for Predicting Fracture Resistance of Fiber Reinforced Concrete," *Cement and Concrete Research,* Vol. 13, No. 6, 1983, pp. 819–829.

1.22 Visalvanich, K.; and Naaman, A. E. "Fracture Model for Fiber Reinforced Concrete," *ACI Journal,* Proceedings, Vol. 80, No. 2, 1982, pp. 128–138.

1.23 Jenq, Y. S.; and Shah, S. P. "Crack Propagation in Fiber Reinforced Concrete," ASCE, *Journal of Structural Engineering,* Vol. 112, No. 1, 1986, pp. 19–34.

1.24 Shah, S. P. "Determination of Fracture Parameters (K_{IC}^s and $CTOD_C$) of Plain Concrete Using Three-Point Bend Tests," RILEM, *Materials and Structures,* Vol. 23, 1990, pp. 457–460.

1.25 Lawrence, P. "Some Theoretical Considerations of Fibre Pull-Out from an Elastic Matrix," *Journal of Material Science,* Vol. 7, 1972, pp. 1–6.

1.26 Laws, V.; Lawrence, P.; and Nurse, R. W. "Reinforcement of Brittle Matrices by Glass Fibers," *Journal of Physics D: Applied Physics,* Vol. 6, 1972, pp. 523–537.

1.27 Gopalaratnam, V. S.; and Shah, S. P. "Tensile Failure of Steel Fiber-Reinforced Mortar," ASCE, *Journal of Engineering Mechanics,* Vol. 113, No. 5, 1987, pp. 635–652.

1.28 Stang, H.; and Shah, S. P. "Failure of Fiber Reinforced Composites by Pull-Out Fracture," *Journal of Materials Science,* Vol. 21, No. 3, 1986, pp. 935–957.

1.29 Morrison, J. K.; Shah, S. P.; and Jenq, Y. S. "Analysis of the Debonding and Pull-Out Process in Fiber Composites," ASCE *Journal of Engineering Mechanics,* Vol. 117, 1991.

1.30 Gopalaratnam, V. S.; and Cheng, J. "On the Modeling of Inelastic Interfaces in Fibrous Composites," in *Bonding in Cementitious Composites,* S. Mindess and S. P. Shah (eds.) Materials Research Society, Boston, Vol. 114, December 1988, pp. 225–231.

Interaction between Fibers and Matrix

The interaction between the fiber and matrix is the fundamental property that affects the performance of a cement-based fiber composite material. An understanding of this interaction is needed for estimating the fiber contribution and for predicting the composite's behavior. A large number of investigators have studied the various aspects of this interaction [2.1–2.45]. A variety of factors are involved; the following are the major parameters affecting the fiber interaction with the matrix.

- Condition of the matrix: uncracked or cracked
- Matrix composition
- Geometry of the fiber
- Type of fiber: for example, steel, polymeric, mineral, or naturally occurring fibers
- Surface characteristics of the fiber
- Stiffness of the fiber in comparison with matrix stiffness
- Orientation of the fibers: aligned versus random distribution
- Volume fraction of fibers
- Rate of loading
- Durability of the fiber in the composite and the long-terms effects

Owing to the interaction that exists between these parameters, the available theoretical models deal with them only to a limited extent.

Physical concepts, experimental results, and prediction models dealing with the various aspects of fiber contribution are discussed in this chapter.

2.1 Fiber Interaction with Homogeneous Uncracked Matrix

This type of interaction occurs in almost all composites during the initial stages of loading. In certain cases, such as highly reinforced thin sheets, the composite may remain uncracked during the service life. However, in most cases, the matrix will crack during the service life. The fiber interaction with the uncracked matrix has therefore limited importance in practical applications. Study of this interaction does yield useful information for understanding the overall behavior of the composite. In addition, even when cracks develop in the composite, the uncracked portions of the structure affect the overall behavior of the structural system.

A simple fiber-matrix system containing a single fiber is shown in Figure 2-1. In the unloaded stage, the stresses in both the matrix and the fiber are assumed to be zero (Figure 2-1a). Applying tensile or compressive load to the composite or subjecting the composite to temperature change results in the development of stresses and deformations that must remain compatible. In the case of a cement matrix, the hydration of cement alone may induce stresses both in the matrix and in the fiber. When the load is applied to the matrix, part of the load is transferred to the fiber along its surface. Because of the difference in stiffness between the fiber and the matrix, shear stress develops along the surface of the fiber. This shear stress helps to transfer some of the applied load to the fiber. If the fiber is stiffer than the matrix, the deformation at and around the fiber will be smaller, as shown in Figure 2-1b and 2-1c. This type of situation arises with steel and mineral fibers. If the fiber modulus is less than the matrix modulus, then the deformation around the fiber will be higher. This occurs in composites with polymeric and some naturally occurring fibers.

Elastic stress transfer exists in an uncracked composite as long as the matrix and the fiber are within elastic stress range. The stress-strain response of the matrix could exhibit nonlinearity and inelastic behavior prior to fracture. Mathematical equations have been developed for both the shear stress at the interface τ and the stress along the fiber. These models are normally referred to as shear lag models and are based on a number of simplifying assumptions. These assumptions include (1) linear elastic behavior of fiber and matrix, (2) a perfect bond between fiber and matrix, (3) the interface being extremely thin and its property the same as the property of

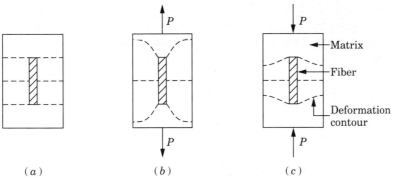

Figure 2-1   Fiber-matrix interaction, uncracked matrix: (a) unloaded; (b) tension; (c) compression.

the matrix elsewhere, and (4) the fibers having straight and regular cross-sections. In the case of multiple fibers, they are arranged in a predictable form and no interaction exists among them.

The distribution of the shear stress $\tau(x)$ at a distance x from the end of the fiber can be expressed as

$$\tau(x) = E_f \epsilon_m \left[\frac{G_m}{2E_f \ln(R/r)} \right]^{1/2} \frac{\sinh \beta_1(l/2 - x)}{\cosh \beta_1 l/2} \tag{2.1}$$

where

$$\beta_1 = \left[\frac{2G_m}{E_f r^2 \ln(R/r)} \right]^{1/2} \tag{2.2}$$

E_m, E_f = Young's modulus for matrix and fiber
G_m = shear modulus of the matrix
l = length of the fiber
R = radius of the matrix around the fiber (interfiber spacing)
r = radius of the fiber
ϵ_m = strain of the matrix

The ratio R/r is expressed as a function of fiber volume fraction and how the fiber is arranged. Expressions have been developed for square and hexagonal arrangements of fibers. Axial stress in the fiber $\sigma_f(x)$ can be computed using the following equation:

$$\sigma_f(x) = E_f \epsilon_m \left[\frac{1 - \cosh \beta_1(l/2 - x)}{\cosh \beta_1 l/2} \right] \tag{2.3}$$

Both the shear stress $\tau(x)$ and the axial stress $\sigma_f(x)$ distributions are nonlinear along the length of the fiber. Note that the Poisson's effects are neglected in these equations.

Even though equations (2.1) through (2.3) are based on very restrictive assumptions stated earlier, they provide some means of computing the stresses in the fibers and their contribution to the composite.

In practical cases, fibers are randomly distributed at least in two dimensions. In the case of fiber-reinforced concrete they are randomly distributed in all three (mutually perpendicular) directions. In addition, most steel fibers and some polymeric fibers have surface or end deformation. In almost all cases there is interaction between fibers, giving rise to the complexity of the problem. Hence, mathematical models for use in practical applications are still in early stages of development.

Based on the composite behavior, it is established that fibers contribute to both composite strength and stiffness. The amount and nature of contribution depend on fiber type, fiber volume fraction, and matrix properties. For example a composite containing a 10 percent volume of steel fibers exhibits a fivefold increase in strength, whereas the increase is negligible at volume fractions of less than 2 percent.

2.2   Fiber Interaction in Cracked Matrix

When the composite containing the fibers is loaded in tension (Figure 2-1b), at a certain stage the matrix cracks (Figure 2-2). Once the matrix cracks, the fiber carries the load across the crack, transmitting load from one side of the matrix to the other. In practice, several fibers will bridge the crack, transferring the load across the crack. If the fibers can transmit sufficient load across the crack, more cracks will form along the length of the specimen. This stage of loading is called the multiple cracking stage (as explained in Chapter 3). In most practical applications, this multiple cracking stage occurs under service load conditions. The fiber interaction characteristics also determine the peak load-carrying capacity of the composite and the postpeak load-deformation behavior (Chapter 3).

The critical issues to be addressed in fiber interaction are: (1) load-slip variation, (2) geometry and orientation effect, (3) how to quantify the pull-out resistance (load) of a single fiber, and (4) interaction of randomly distributed fibers, so as to evaluate the behavior of multiple fiber pull-out. Experimental techniques used to study the fiber pull-out of single and multiple fibers, interpretation of the experimental results, and mathematical modeling of the fiber-matrix interaction are discussed in the subsequent sections. The effect of the interaction on tension behavior of the composite is discussed in Chapter 3.

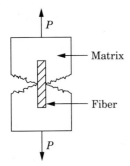

Figure 2-2
Fiber-matrix
interaction,
cracked
matrix.

2.2.1 Experimental techniques for evaluating fiber-matrix bond

Fiber-matrix bond behavior can be studied using either direct or indirect tests. For the indirect tests, the composite is tested in tension or bending, and the fiber contribution is evaluated. Extensive mathematical analysis is used to separate the resistance provided by the fiber from the matrix resistance. The results obtained in this process are highly dependent on the mathematical model used for the analysis.

In the direct tests, either a single fiber or an array of parallel fibers are pulled out from the matrix. The test results can be used to estimate the interfacial properties, average bond strength, and the load-slip behavior. Figure 2-3 shows some typical test setups used for studying the fiber pull-out problem. The average bond stress that the fiber can sustain can be obtained using the setup shown in Figure 2-3a. The embedment length is kept short to provide easy pull-out. The test results give the load at which the fiber starts to pull out and the resistance provided by the fiber during the progressive pull-out. The type of support provided for the matrix, Figure 2-3a parts i and ii, results in different boundary conditions. The case shown in part i provides some compression around the fiber. This setup still provides a better simulation of a fiber embedded in composite since the support system interference is less than that for the setup shown in part ii.

The setups shown in Figure 2-3b, where longer fiber lengths are used, could provide information on the debonding of fibers at the loaded end. If the embedment length is too long, the fiber might fracture rather than pull out. In the setups shown in Figure 2-3a and b, some influence of the supports always exists.

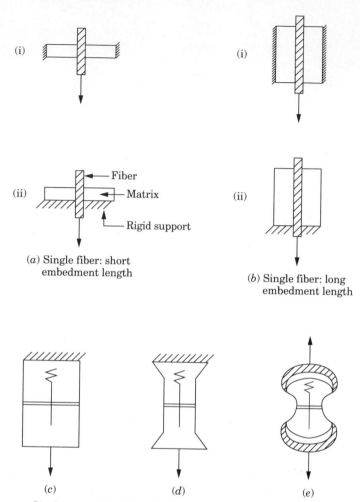

(a) Single fiber: short
embedment length

(b) Single fiber: long
embedment length

Single or multiple fibers for short or long embedment length

Figure 2-3 Typical fiber pull-out test setups.

The setups shown in Figure 2-3c, d, and e produce less of a confining effect and less interference during the fiber pull-out. These tests also provide a better simulation of a fiber composite subjected to tension. Moreover, it is possible to use multiple fibers, and the fibers can be embedded in an angle with respect to the pulling direction.

A test setup was also developed to determine the critical fiber length, as shown in Figure 2-4 [2.36]. The critical fiber length is the length of embedment that provides for maximum pull-out load without fracturing the fiber. In most civil engineering applications, pulling out during the failure of the composite is preferred since this failure mode provides more ductility.

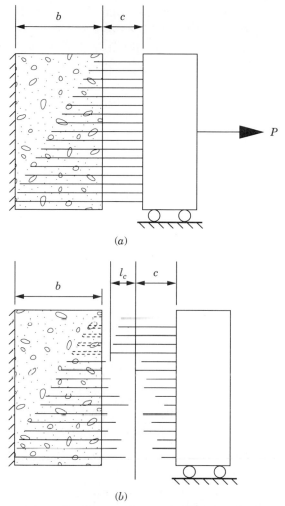

Figure 2-4 Bond strength test (critical length l_c method) (a) before pull-out; (b) after pull-out.

Test setups have also been developed for evaluating the pull-out be-havior at high rates of loading [2.33, 2.37]. The results obtained from these tests are useful for evaluating the fiber composites subjected to dynamic loading.

The following points should be taken into consideration while se-lecting a set up for pull-out tests.

- The setup should ideally simulate the random fiber located in the composite. In most cases it is impossible to obtain a perfect simula-tion. However, attempts should be made to avoid boundary conditions around the fiber and the fiber-matrix junction that are quite different from what is happening in the cracked composite.

- In most cases, the concrete (matrix) around the fiber is in tension when the composite is subjected to tension. Pull-out tests that simulate this condition are preferable.

- The setup should lend itself for the use of multiple and inclined fibers.

- The setup should provide reproducible results with reasonable variance. The coefficient of variation for pull-out tests is normally high (in the range of 20 to 50 percent).

2.2.2   Typical experimental results

Typical results of fiber pull-out are presented in Table 2.1 [2.12, 2.38–2.43]. The bond strengths reported in the table were calculated assuming uniform bond stress along the fiber embedment length. A typical pull-out load-versus-slip relationship for a single aligned fiber is shown in Figure 2-5. The load-slip relationships for multiple-fiber and inclined-fiber arrangements are shown in Figure 2-6.

The study of Table 2.1, Figures 2-5 and 2-6, and the discussions provided by various investigators lead to the following observations:

TABLE 2.1   Typical Pull-Out Test Results

	Matrix properties			Fiber properties			Pull-out test results	
Ref.	S/C	W/C	Casting direction*	Type	Embedded length (mm)†	Dia (mm)	Peak stress (Pa)‡	Slip at peak stress (mm)
[2.38]	2.5	0.6	PD	Brass-coated	12.7	0.25	2.6	0.76
[2.39]	3.0	0.5	PD	High-tensile	10.2–13.7	0.38	4.0–4.2	—
[2.40]	2.0	0.5	PD	strength	50.8		1.3	—
	2.0		PL				2.3	—
	3.0		PD				1.4	—
	3.0		PL				2.2	—
	4.0		PD				1.6	—
	4.0		PL				1.8	—
[2.43]	0	0.31	PD	Coil wire	30.0	0.64–0.85	1.5–2.0	—
[2.12]	2.5	0.6	PD	Arbed	12.5	0.38	0.45	0.20
[2.41]	—	0.55	PD		12.5	0.38	2.00	0.20
			PD		40.0	0.50	0.95	0.25
[2.42]	0	0.30	PL	Low-carbon	17.5	0.38	2.60	—

*Casting direction: with respect to fiber, PD = perpendicular and PL = parallel; S/C = sand/cement, W/C = water/cement
†1 in. = 25.4 mm.
‡1 psi = 6.895 kPa.

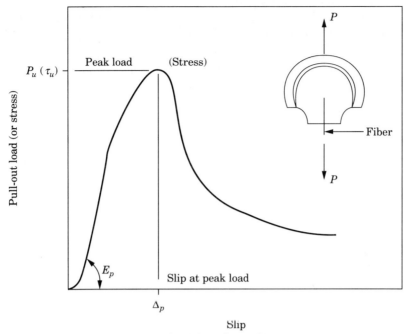

Figure 2-5 Typical fiber pull-out load-slip relationship.

- Because of the random distribution of fibers in the cement-based composite, an understanding of the pull-out performance of inclined fibers is needed for predicting the composite behavior. The two main contributing factors to pull-out resistance are shear resistance at interface and fiber dowel action. The interfacial shear strength primarily depends on the properties of the matrix, fiber geometry properties, and the number of fibers per unit area. The dowel action (fiber bending effect and friction effect) depend on the angle of inclination and the properties of the fiber. The action damages the matrix at the crack and hence multiple fibers are not as effective as individual fibers.

- For steel fibers, the additional work required in bending the fiber can be beneficial. For polymeric fibers the bending effect is often negligible. However, the increased frictional stresses owing to inclination can be substantial. For glass fibers, the maximum pull-out load can be lower than that of aligned fibers because the fibers are very brittle and hence can easily break.

- The postpeak performance of the inclined fiber is quite different from the aligned fiber. In the case of steel and polymeric fibers, the inclined fiber can sustain much higher energy than the aligned fiber (Figure 2-6).

- Multiple inclined fibers can sustain only a fraction of the pull-out load sustained by a single fiber (Figure 2-7).

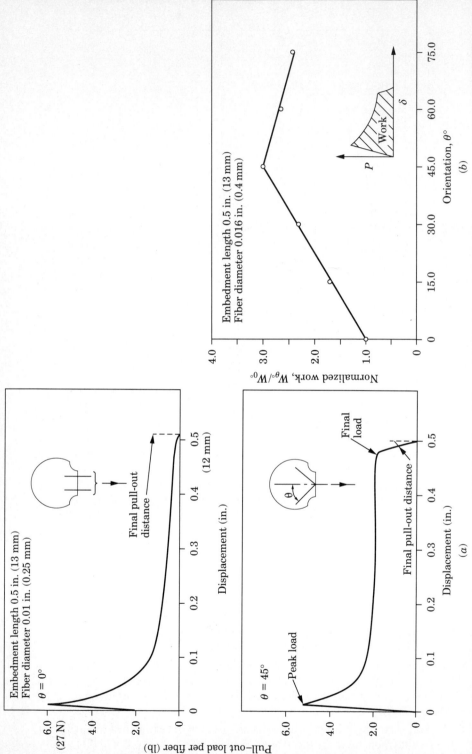

Figure 2-6 Experimental results from pull-out tests [2.8]. Smooth steel fibers of circular cross-section are pulled out from mortar matrix briquettes: (a) Typical load-displacement behavior for aligned fiber and inclined fiber pull-out; (b) Pull-out work requirement.

- The test data show significantly higher variation than other standard test data, such as the tension test data of composites.

2.3 Interpretation of Test Data and Analytical Models

The mechanics of bond failure at the fiber-matrix interface is an important parameter in understanding the load-slip response. The primary issue is the role that the fiber-matrix interface plays after the matrix

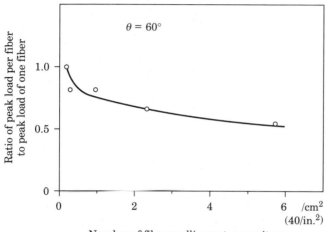

(a) Pull-out strength

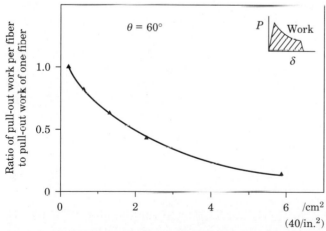

(b) Pull-out work

Figure 2-7 Effect of the number of inclined fibers on their pull-out behavior [2.8].

cracks. Let us examine the case of the single fiber pull-out test, shown schematically in Figure 2-8a. When the load is small, the bond behavior can be assumed to be elastic. The shear stress distribution can be computed using the theory of elasticity, by solving the equations of equilibrium, compatibility, and boundary conditions. Assuming that the influence of fiber extends up to a radius of half the fiber spacing, an equation for the elastic bond stress τ_e can be written as the following [2.46]:

$$\tau_e(x) = -\frac{\beta}{2\pi r}(C_1 \sinh \beta x + C_2 \cosh \beta x) \tag{2.4}$$

where
$$\beta^2 = \frac{2\pi G_m}{\ln(\frac{1}{2}\sqrt{\pi/V_f})A_f E_f} \tag{2.5}$$

$$C_1 = -\alpha P \tag{2.6}$$

$$C_2 = \frac{P(1 - \alpha(1 - \cosh \beta l/2))}{\sinh \beta l/2} \tag{2.7}$$

$$\alpha = \frac{A_f E_f}{A_m E_m} \tag{2.8}$$

$A_f,\ A_m$ = area of fiber and matrix

G_m = shear modulus of the matrix

x = length measured from the embedded end of the fiber (Figure 2-8)

The shear stress distribution expressed using equation (2.4) results in a maximum shear stress at the cracked surface of the matrix. When this shear stress exceeds shear strength of fiber-matrix interface τ_s, debonding occurs. The debonding, which starts at the crack surface, progresses along the fiber (Figure 2-8b). In the debonded zone, frictional resistance τ_i provides some resistance to pull-out. This frictional resistance, which is less than the matrix-bond strength, is often assumed to be constant with respect to slip (Figure 2-9). For the partially debonded case, the pull-out load P can be expressed as

$$P = \frac{2\pi r \tau_s}{(1-\alpha)} \tanh \frac{\beta m l}{2} + \frac{\pi r l \tau_i(1-m)}{1-\alpha} \tag{2.9}$$

where $(1 - m)l/2$ is the debonded length. In the region 0 to $ml/2$ the bond is elastic, whereas in the region $ml/2$ to $l/2$ the resistance is purely frictional. Equation (2.9) is valid only for steel fibers.

If the frictional strength τ_i is less than bond strength τ_s, the pull-out load maximizes only after a stable debonding process, as shown in Figure 2-10. This indicates that a pull-out test exhibits a region

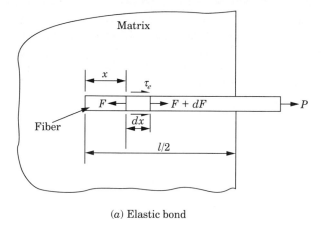

(a) Elastic bond

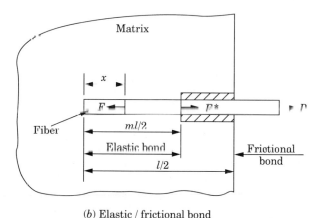

(b) Elastic / frictional bond

Figure 2-8 Schematic of the fiber pull-out problem solved showing definitions of the model parameters.

of stable debonding prior to the point at which the ultimate strength is reached. The average composite stress-strain curve becomes nonlinear primarily because of this stable debonding process, which takes place prior to pull-out of the fibers.

Generally, bond strength is calculated assuming a uniform interface bond strength $\bar{\tau}$. The maximum pull-out load is divided by the fiber contact area to obtain this uniform interface bond strength. It has been demonstrated that this approach toward the characterization of the bond strength depends on both embedment length and fiber diameter [2.46]. An alternative approach is to use a two-parameter model using τ_s and τ_i as the material properties. This approach has been shown

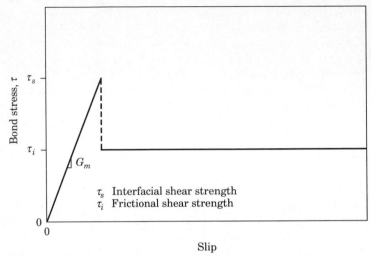

Figure 2-9 Schematic of the assumed interface property. A purely elastic behavior is evident until the shear strength of the interface τ_s is reached, followed by a constant frictional bond of strength τ_i.

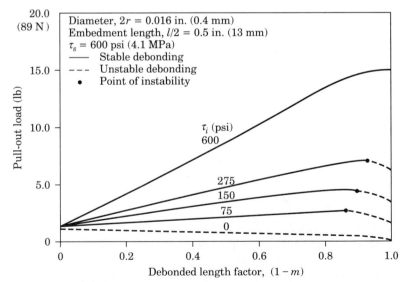

Figure 2-10 Model-predicted pull-out load versus the debonded length factor $(1 - m)$, showing stable and unstable debonding regimes depending on the ratio τ_s / τ_i assumed [2.46].

to provide a good prediction of the experimental results (Figure 2-11) [2.46]. The parameters τ_s and τ_i could perhaps be considered as basic material properties.

The pull-out load calculated using the behavior explained so far can be used to predict the fiber-composite behavior in tension. The development of procedures for such a prediction is discussed in Chapter 3.

Perhaps part of the scatter observed in the experimental results may be attributed to the various methods of analysis that so far have used a stress-based approach. The stress criterion is based on the maximum interfacial bond stress that can be developed (bond strength) for a given fiber and matrix, beyond which debonding initiates.

Models based on fracture mechanics are being investigated. In this approach, the process of debonding and pull-out is modeled as development of an interfacial crack. Criteria for crack propagation based on energy considerations and fracture mechanics principles are discussed in detail in Chapter 3. The energy approach seems to be more geometry-independent than the stress approach.

2.4   Composition of the Matrix

The composition of the portland cement–based microstructure has a significant effect on the properties of the composite. The hardened cement paste is a solid that contains pores of varying sizes and a

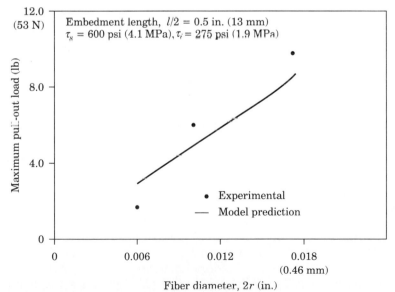

Figure 2-11   Experimental and model-predicted variation of the maximum pull-out load with respect to the fiber diameter [2.46].

microstructure that exhibits significant volume change, that is, creep and shrinkage, owing to moisture migration. The hydration of the cement presents an alkaline environment with a pH in the range of 12 to 12.5. The durability of many fiber systems in this highly alkaline environment must be carefully studied, because the reinforcing effects of fibers may degrade in time. The main products of the hydration of portland cement are calcium silicate hydrate, C-S-H, and calcium hydroxide, CH. The calcium silicate hydrate has a poor crystalline microstructure, modeled as thin plates, which are stacked in a random order with a spacing of approximately 11 Å. The water may be present in-between these layers. In addition, larger pores exist, which are referred to as capillary pores. The C-S-H has a significantly high surface area $(200\text{m}^2/\text{g})$. Interlayer bonding in-between the C-S-H sheets creates the strength of C-S-H. The calcium hydroxide, which is a product of the hydration reaction of calcium silicates with water, consists of large hexagonal crystals, approximately 0.20–1 μm in size. The distribution, size, and amount of calcium hydroxide is significantly dependent on the constituents of the matrix and the use of pozzolans, such as silica fume, fly ash, or mica. It is the reaction of calcium hydroxide with the pozzolans that generates more cementitious materials such as the C-S-H. The amount of C-S-H increases with the duration of curing; the amount of calcium hydroxide, however, is dependent on the presence of pozzolans.

The interface between the matrix and the fibers has been characterized as having a microstructure that is different from the microstructure of the rest of the matrix. The interface zone exists up to 50 μm from the fiber surface [2.48]. This zone consists of a duplex film roughly 1–2 μm thick that surrounds the fiber, a zone that is quite rich in large CH crystals of up to 30 μm, and a zone that is quite porous. The existence of the porous zone has been verified using a microhardness test. The contribution of the interfacial region to the mechanical performance of the composite is made through the process of debonding and pull-out. Furthermore, a weak interface could debond and deflect a propagating matrix crack as it approaches the fiber. This mechanism, known as the Cook-Gordon cracking mechanism [2.49], results in a change in the crack path leading to crack arrest or toughening of the composite.

The bond of fiber to matrix is quite important if a composite is to perform in an optimum manner. Because there is no chemical reaction, the bonding of polymeric fibers such as polypropylene or polyvinyl alcohol (PVA) to the inorganic microstructure of the cement paste is primarily by mechanical interlock. The interfacial bond strength has been measured at about 1 MPa ($\sim$150 psi), which is quite low. New techniques have been developed for changing the surface microstructure of the fibers such as the use of electrical charge to cause surface roughness [2.50]. Addition of calcium carbonate, $CaCO_3$, or calcium sulfate, $CaSO_4$, during polymeric fiber production has also resulted in significant bond improvements.

A relatively high bond strength does not necessarily provide for a better composite performance. If a perfect bond exists, the fibers that bridge the wake of a propagating crack would not undergo the debonding process. The crack opening results in significant localized strain development in the fibers, resulting in their fracture as opposed to pull-out. Although the strength of the composite might increase slightly, the failure by fracture of fibers reduces the fracture toughness of the composite significantly. This type of failure mechanism is present in many cementitious composites that exhibit time-varying mechanical properties. Embrittlement can develop for both glass and wood fiber–reinforced concrete under high-temperature and high-humidity conditions and is described in detail in Chapter 13.

It should be mentioned that improved shear-lag and cohesive crack models are being developed. These models include some of the effects (such as nonuniform interfacial traction) neglected in earlier models. The description of these models can be found in Reference [2.51].

2.5   References

2.1   Cox, H. L.: "The Elasticity and Strength of Paper and Other Fibrous Materials," *British Journal of Applied Physics*, Vol. 3, 1952, pp 72–79.

2.2   DeVekey, R. C.; and Majumdar, A. J. "Determining Bond Strength in Fiber Reinforced Composites," *Magazine of Concrete Research*, Vol. 20, No. 65, 1968, pp. 229–34.

2.3   Creszczuk, L. B. "Theoretical Studies of the Mechanics of the Fiber Matrix Interface in Composites," ASTM-STP452, 1969, pp. 42–58.

2.4   Kelly, A. "Interface Effects and the Work of Fracture of a Fibrous Composite", *Proceedings of Royal Society of London*, Series A.319, 1970, pp. 95–116.

2.5   Lawrence, P. "Some Theoretical Considerations of Fiber Pull-Out from an Elastic Matrix," *Journal of Material Science*, Vol. 7, No. 1, 1972, pp. 1–6.

2.6   Takaku, A.; and Arridge, R. G. C. "The Effect of Interfacial, Radial, and Shear Stress on Fiber Pull-Out in Composite Materials," *Journal of Physics D: Applied Physics*, Vol. 6, 1973, pp. 2038–2047.

2.7   Kelly, A.; and Zweben, C. "Poisson Contraction on Aligned Fiber Composites Showing Pull-Out," *Journal of Material Science*, Vol. 11, 1976, pp. 582-587.

2.8   Naaman, A. E.; and Shah, S. P. "Pull-Out Mechanism in Steel Fiber Reinforced Concrete," ASCE, *Journal of Structural Division*, Vol. 102, No. 8, 1976, pp. 1537–1548.

2.9   Pinchin, D. J.; and Tabor, D. "Interfacial Phenomena in Steel Fiber Reinforced Cement. II. Pull-Out Behavior of Steel Wires," *Cement and Concrete Research*, Vol. 8, No. 2, 1978, pp. 139–50.

2.10 Pinchin, D. J.; and Tabor, D. "Inelastic Behavior in Steel Wire Pull-Out from Portland Cement Mortar," *Journal of Material Science*, Vol. 13, No. 6, 1978, pp. 1261–1266.

2.11 Pinchin, D. J.; and Tabor, D. "Interfacial Contact Pressure and Frictional Shear Transfer in Steel Fiber Cement," *RILEM Conference on Testing and Test Methods of Fiber Cement Composites*, 1978, Construction Press, U.K., pp. 337–44.

2.12 Stroeven, P.; de Haan, Y. M.; Bouter, C.; and Shah, S. P. "Pull-Out Tests of Steel Fibers," *RILEM Conference on Testing and Test Methods of Fiber Cement Composites*, 1978, Construction Press, U.K., pp. 345–53.

2.13 Bowling, J.; and Groves, G. W. "The Debonding and Pull-Out of Ductile Wires from Brittle Matrix," *Journal of Material Science*, Vol. 14, 1979, pp. 431–442.

2.14 Bartos, P. "Analysis of Pull-Out Tests on Fibers Embedded in Brittle Matrices," *Journal of Material Science*, Vol. 15, 1980, pp. 3122–3128.

2.15 Phan-Thien, N. "A Contribution to the Rigid Fiber Pull-Out Problem," *Fibre Science and Technology*, Vol. 13, 1980, pp. 179–186.

2.16 Phan-Thien, N.; and Goh, C. J. "On the Fiber Pull-Out Problem," *Journal of Applied Mathematical Mechanics*, ZAMM, Vol. 61, 1981, pp. 89–97.

2.17 Bartos, P. "Bond in Fiber Reinforced Cement Concretes: A Review Paper," *International Journal of Cement Composites*, Vol. 3, No. 3, 1981, pp. 159–232.

2.18 Phan-Thien, N.; Pantelis, G.; and Bush, M. B. "On the Elastic Fiber Pull-Out Problem: Asymptotic and Numerical Results," *Journal of Applied Mathematical Physics*, ZAMP, Vol. 33, 1982, pp. 251-265.

2.19 Laws, V.; "Micromechanical Aspects of the Fiber-Cement Bond," *Composites*, Vol. 13, 1982, pp. 145–151.

2.20 Atkinson, C.; Avila, J.; Betz, E.; and Semlson, R. E.; "The Rod Pull-Out Problem, Theory and Experiment," *Journal of Mechanics and Physics of Solids*, Vol. 30, No. 3, 1982, pp. 97–120.

2.21 Gray, R. J. "Analysis of the Effect of Embedded Fiber Length on the Fiber Debonding and Pull-Out from an Elastic Matrix, Part 1. Review of Theories," *Journal of Material Science*, Vol. 19, No. 3, 1984, pp. 864–870.

2.22 Gray, R. J. "Analysis of the Effect of Embedded Fiber Length on the Fiber Debonding and Pull-Out from an Elastic Matrix, Part 2. Application to Steel Fiber-Cementitious Matrix Composite System," *Journal of Material Science*, Vol. 19, 1984, pp. 1680–1691.

2.23 Stang, H. "The Fiber Pull-Out Problem: An Analytical Investigation," Series R, No. 204, Tech. Univ. of Denmark, 1985.

2.24 Stang, H.; and Shah, S. P. "Failure of Fiber-Reinforced Composites by Pull-Out Fracture," *Journal of Material Science*, Vol. 21, No. 3, 1986, pp. 953–957.

2.25 Bien, J. "Holographic Interferometry Study of the Steel Concrete Bond in Pull-Out Testing," Report 1-86-9, Delft University of Technology, Stevin Laboratory, The Netherlands, 1986.

2.26 Stang, H.; and Shah, S. P. "Fracture Mechanical Interpretation of the Fiber/Matrix Debonding Process in Cementitious Composites," *Fracture Toughness and Fracture Energy of Concrete*, Elsevier, 1986, pp. 513–523.

2.27 Steif, P. S.; and Haysan, S. F. "On Load Transfer between Imperfectly Bonded Constituents," *Mechanics of Materials*, Vol. 5, 1986, pp. 375–382.

2.28 Gopalaratnam, V. P.; and Shah, S. P. "Tensile Failure of Steel Fiber Reinforced Mortar," ASCE, *Journal of Engineering Mechanics*, Vol. 113, No. 5, 1987, pp. 635–652.

2.29 Bien, J.; and Stroeven, P. "Holographic Interferometry Study of Debonding Between Steel and Concrete," *Engineering Applications of New Composites,* Omega Scientific, 1988, pp. 213–218.

2.30 Morrison, J. K.; Shah, S. P.; and Jenq, Y. S. "Analysis of Fiber Debonding and Pull-Out in Composites," ASCE, *Journal of Engineering Mechanics Division*, Vol. 114, No. 2, 1988, pp. 277–294.

2.31 Nammur, G.; and Naaman, A. E. "Bond Stress Model for Fiber Reinforced Concrete Based on Bond Stress-Slip Relationship," *ACI Materials Journal*, Vol. 86, No. 1, 1989, pp. 45–57.

2.32 Banthia, N.; Pigeon, M.; and Trottier, J. F. "Steel Fiber Cementitious Matrix Bond Studies at High Stress Rates," *Experimental Techniques*, Vol. 13, 1989, pp. 19–22.

2.33 Banthia, N.; Trottier, J. F.; and Pigeon, M. "Fiber Pull-Out Mechanisms: Effects of Fiber Geometry, Loading Rate, and Sub-Zero Temperatures," *Fiber Reinforced Cements and Concretes: Recent Developments*, Elsevier, U.K., 1989, pp. 136–145.

2.34 Banthia, N; and Trottier, J. F. "Deformed Steel Fiber-Cementitious Matrix Bond Under Impact," *Cement and Concrete Research*, Vol. 21, 1991.

2.35 Stang, H.; Li, Z.; and Shah, S. P. "The Pull-Out Problem: The Stress versus Fracture Mechanical Approach," *ASCE, Journal of Engineering Mechanics Division*, ASCE, Vol. 116, No. 10, 1990, pp. 2136–50.

2.36 Wang, Y.; Backer, S.; and Li, V. C. "An Experimental Study of Synthetic Fiber Reinforced Cementitious Composites," *Journal of Material Science*, Vol. 22, 1987, pp. 4281–91.

2.37 Gokoz, U. N.; and Naaman, A. E. "Effect of Strain Rate on the Pull-Out Behavior of Fibers in Mortar," *International Journal of Cement Composites*, Vol. 3, No. 3, 1981, pp. 187–202.

2.38 Naaman, A.; and Shah, S. "Bond Studies on Oriented and Aligned Steel Fibers," *RILEM Fibre Reinforced Cement and Concrete*, Construction Press, U.K., 1975, pp. 171–178.

2.39 Maage, M. "Fibre Bond and Friction in Cement and Concrete," *RILEM Conference on Testing and Test Methods of Fibre Cement Composites*, Construction Press, U.K., 1978, pp. 399–408.

2.40 Gray, R.; and Johnston, C. "The Measurements of Fibre-Matrix Interfacial Bond Strength in Steel Fiber-Reinforced Cementitious Composites," *RILEM Conference on Testing and Test Methods of Fibre Cement Composites*, Construction Press, U.K., 1978, pp. 317–328.

2.41 Burakiewicz, A. "Testing of Fiber Bond Strength in Cement Matrix," *RILEM Conference on Testing and Test Methods of Fibre Cement Composites*, Construction Press, U.K., 1978, pp. 355–365.

2.42 Tattersall, G.; and Unbanowicz, C. "Bond Strength in Steel-Fibre-Reinforced Concrete," *Magazine of Concrete Research*, Vol. 26, 1974, pp. 105–113.

2.43 Hughes, B.; and Fattuhi, N. "Fibre Bond Strengths in Cement and Concrete," *Magazine of Concrete Research*, Vol. 27, 1975, pp. 161–166.

2.44 Maage, M. "Steel Fibre Bond Strengths in Cement Based Matrices Influenced by Surface Treatments," *Cement and Concrete Research*, Vol. 7, 1977, pp. 703–710.

2.45 Giaccio, G.; Giovambattista, A.; and Zerbino, R. "Concrete Reinforced with Collated Steel Fibers: Influence of Separation," *ACI Journal*, Vol. 83, 1986, pp. 232–235.

2.46 Gopalaratnam, V.; and Shah, S. P. "Failure Mechanisms and Fracture of Fiber Reinforced Concrete," *Fiber Reinforced Concrete Properties and Applications*, SP-105, American Concrete Institute, Detroit, Michigan, pp. 1–25.

2.47 Soroushian, P.; and Bayasi, Z. "Prediction on the Tensile Strength of Fiber Reinforced Concrete: A Critique of the Composite Material Concept," *Fiber Reinforced Concrete Properties and Applications,* SP-105, American Concrete Institute, Detroit, Michigan, 1987, pp. 71–84.

2.48 Bentur, A.; Diamond, S.; and Mindess, S. "The Microstructure of the Steel Fiber-Cement Interface," *Journal of Material Science*, Vol. 20, 1985, pp. 3610–3620.

2.49 Cook, J.; and Gordon, J. E. "A Mechanism for the Control of Crack Propagation in All Brittle Systems," *Proc. Royal Soc.*, 282A, pp. 508–520.

2.50 Krenchel, H.; and Jensen, H. W. "Organic Reinforcing Fibers for Cement and Concrete," Proceedings, Symposium on Fibrous Concrete (CI80), Construction Press, U.K., 1980, pp. 87–94.

2.51 Shah, S. P.; and Ouyang, C. "Mechanical Behavior of Fiber-Reinforced Cement-Based Composites," *Journal of American Ceramic Society,* Vol. 74, No. 11, 1991, pp. 2727–38 and 2947–2953.

3

Basic Concepts and Mechanical Properties: Tension

3.1 Basic Concepts: Strong Brittle Fibers In Ductile Matrix

The development of fiber-reinforced composite materials stems largely from research by Griffith, who used high-strength glass fibers in order to show that the apparent strength of brittle materials increases many-fold as the size of internal flaws (which inherently exist in materials) is reduced. In a paper published in 1921, he reported that the strength of glass fibers increased manyfold as their diameters decreased. The tensile strength of these fibers was very high, but they were brittle and notch-sensitive. Any microdefect could lead to sudden failure at stresses much lower than their normal tensile strength. This sensitivity for defects made it almost impossible to manufacture longer or thicker fibers. In addition, in order to capitalize on the high stiffness and strength of the fibers in a structural member, it was necessary to use them in a manner that would provide for the various geometries and loading conditions. This led to the embedding of these fibers in a ductile matrix to develop a high-strength and notch-insensitive composite. A polymeric matrix such as polyester resin was used as a matrix leading to a very successful composite known as fiberglass. This concept is still being used to manufacture a variety of fiberglass products including fiberglass reinforcing bars.

In fiberglass types of composite, high-strength brittle fibers are embedded in a ductile matrix, having a fiber volume fraction of up to 40 percent. The relative stress-strain curves in tension for glass fibers and an epoxy matrix are shown in Figure 3-1. Since the surface area

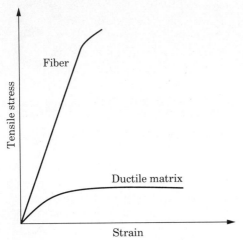

Figure 3-1   Tensile stress-strain curves for fibers with a ductile matrix.

of the fibers is large, it is possible to obtain perfect bond between the fibers and the matrix. This bond is strong enough to make the composite stronger and stiffer than the matrix and more ductile compared to the fiber behavior. The failure strain of the matrix, which is larger than the failure strain of the fiber, permits the use of the full potential strength of the fibers.

Composites made of high-strength carbon (graphite) fibers and a resin matrix, which combine strong and stiff fibers and a ductile matrix, provide high performance for aerospace applications. Graphite is another example in which high strength is being successfully utilized.

3.2   Strong Fibers in a Brittle Matrix

The primary reason for using strong fibers in a relatively weak (in tension) brittle matrix is to improve the ductility of the matrix. The fibers contribute to the increase in strength as well. However, in many applications involving this type of combination, the fiber volume fractions are kept relatively low ($< 1\%$), resulting in an insignificant increase in strength. In these composites since the ultimate strain capacity of the matrix is lower than the strain capacity of the fibers, the matrix fails before the full potential capacity of the fiber is achieved. The fibers that bridge the cracks formed in the matrix contribute to the energy dissipation through processes of debonding and pull-out. In applications where higher volume fractions of fibers ($>5\%$) are used, there is a significant increase in tensile strength. In this chapter, the responses of composites made with both high and low volume fractions of fibers in a brittle matrix are studied.

The matrix typically consists of a portland cement composite or other building materials such as gypsum. The portland cement composite could consist of cement paste, mortar (cement plus fine aggregate), or concrete (cement plus fine and coarse aggregate). In the construction practice, various admixtures such as high-range water-reducing and air-entraining admixtures are also being used to improve the overall strength and durability of the composite.

The fiber could be metallic, mineral, polymeric, or naturally occurring. The metallic fibers have typically high modulus and high strength. Their behavior is also ductile. Mineral fibers (typically glass) have moduli higher than the cement product but lower than steel. Their failure is relatively brittle in nature. Polymeric fibers are strong and ductile, but their moduli are normally lower than those of cement composites. Certain polymeric fibers such as Kevlar have higher moduli. Organic fibers are relatively strong, but their bond characteristics are not as good as metallic or mineral fibers. The overall behavior of a composite will depend on the type of fiber used. However, the following general concept applies to all fiber types.

The relative stress-strain curves in tension for a strong fiber and a weak brittle matrix are shown in Figure 3-2. In this case, the matrix will fracture (crack) long before the fiber reaches its tensile strength

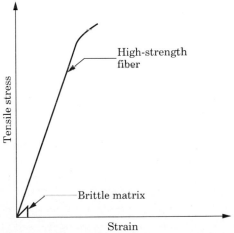

Figure 3-2 Tensile stress-strain curves for fibers with a brittle matrix.

since the fracture strain for the matrix is very low compared to the fracture strain of fiber. Once the matrix cracks, further behavior of the composite could be one of the following.

1. The composite will fracture immediately after the fracture of the matrix. Figure 3-3a shows this type of behavior. Very low fiber volume fraction could lead to this type of failure.

2. After the matrix cracks, the load-carrying capacity could drop but the composite could continue to resist loads that are lower than the peak load (Figure 3-3b). When the matrix cracks, the load is transferred from the composite (matrix plus fibers) to the fibers at the crack interface. Hence, further load-carrying capacity comes from the fibers transferring the load across the crack. As the deformation increases, fibers pull out of the matrix, resulting in lower and lower load-carrying capacity. This type of composite does not provide an increase in strength (over the matrix strength) but provides ductile behavior. The area under the stress-strain curve is an indication of the ductility or toughness of the composite.

3. If the volume fraction of fibers is large enough, after the matrix cracks, the fibers will start carrying the increased loads. If there are sufficient fibers across the crack, then they will continue to resist higher loads than the cracking load, Figure 3-3c. The stiffness of the stress-strain curve will drop because of the loss of matrix contribution. The slope of the postcracking response would depend on the volume fraction of fibers and their bonding capacity to the matrix. As the load increases, more cracks will form along the length of the specimen. Eventually, when fibers start to pull out from the matrix, the slope of the stress-strain curve will reach zero and the load-carrying capacity will start to drop. This type of failure provides for optimum

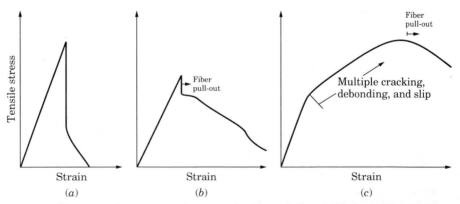

Figure 3-3 The composite stress-strain curves for fiber-reinforced brittle matrices: (a) Low fiber volume fractions; (b) Intermediate fiber volume fractions; (c) High fiber volume fractions.

use of the fiber and matrix properties. Most of the fiber composites used in the field follow failure patterns shown in Figure 3-3c and b. Various aspects of these stress-strain behaviors and their prediction and design considerations are presented in subsequent sections.

3.3 Tensile Behavior
of Fiber-Cement Composites

As mentioned in the previous section, stress-strain (load-deformation) behavior can be divided into three regions identified as initial elastic range, inelastic (multiple cracking) range, and postpeak range (Figure 3-4). All three regions may not be present for all fiber types and volume fractions. Figures 3-5, 3-6, and 3-7 show the experimentally obtained stress-strain response of fiber-reinforced concrete containing steel, glass, and polypropylene fibers [3.1, 3.2]. From these figures, it can be seen that some fiber types and volume fractions result in more pronounced inelastic regions than others.

The linear range is present for all the composites shown in Figure 3-4 and may be characterized by means of Young's modulus of the composite. The determination of this parameter is significantly dependent on the means of measuring deformations or strains. It has been established that, depending on the testing procedure, the definition of the first-crack load could significantly change the extent of the elastic region [3.3]. Figure 3-8 clearly shows the influence of the type of measurement on the identification of the elastic region [3.4]. The strains are much more sensitive to type of measurement, and the variation can range by a significant amount. If the spurious deformations (deformation at the grips, support settlement) are included in the specimen response, there could be a significant overestimation of the composite response. This is a nonconservative approach that might result in the composite design based on linear elasticity being used

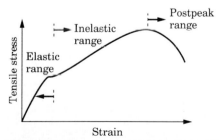

Figure 3-4 The three ranges of tensile stress-strain behavior for high fiber volume composites.

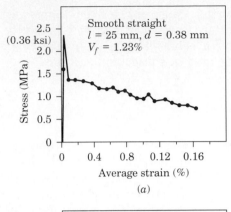

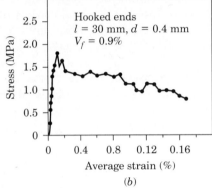

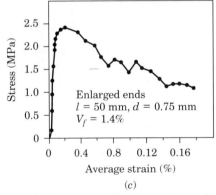

Figure 3-5 Experimentally observed tensile behavior of FRC: Steel fibers (1 in. = 25.4 mm)[3.1, 3.2].

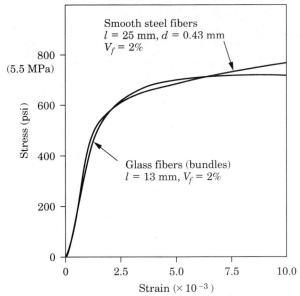

Figure 3-6 Experimentally observed tensile behavior of FRC: Steel and glass fibers (1 in. = 25.4 mm) [3.1, 3.2].

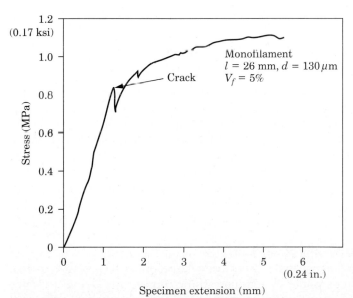

Figure 3-7 Experimentally observed tensile behavior of FRC: Polypropylene fibers (1 in. = 25.4 mm) [3.1, 3.2].

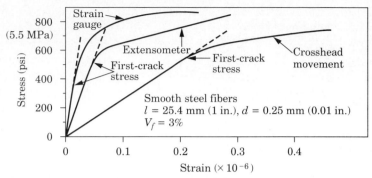

Figure 3-8 Tensile stress-strain response of FRC: Comparison of three types of strain measurement [3.4].

in the inelastic range. Theoretical derivation of the elastic modulus of a composite based on the rule of mixtures is described in Section 3.5. If the fiber volume fraction is relatively high, then fibers will carry more and more load, leading to the multiple cracking stage. This situation is described further in detail in Section 3.7. A number of researchers have attempted to predict this response. Shear lag theory [3.5], micromechanics-based approaches [3.6–3.9], and fracture mechanics [3.10–3.12] have been successfully used. These approaches are presented in Sections 3.10 and 3.11.

3.4   Experimental Evaluation of Conventional Fiber-Cement Composites

In general, for composites with less than 2% fiber content (by volume), the stress-strain response can be approximated by three segments. The first segment, representing precracked regions, can be defined by the modulus of elasticity of the matrix. The second segment can be taken as a zone of nonlinear deformation between the first cracking of the matrix and the ultimate tensile strength of the composite. For most cases, this segment is relatively small. The third segment of the curve representing the postpeak (reserve) strength can be attributed to the fiber pull-out resistance. Depending on the fiber volume fraction, fiber composition, and fiber geometry, this reserve strength can vary from negligible to 40% of peak strength.

Recent advances in instrumentation have made it possible to test tension specimens under closed-loop stable conditions. Using this type of test setup, postpeak responses can be obtained even for plain concrete and FRC [3.13, 3.14]. Figure 3-9 represents the response of plain mortar and steel fiber–reinforced concrete tested in uniaxial tension; the stress-displacement responses for various fiber volume fractions are also shown [3.14]. These responses were obtained using notched specimens loaded such that the notch mouth opening displacement increased monotonically. The following discussion provides an in-depth evaluation of the stress-strain response as reported in References [3.13] and [3.15].

3.4.1   Tensile stress-displacement response

Plain mortar matrix specimens exhibit linear elastic behavior up to about 50% of their tensile strengths. The onset of inelastic behavior prior to the peak load suggests the initiation of the microcracking process, although no cracks are detected prior to the peak loads (optically, at a 100% magnification). Almost immediately after peak loads, deformations become localized (nonunique local strains) with the eventual widening of a single crack. The unreinforced matrix exhibits significant traction capacity at displacements several times those observed at the peak load (Figure 3-9). More detailed information on the fracture characteristics of the unreinforced mixes can be found in Reference [3.14.]

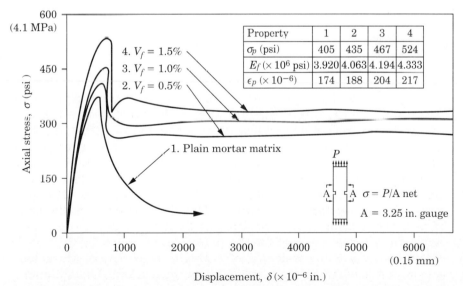

Property	1	2	3	4
σ_p (psi)	405	435	467	524
E_f ($\times 10^6$ psi)	3.920	4.063	4.194	4.333
ϵ_p ($\times 10^{-6}$)	174	188	204	217

4. $V_f = 1.5\%$
3. $V_f = 1.0\%$
2. $V_f = 0.5\%$

1. Plain mortar matrix

$\sigma = P/A$ net

A = 3.25 in. gauge

Displacement, δ ($\times 10^{-6}$ in.)

Figure 3-9   Tensile stress-strain response: Comparison of plain and fiber-reinforced concrete (1 psi = 6.895×10^{-3} MPa) [3.14].

Steel fiber–reinforced composite (SFRC) behavior is linearly elastic up to about 80% of the matrix tensile strength. Nonlinear deformations take place beyond the linear elastic limit. Composite peak stress and corresponding displacements (or strains) are larger than the corresponding values for the unreinforced matrix mix ($\leq$ 25–30% for the 1.5% SFRC mix). After the composite peak stress, the load-carrying capacity abruptly drops to a stress level termed by some investigators as the postcracking strength. One single crack becomes visible at the critical section. With further increases in displacement, the load-carrying capacity gradually drops (in a linear fashion) with increases in the displacement. A frictional type of pull-out resistance (because of interfacial bond slip) is perhaps responsible for the linear $\sigma - w$ relation with a negative tangent modulus.

3.4.2 Localization of deformation

Tests on notched rectangular prismatic SFRC specimens show that the fracture of such composites (using short, random steel fibers) is due to the formation of a single major crack near the peak stress level. Strains at various locations on the specimen were monitored during a monotonic test conducted on an SFRC specimen ($V_f = 1\%$). The behavior of the fracture zone and the zones surrounding it were studied, and the results of this test are illustrated in Figure 3-10. Although crack widths could be measured only after peak load levels, the strain reading at location 6 (Figure 3-10) was significantly lower than those at locations 1 to 5 at the peak stress, suggesting the formation of a major crack at the critical section. Rapid increases in strain values at this section and the unloading of gauges 6 and 7 were observed, similar to the softening behavior of the plain matrix [3.14]. However, the unloading of gauges 6 and 7 was not as significant as that observed for the plain matrix, probably because the composite, unlike the plain matrix, carries substantial loads even beyond its peak stress. Additionally, it is believed that microcracks that form in the zones adjacent to the major crack are restrained from closing up because of the presence of fibers, unlike in the unreinforced matrix. This is evidenced in the size of permanent deformations (strains) at locations 6 and 7—they are larger than those observed for the plain matrix. Final fracture still occurs because of the widening of a single major crack, which has a more tortuous profile as a result of the presence of fibers. Confirmation of this behavior was also obtained from optical crack-width measurements.

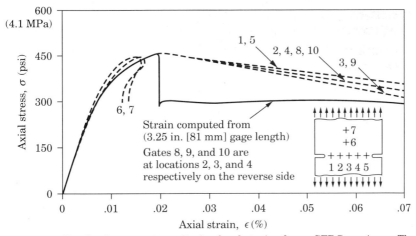

Figure 3-10   Results from a test monitoring local strains for an SFRC specimen. The net axial stress at the critical section is plotted versus the strain computed from w and is compared with the local strain behavior. Dashed lines are used for the local strain behavior, measured only at discrete points [3.14].

3.4.3   Optical crack-width measurements

Microscopic measurements indicate that (1) a single crack of measurable width becomes visible across the critical section almost immediately after peak load and (2) the postcracking resistance observed in SFRC is due to the widening of a single crack. Average crack widths computed from the displacement measurements w_δ defined in the inset of Figure 3-11 matched optically measured crack widths ω_o. The line connecting all the points $(\omega_\delta, \omega_o)$ is further away from the 45° line for SFRC than for the plain matrix, again confirming larger permanent deformations in the noncritical zones for SFRC specimens. It has been shown [3.14] that the softening behavior of the unreinforced matrix can be uniquely represented by an exponential stress–crack width relation.

3.4.4   Residual deformation

Unloading-reloading tests conducted in the postcracking regime showed little degradation in the apparent material stiffness compared with that observed for plain matrix, even at large unloading displacements (Figure 3-12). This results from the interfacial failure of the fibers and the irreversible nature of fiber slip across the crack. The frictional restraint to crack closing provided by the fibers causes the residual deformations to be very large.

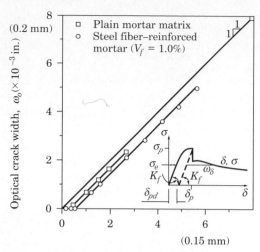

Figure 3-11 Comparison of crack widths measured using optical means and computed from stress-strain response. The inset illustrates the method used to compute the crack width from displacement measurement [3.14].

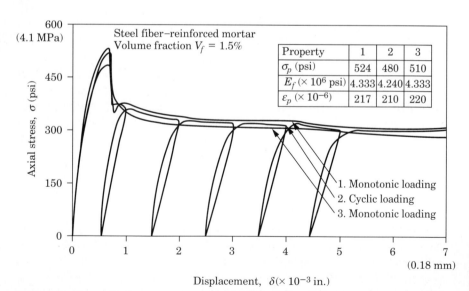

Figure 3-12 Stress-displacement response of FRC: Comparison of monotonic and cyclic stresses [3.14].

3.5   Elastic Response in Tension

3.5.1   Composite Stiffness

In the elastic range, the primary variables involve the moduli of elasticity of the matrix and the fibers. The rule of mixtures can be used as a means of obtaining the composite response. Experimentally, the slope of the stress-strain curve up to the first cracking point can be used as the composite stiffness. Let us represent the moduli of elasticity of the matrix and the fibers as E_m and E_f and the volume fractions of each phase as V_m and V_f. A parallel model is assumed, indicating that the strains in the matrix, the fibers, and the composite are equal; that is, there is no slip. Furthermore, the load being carried by the composite is the summation of the loads being carried by the two phases. These equations are represented as

$$\epsilon_c = \epsilon_m = \epsilon_f \tag{3.1}$$

and

$$\sigma_c A_c = \sigma_f A_f + \sigma_m A_m \tag{3.2}$$

where A_c, A_f and A_m are areas of composite, fiber, and matrix respectively. Then the modulus of the composite E_c can be expressed as

$$E_c = E_f V_f + E_m V_m \tag{3.3}$$

since

$$V_f + V_m = 1 \tag{3.4}$$

Equation (3.3) can also be written as

$$E_c = V_f E_f + (1 - V_f) E_m \tag{3.5}$$

Equation (3.5) is valid only if the fibers are continuous and aligned in the loading direction and if there is perfect bond between the matrix and the fibers. As the volume fraction of fiber V_f increases from 0 to 1, the composite modulus will change from E_m to E_f in proportion to V_f. More detailed information on various types of fibers is presented in Chapter 5. Moduli of cement-based matrices vary from 20 to 30 GPa. For steel fiber–reinforced concrete, the modulus of the composite E_c can be expected to be greater than the modulus of the matrix E_m. For concrete reinforced with synthetic fibers with modulus E_f smaller than E_m, the modulus of the composite E_c shows a decrease.

If the fibers are discrete (discontinuous) and randomly distributed, the term representing the modulus of the fibers E_f in equation (3.3) could be replaced with an equivalent modulus E_f^*. The modulus of the composite E_c would depend on lengths, geometry, and the volume fraction. The equivalent modulus could be as low as one-third of E_f. It should be noted that for many fiber-reinforced concrete applications, the volume fraction of fibers is normally less than 2%. Hence, for these composites the modulus E_c is about the same as the matrix modulus E_m. In the case of manufactured products such as glass fiber–reinforced panels, the fiber volume fraction could be relatively high, resulting in a higher composite modulus E_c.

Once the composite modulus E_c is known, the stress-strain relation up to the first crack can be written as

$$\sigma_c = E_c \epsilon_c \tag{3.6}$$

where σ_c and ϵ_c are stress and corresponding strain values of the composite. Since it is assumed that there is no slip between the matrix and the fiber, the strain values of the composite ϵ_c, matrix ϵ_m, and fiber ϵ_f are all the same.

3.5.2   Fiber orientation and length efficiency factors

Modifications to the rule of mixtures are introduced through the definition of fiber efficiency factors. The most common efficiency factors are for orientation and finite length. The efficiency factors for length have been reported in References [3.16–3.18]. Krenchel has derived the fiber orientation efficiency as 3/8 for a random two-dimensional arrangement of fibers [3.16]. This efficiency factor may change as the matrix cracks and the fibers bridging the crack tend to align themselves and thus carry loads. Aveston, Mercer, and Sillwood have used a value of 1/2 [3.18]. Laws, Lawrence, and Nurse take into account the contribution caused by the length effects and by the effect of bond strength on the increase in the fiber alignment efficiency factor [3.19].

The length effects of short fibers can be viewed in terms of the parameter called critical length. Stress transfer from the matrix to the fiber will result in shear forces at the interface between fiber and matrix. As the length of the fiber decreases, its efficiency in carrying loads decreases since the ends of the fiber are partially loaded. The minimum length of the fiber required to achieve the full strength capacity of the fiber is referred to as the critical length. Composites with fiber lengths lower than l_c fail without fiber fracture during the fiber pull-out since the fiber is not long enough to generate fracture. It can be shown that the bond strength controls the critical length. Figure 3-13 represents a simple approach toward the shear stress transfer as affected by the fiber length. In composites with $l \leq l_c$, (Figure 3-13a, b), the fiber strength cannot be achieved, and the failure is governed by debonding, or matrix fracture. In the composites with $l > l_c$, fiber strength can be attained (Figure 3-13c) with the minimum fiber length defined as the critical length. It can be shown that the critical length is inversely proportional to the average bond strength, that is,

$$l_c = \frac{2\sigma_{fu}A}{p\tau} \tag{3.7}$$

where τ and σ_{fu} represent the average bond strength of the interface and the ultimate strength of the fiber respectively. The variable p represents the perimeter of the fiber. Recent approaches (for example,

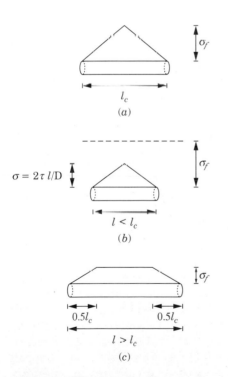

Figure 3-13 Fiber length effects in the transfer of shear stresses and the critical length.

Reference [3.20]) for fiber efficiency have been developed based on the energy dissipation of the fibers. Use of the pull-out slip response as affected by orientation and length effects has also been used as a measure of the efficiency of the fibers in carrying load [3.21].

3.5.3   Critical fiber volume fraction

The tensile response of fiber-cement composites significantly depends on the volume fraction of the fibers used. The response of a fiber-reinforced composite up to the cracking strength of the matrix can be described using equation (3.6). Beyond this point, assuming that the matrix does not contribute any further, the strength of the composite is a function of the strength and volume fraction of the fibers, that is,

$$\sigma_{cu} = V_f \sigma_{fu} \tag{3.8}$$

At a fiber loading up to the critical volume, the release of a matrix-cracking load onto the fibers causes the exhaustion of fiber strength; failure is thus by the formation of a single crack. For high volume fractions, the fibers are able to carry load in excess of matrix-cracking load; thus the ultimate strength of the composite is higher than the matrix strength, and distributed cracking can exist. The transition from single cracking to multiple cracking takes place above a certain fiber volume fraction, referred to as the critical volume fraction V_{cr}. It has been derived by setting the composite load-carrying capacity equal to the fiber load-carrying capacity [equations (3.5) and (3.8)] equal to each other and solving for V_f.

$$V_{cr} = \frac{\sigma_{mu}}{\sigma_{mu} + (\sigma_{fu} - \sigma'_{fu})} \tag{3.9}$$

where σ'_{fu} represents the stress in the fiber at the strain level corresponding to the matrix-cracking strain. Note that this derivation assumes linear elastic behavior both for fibers and matrix, no debonding, brittle response of matrix, perfectly aligned and continuous fibers, in addition to a perfect bond. Typical V_{cr} values for aligned steel, glass, and polypropylene range from 0.3% to 0.8% [3.22]. For chopped and randomly oriented fiber composites the critical fiber volume fraction is significantly higher than these values owing to the efficiency factors.

3.6   Prediction of Composite Strength Based on Empirical Approaches

Both the matrix and the fibers contribute to the tensile load-carrying capacity of the composite. However, at the point of ultimate strength of the composite, also designated as failure stress (and represented by the peak of the stress-strain curve), the relative proportion of the load being carried by each phase is not known. If the fibers are uniformly distributed in all three dimensions and the bond between the fibers and the matrix is strong enough to induce fractures in all the fibers, the following theoretical expression can be written for the composite's strength. Let

$$\sigma_{mu} = \text{matrix tensile strength}$$

$$\sigma_{fu} = \text{fiber tensile strength}$$

$$\sigma_{cu} = \text{composite tensile strength}$$

V_f and V_m are fiber and matrix volume fractions defined in the previous section. At a given cross-section, assuming that the two phases are behaving independently of each other, the contribution of the strengths of the matrix and the fibers will be $V_m\sigma_{mu}$ and $V_f\sigma_{fu}$ respectively. Hence

$$\sigma_{cu} = V_f\sigma_{fu} + V_m\sigma_{mu} \tag{3.10}$$

Equation (3.10) is not directly applicable to FRC for the following three reasons:

1. For short randomly oriented fiber composites, the fibers place themselves in different orientations depending on fabrication and dimension of the member. In any given direction, the contribution of the fibers will be less than the total volume fraction V_f.

2. The tensile strength of fibers is generally high compared to the tensile strength of concrete or mortar. Typical strengths of fibers include the following: steel 100–200 ksi (689–1378 MPa), polymeric (nylon, polypropylene, polyethylene, polyester) 100–150 ksi (689–1034 MPa), glass 100–300 ksi (689–2067 MPa), and natural (sisal, coconut, wood) 60–100 ksi (413–689 MPa). In comparison, the tensile strength of concrete varies from 0.4 to 1 ksi (2.8–7 MPa). In addition, the strains that are needed to develop the tensile strengths are quite different for matrix and fibers. Hence, when the matrix cracks, the fibers have not developed their full strength, thus reducing their contribution to the composite strength. For stiff fibers such as steel and glass, the strains needed to develop full strength are relatively low compared

to polymeric fibers. Hence, for the same volume fraction and geometry of fibers, stiff fibers can provide better composite strength.

3. In most cases, the bond between fibers and matrix is not strong enough to develop the full fiber strength. Hence, a major fraction of the fibers pull out as opposed to fracture. The debonding and pull-out process is a major contributor to the ductility observed in the stress-strain behavior needed for structural components. The pulling out of fibers does reduce the strength of the composite at the expense of a more ductile performance. The factors that affect the bond are fiber length/diameter (l/d) ratios, commonly referred to as aspect ratios; fiber shapes, such as smooth, deformed, enlarged ends, or hooked ends; and the fiber type, such as steel, polymeric, or naturally occurring. Higher aspect ratios provide better performance; however, the aspect ratio has to be limited for practical purposes such as mixing and consolidation. Fibers that provide some kind of mechanical anchorage (hooked or enlarged ends) fare better than smooth fibers.

Since the failure caused by fiber pull-out is so common, the contribution from fibers is normally expressed as a function of bond strength (between fibers and matrix), rather than the fiber tensile strengths. The higher strains needed to fracture fibers (as explained in reason 2) is another reason that the fiber strengths are not utilized in the calculation of composite strengths.

Based on the aforementioned discussions, equation (3.10) can be modified as

$$\sigma_{cu} = \alpha V_m \sigma_{mu} + \beta \tau V_f \frac{l}{d} \tag{3.11}$$

where
$\quad l$ = length of the fiber
$\quad d$ = diameter of the fiber
$\quad \alpha, \beta$ = empirical constants
$\quad \tau$ = average bond strength

The coefficient α represents the strain-softening effect exhibited by concrete in the postcrack loading region. The constant β accounts for the effects of the spatial distribution of fibers and other simplifying assumptions such as those involved in defining average bond strength from a single fiber pull-out test. The constants α and β are determined using test results for a given fiber type and matrix composition.

To predict the postpeak, strain-softening response of composites, approaches based on a cohesive crack have been used. In such models, relationships between load and crack-opening displacement are derived from fiber pull out–slip response.

3.7   Experimental Evaluation of High–Volume Fraction Fiber Composites

A possible means of increasing the tensile-strength capacity and dramatically increasing the tensile-strain capacity of cementitious materials is by addition of fibers sufficiently higher than the critical volume. Currently, the standard procedure for FRC materials is limited to low-volume fiber contents (approximately 1% by volume), for which the contribution of fibers is mostly apparent in the postcracking response. That is, fibers do not enhance the tensile-strain capacity of the cementitious matrix. Recent research, however, indicates that as the volume fraction of fibers increases (up to 12–13%) and as fibers become more uniformly dispersed, they can hinder the growth of microcracks, suppress localization, and consequently substantially increase fracture strength and strain capacities of the matrix [3.23–3.25].

Several points must be considered in order to investigate such properties. The rule of mixtures has been widely used as a method of characterizing the response of multiphase materials. Properties of the constituents in a composite material are assumed to be independent of each other when using the rule of mixtures. In the presence of interactions such as the toughening of the matrix by fibers, traditional approaches [3.26, 3.28] would indicate a significant improvement in the overall contribution of the matrix, and a significant increase in the strength of the matrix.

Understanding how and why fibers alter fracture strength and strain capacities of a matrix may improve the design procedures of high-performance cement-based composites. Since the failure mechanisms in these composites are complex, test methods are required that can characterize the entire loading history. The following section deals with the following topics:

1. Interaction between the fibers and a quasi-brittle matrix and how the composite response is affected.

2. Experimental methodology to evaluate the evolution and accumulation of damage parameters associated with such failure mechanisms.

3. The theoretical formulations to model and predict the interaction phenomena and stress-strain responses of these composites.

Experimental evaluation of the toughening mechanisms of composites was achieved using the techniques of quantitative image analysis, laser holographic interferometry, and acoustic emission, which are described in the following subsections.

3.7.1 Interaction effects of fibers and matrix

The beneficial interaction between fibers and matrix is illustrated in Figure 3-14, in which the stress-strain curves of fiber–reinforced specimens subjected to uniaxial tension are shown. Fibrillated continuous uniaxial polypropylene fiber–reinforced concrete specimens with volume fractions of the fibers in the range of 8–13% were studied. In order to achieve such high volume fractions, special manufacturing techniques were used [3.26]. The stress-strain curves of these composites can be divided into two parts, an initially stiff region followed by a substantially less stiff region. The endpoint of the initial, stiff part of the curve is often termed the bend-over point (BOP). The stress-strain curves were observed to be essentially linear up to this point. The stress-strain responses of the matrix calculated from the response of the composite specimen are shown in Figure 3-15. In order to calculate the matrix stress-strain

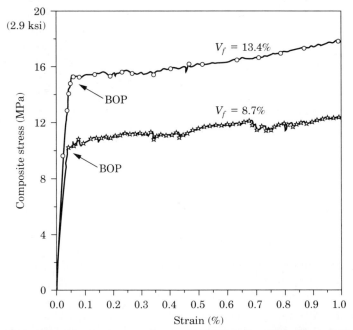

Figure 3-14 Stress-strain response of fibrillated polypropylene-fiber cement–based composites with fiber volume fractions of 8.7 and 13.4% [3.26, 3-27].

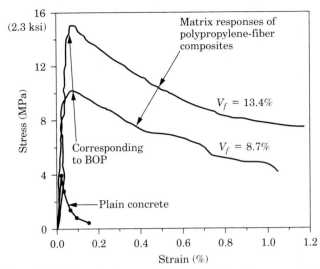

Figure 3-15   Stress-strain response of the matrix phase computed from the composite response [3.26].

curve, the contribution of the fibers (which was experimentally and independently determined, as explained in Reference [3.26]) subjected to an identical elongation was subtracted from that of the composite. The peak value of the matrix stress is observed to occur at about the BOP and depends on the volume fraction of fibers. This implied that macroscopically linear behavior of a matrix up to a stress level of 15 MPa (~2.1 ksi) may be the result of the suppression of localized cracking in the presence of fibers. Note that even at a strain level of 2%, the matrix capacity is almost 8 MPa (~1.2 ksi). Such behavior has been shown to exist for composites made with steel, glass, and polymeric fibers [3.26–3.28].

3.7.2   Quantitative image analysis

To observe accurately the internal microdamage, fiber-reinforced specimens were strained to a predetermined value. To avoid crack closures, these strains were mechanically frozen by gluing steel as indicated in Figure 3-16. The specimens were then vacuum-impregnated with epoxy containing fluorescent dye and prepared for thin-sectioning [3.23, 3.24].

The optical microscopic image was analyzed using a digital image analysis system consisting of a stereomicroscope, a high-resolution video camera, and an image analysis system. An image was represented with 512 × 512 pixels (Figure 3-17), and each pixel had an

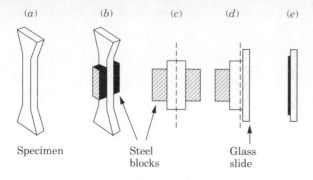

(a) Specimen   (b) Steel blocks   (c)   (d) Glass slide   (e)

(a) Load the specimen to a specified strain value.

(b) Attach steel blocks while under load.

(c) Vacuum-impregnate with epoxy dye.

(d) Attach to glass slide; cut, grind, and polish.

(e) View glass slide under UV fluorescence.

Figure 3-16 The procedure for the mechanical freezing of the microcracking and the thin-sectioning procedure [3.23, 3.24].

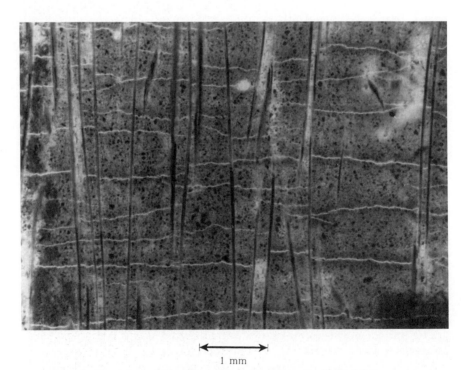

1 mm

Figure 3-17 A gray image of a specimen loaded to 1% strain [3.23, 3.24].

intensity value within the range of 0 to 255. To measure a feature of an image (e.g., microcracking), the gray-level image was initially enhanced by edge-enhancement filters. Since the microcracks have a higher pixel intensity than other components of the image, a binary image isolating the microcracks was created by setting a threshold intensity value. A binary image is simply an image in which each pixel is either black or white. The unwanted features within the binary image (i.e., pores) were then removed by means of filtering, thus leaving the microcracks as the only features present. Quantitative measurements of the microcracks were then obtained. A gray image of a thin section cut from a specimen of V_f = 12%, loaded up to 1% strain, is shown in Figure 3-17. The cracked state of the specimen can clearly be identified in the figure. Crack density, defined as the total crack perimeter per unit area of test section, can be calculated from such an image by developing operator-independent computer algorithms. Figure 3-18 represents the distribution of crack density plotted as a function of location on the section for a specimen subjected to 0.1% strain. The presence of highly damaged areas in the form of crack bands is quite evident. The crack density across the band is not homogeneous, which suggests that the cracks are propagating from one side of the specimen toward the other. The mean value of crack density S_v, defined as the center of mass for the microcrack surface density, is plotted as the function of strain level in Figure 3-19. Note that S_v increases with increasing strain level and is a measure of internal damage evolution caused by the applied loads.

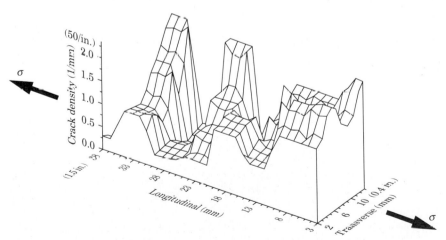

Figure 3-18 Spatial distribution of the crack density at 0.1% strain [3.23, 3.24].

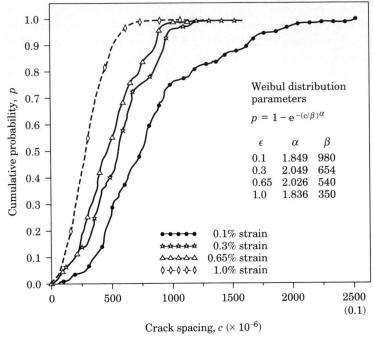

Figure 3-19 Crack density as a function of the applied strain [3.23, 3.24].

3.7.3 Laser holographic interferometry

The experimental setup for laser holographic interferometry is shown in Figure 3-20. Single-beam reflection holograms (Lippmann-Denisyuk) were used. A 10 mW He-Ne laser of wavelength 633 nm (red) and coherence length of approximately 80 mm was used. The light from the laser source is directed by the mirrors and used to illuminate the specimen. This light is referred to as the reference beam.

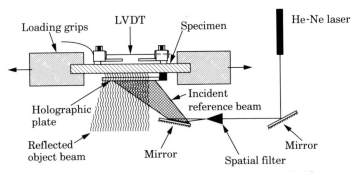

Figure 3-20 Reflection holographic interferometry setup [3.25].

A photographic plate is also placed in the path of the reference beam. The reference beam is reflected from the surface of the specimen and is then referred to as the object beam. The interference obtained from the reference beam and the object beam is recorded as a hologram on the photographic emulsion. An incremental displacement is applied to the specimen between the two exposures. Since interference takes place between the wave fronts reconstructed by the two holograms of the same object at different states of deformation, interferometric fringes are obtained by means of double-exposure holography. The interferometric fringes include all the displacement sustained between exposures. By analyzing the fringes recorded, both crack length and opening can be obtained [3.25]. With laser holography, it is possible to observe cracks as fine as one-quarter of a micrometer.

Crack propagation at different loading stages is indicated in Figure 3-21 for a composite with a 12% volume fraction of fibers. Cracks initiate at one side of the specimen at early stages of loading (point a in Figure 3-21a and d). These cracks tend to localize, and a critical crack propagating through the entire specimen is observed at point b, (the BOP, point b in Figure 3-21d). After the BOP, periodic cracks along the specimen length are developed (point c in Figure 3-21d). This is in agreement with variations in the microcrack surface density across the specimen. Crack spacing decreases with increasing value of strain. It is noted that the strain softening occurs after the BOP for plain concrete matrix, whereas the stress increases with increasing strain after the BOP when the matrix is reinforced by relatively high fiber volume fraction. This fact suggests that damage localization after the BOP in concrete can be reduced or eliminated by the use of fibers.

3.7.4   Acoustic emission

Techniques of acoustic emission (AE) are based on monitoring the stress waves that are generated by the rapid local redistribution of stresses that accompanies many damage mechanisms. The elastic waves, which originate at the source of activity, radiate away to the external surface of the specimen and can be detected by sensors attached to the body. If several sensors are used, then analysis of the waveforms can be used to locate the source of the event (e.g., microcracking) as well as to determine the nature of the damage mechanism (e.g., opening or sliding mode of cracking). Accurate analysis of an acoustic event may lead to distinguishing among matrix cracking, fiber debonding, and other mechanisms of damage. Acoustic emission waveforms were acquired by using piezoelectric transducers, in a digitization system with a 32 MHz frequency as shown in Figure 3-22. The details of acoustic emission analysis are given in Reference [3.24] and in Reference [3.29], for plain concrete.

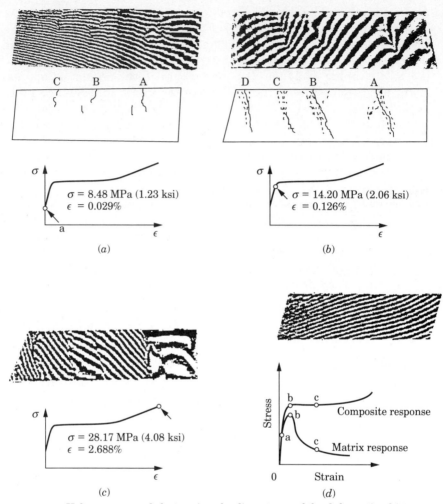

Figure 3-21 Holograms recorded at various loading stages of the deformation history.

Figure 3-23 shows the stress-strain response and the corresponding cumulative AE event count for specimens reinforced with two different volume concentrations of fibers. The higher the fiber content, the higher the AE event count will be. This would mean that the number of microcracks increases with an increasing volume of fibers. The maximum AE event rate seems to occur around the BOP. This rate decreases monotonically as the specimens are strained beyond the BOP. The relationship between the BOP and the AE event rate is shown in Figure 3-24. Note that the higher the volume fraction of fibers, the higher the stress level at the BOP will be, and the higher the AE rate.

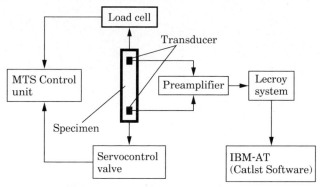

Figure 3-22 The experimental setup used for acoustic emission experiments [3.24].

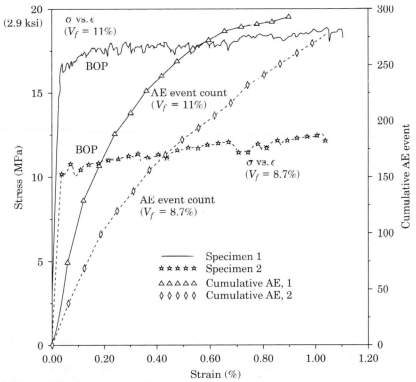

Figure 3-23 Stress-strain response and the cumulative acoustic emission events [3.24].

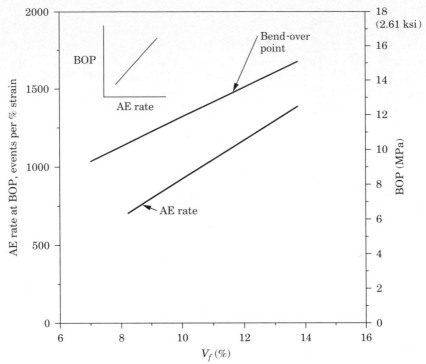

Figure 3-24   Effect of fiber volume fraction on the acoustic emission rate and the BOP [3.24].

Although the composite behaves macroscopically elastic prior to the BOP, microcracking exists in the linear region, and its rate is directly proportional to the volume fraction of fibers.

3.8   Fracture Mechanics Approach

3.8.1   Historical perspective

In the late 1800s theoretical formulations to predict the strengths of solids were based on interatomic forces. For a body in an unstressed state, rows of atoms maintain a specific distance from each other. This spacing is determined by the equilibrium that exists between the forces of attraction and repulsion. Theoretical strength can be estimated as the magnitude of applied force required to separate two adjacent layers of atoms from each other. Using the interatomic forces, the theoretical strengths of solids were determined as $E/10$, where E represents Young's modulus of the material. This magnitude, however, is two orders of magnitude larger than the experimental results for strength. It is clear that theories at the atomic level were unable to accurately predict the strength of bulk materials.

The phenomenon of fracture has been observed by all of us. Most simply it pertains to the separation of one solid into at least two through the creation of new surfaces. The explanation for the discrepancies between the theoretical and experimental results of the strengths of solids was provided by the work of Griffith in 1920 [3.30]. It was postulated that regardless of the processing techniques, materials contain inherent flaws. The stress concentration at the tip of these flaws results in stresses that are orders of magnitude greater than the stresses away from the crack tip. Hence failure occurs at the tip, resulting in the reduction of the overall strength of the material. The formation of new surfaces caused by fracture requires energy dissipation. Furthermore, as the crack forms, the material in the vicinity of the crack tip undergoes elastic recovery, resulting in a decrease in the strain energy of the system. A condition of equilibrium would exist if the rate of the release of strain energy is equal to the rate of the consumption of energy caused by crack surface creation. The Griffith criterion postulates that a crack will propagate when the decrease in the elastic strain energy is at least equal to the energy needed to create new surfaces associated with the crack. Using the elasticity solutions of Inglis, it was shown that the far-field stress (the stress away from a crack tip) required to cause the fracture of a material containing a flaw of size $2c$ is given by the equation

$$\sigma = \left(\frac{2E\gamma}{\pi c}\right)^{\frac{1}{2}} \tag{3.12}$$

where E and γ represent Young's modulus and the surface energy of the material [3.31]. Metals exhibit a significant degree of plastic yielding prior to fracture, the surface energy is therefore a minor component of energy dissipation. The strain energy release rate is therefore defined as the source of total energy required to cause the extension of the crack, that is, the energy required to increase the crack by some unit area. Once the strain energy release rate reaches a critical value, usually denoted as G_{IC}, an instability condition is reached and crack propagation occurs. This is shown by the straight line in Figure 3-25. The crack is stationary, until the strain energy release rate reaches G_{IC}, at which time the crack propagation occurs. Note that the single parameter G_{IC} is required to characterize the fracture toughness, and it can be measured using the peak load of a notched specimen tested under mode I condition. Quasi-brittle materials such as concrete exhibit energy dissipation owing to frictional sliding, aggregate interlock, and fracture surface tortuosity. After a crack is initiated, it propagates under stable conditions, and, as the aforementioned processes evolve, the energy requirement for the propagation of the crack increases. The material thus exhibits a certain degree of toughening

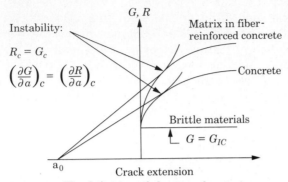

Figure 3-25 The definition of the G_{IC}, the strain energy release rate, and R, the fracture resistance curve.

owing to the formation of the zone commonly referred to as the fracture process zone. The stable crack growth and the increased toughness of the material are exhibited by means of an R-curve that correlates the increased toughness with the stable crack growth. Note that the condition for stable crack growth is defined as

$$G = R \qquad \frac{\partial G}{\partial a} < \frac{\partial R}{\partial a} \qquad (3.13)$$

The condition for crack instability can be defined as

$$G = R \qquad \frac{\partial G}{\partial a} = \frac{\partial R}{\partial a} \qquad (3.14)$$

The considerations required in obtaining both G and R are provided in the subsequent sections.

3.8.2   The compliance approach

A possible means of determining G, the strain energy releaserate, is by means of the compliance approach. Consider the case inwhich a notched specimen exhibits infinitesimal crack growthunder constant displacement, as shown in Figure 3-26a. Figure 3-26b is the load-deformation response of the specimenbefore and after the crack propagation. The shaded area represents the energy released because of the incremental crack growth. Equivalently, for an infinitesimal crack growth under a fixed load condition (Figure 3-26c), the load-deformation curve can be used to obtain the energy released, represented by the shaded area. Under both cases, the displacement of the

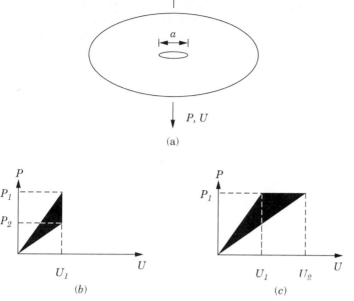

Figure 3-26 Specimen response representation for compliance approach: (a) definition of the compliance for a notched specimen; (b) crack growth under constant displacement; (c) crack growth under constant applied load.

body loaded with a single force P, in the same direction as the force is called U and is governed by

$$U = CP \qquad (3.15)$$

where C is the compliance of the body. The compliance is, among other parameters, a function of the crack size

$$C = C(a, \ldots) \qquad (3.16)$$

Note that in both instances shown in Figure 3-26 the energy released is equal, and the compliance of the specimen as defined by the inverse of the slope of the load elongation curve is increased. The strain energy release rate can be obtained as

$$\frac{1}{2}\frac{\partial C}{\partial a}P^2\, da = \text{energy released per unit of crack length} = G \quad (3.17)$$

Linear elastic fracture mechanics (LEFM) introduces the equations that describe the magnitudes and distribution of the stresses around the tip of a crack as a function of applied stresses, crack size and shape, and a parameter called the stress intensity factor K. This parameter describes the stress field at the vicinity of the crack, and, like the strain energy release rate, crack extension is imminent when it reaches a critical value K_c. Under plane strain conditions and using LEFM, it can be shown that

$$G_I = \frac{\pi K_I^2}{E(1 - \nu^2)} \tag{3.18}$$

where the subscript I represents an in-plane tensile opening mode, and ν represents Poisson's ratio.

3.9 Applications Based on Linear Elastic Fracture Mechanics

The following section presents the use of a single-parameter fracture mechanics approach in order to obtain the material parameters for various strength parameters. The fiber pull-out problem is initially discussed, and the parameter for interfacial strength is used in the subsequent section to predict the tensile strength of a cement-based composite.

3.9.1 Single-fiber pull-out problem

In the classical fiber pull-out test, shown in Figure 3-27, the debonding zone b is considered stress-free, at least compared with the bonded part of the interface. This means that the debonding zone can be characterized as an interfacial crack with a crack length $b(b < l)$ and a total surface of $2(2\pi ab)$. The fiber is assumed to be circular with diameter $2a$ and length l and linearly elastic with Young's modulus E_f

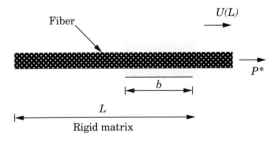

Figure 3-27 The classical fiber pull-out problem.

and Poisson's ratio ν_f. The matrix is also considered to be linearly elastic with the corresponding properties E_m and ν_m.

Using the fracture mechanics approach, the failure load of the composite is estimated using fiber pull-out strengths. When fiber debonds, the debonding zone is treated as an interfacial crack. Failure modes such as fiber yielding, fiber fracture, and further transverse matrix cracking are disregarded. It is assumed that a transverse crack has already been formed and that the debonding starts at the transversely cracked matrix surfaces.

A simple Griffith-type criterion for the crack growth load P_{cr} can then be written as

$$\frac{1}{2}\frac{\partial C}{\partial b}P_{cr}^2 = \frac{d\Omega}{db}\gamma_i \qquad 3.19$$

where γ_i is the interfacial specific work of fracture and Ω represents the crack surface. This criterion can be used to determine γ_i from the single-fiber pull-out test and subsequently used to determine the failure load of FRC composites.

The maximum pull out will be reached when the interfacial crack growth criterion of equation (3.19) is satisfied. That is, for the pull-out problem, using b as a crack length, we obtain

$$\frac{1}{2}\frac{\partial C}{\partial b}P_{cr}^2 = (2\pi a)2\gamma_i \qquad (3.20)$$

where γ_i is the specific interfacial work of fracture.

No exact solution is available for the $C = C(b, \ldots)$. However, partly based on approximate analysis using an asymptotic analysis as well as on finite element analysis, for steel fibers in a portland cement matrix, equation (3.21) can be used to determine the maximum pull-out load [3.32]. A more exact method of calculating compliance based on a shear-lag model is discussed in References [3.33, 3.34]. For simplicity of presentation, the effect of friction (after debonding) is ignored

$$\frac{1}{2}\frac{P_{cr}^2}{E_f\pi a^2} = (2\pi a)2\gamma_i \qquad (3.21)$$

Using equation (3.21), the specific interfacial work of fracture from a single-fiber pull-out can be determined. It is suggested that γ_i is a more accurate measure of bond strength than usually reported average bond strength (τ). An alternative approach in the modeling of pull-out may be achieved by modeling the interface as an elastic-plastic interfacial layer of zero thickness with properties significantly different from the bulk of matrix or fiber. The strain energy release rate for this geometry has been derived elsewhere [3.33]. The experimental procedure to obtain the interfacial parameters is provided in Reference [3.34].

3.9.2   Uniaxial tensile specimen

To predict the tensile strength of fiber–reinforced cement-based composites whose ultimate strength is dictated by the strength of the relatively weak fiber-matrix interface, the analyses described for the single-fiber pull-out case have been applied. For the specimen considered (Figure 3-28), it is assumed that (a) a macroscopic transverse matrix crack is formed, (b) interfacial cracks have initiated, and (c) the fibers are continuous (i.e., sufficiently long) and are aligned in the direction of the applied tensile stress. Only the case of steel fibers is presented, and friction is ignored.

The length of the specimen is $2l$ and the length of the uncracked (bonded) region is $2(l - b)$. Let the cross-section of the specimen be A and the volume fraction of fibers V_f. The compliance of the specimen shown in Figure 3-28 can be expressed as

$$C = 2\left[\frac{b}{V_f E_f A} + \frac{l - b}{E_c A} + H\right] \qquad (3.22)$$

where the first term describes the compliance of the debonded zone and the second term describes the compliance of the bonded region, with E_c defined as Young's modulus of the composite material as derived in equation (3.2). The third term H describes the compliance of the zone where the interfacial cracks end, and for simplicity it is assumed to be independent of b. The total interfacial crack surface can be written as

$$\Omega = 4bAV_f\frac{2}{a} \qquad (3.23)$$

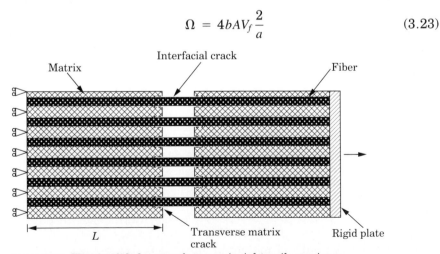

Figure 3-28   The simplified approach to a uniaxial tensile specimen.

Using the criterion proposed in equation (3.7), the following relationship for the interfacial crack growth and the tensile stress (σ_{cr}) is obtained

$$\frac{1}{2}2A\left[\frac{1}{V_fE_f} - \frac{1}{E_c}\right]\sigma_{cr}^2 = 4AV_f\frac{2}{a}\gamma_i \tag{3.24}$$

or

$$\sigma_{cr} = 2V_f\left[\frac{E_cE_f}{E_c - V_fE_f}\frac{2\gamma_i}{a}\right]^{\frac{1}{2}} \tag{3.25}$$

Note that if σ_{cr} is less than the tensile strength σ_u of the unreinforced matrix (which is approximately the same as the limit of proportionality for the commonly employed FRC composites), then the stress-strain curve labeled A in Figure 3-29 is observed for the composites. On the other hand, if σ_{cr} calculated from equation (3.25) is greater than σ_u, the stress-strain curve labeled B in Figure 3-29 is predicted from the proposed analysis. Curves of both these types have been reported for FRC composites, as shown in Figures 3-5 through 3-7. A more rigorous application of fracture mechanics for composites with high volume fractions of fibers is described in Section 3.11.

3.9.3 Comparisons with experimental results

The applicability of the above analysis predicting a composite's tensile strength was verified using experimentally observed data for steel fiber–reinforced concrete. The pull-out tests were conducted using straight and smooth steel fibers [3.35]. The average peak pull-out loads P_{cr} for fibers with an embedment length of about 13 mm and diameters of 0.4, 0.25, and 0.15 mm were observed to be 42.0, 26.5, and 6.2 N. For these values of P_{cr}, using equation (3.21), γ_i values of 13.3, 21.7, and 5.5 J/m^2 are obtained. Taking these values of γ_i, and for $V_f = 0.01$,

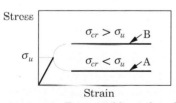

Figure 3-29 Two possible modes of failure for fiber-reinforced composites.

σ_{cr} values of 3.51, 5.67, and 3.68 MPa are obtained by using equation (3.25). These values of tensile strength for cement-based matrix reinforced with aligned fibers appear to be reasonable. Also note that the γ_i value of 6.22 J/m^2 for the steel fiber–matrix interface seems to be reasonable when compared with the values of 20 and 60 J/m^2 observed for unreinforced matrix in mode I crack propagation [3.36, 3.37].

Equation (3.25) points toward a simple and approximate method of determining the failure stress for fiber-reinforced cement composites. The value of γ can be determined from a single-fiber pull-out test. It has been shown that when the effect of friction is included, a single value of γ_i is obtained regardless of the embedment length of fibers. However, formulations based on average bond strength have a higher dependency on the embedment length [3.34].

3.10   Nonlinear Fracture Mechanics

Application of LEFM to cement-based composites was first attempted by Kaplan [3.38]. Shah and McGarry realized that because of the existence of a relatively large fracture process zone, LEFM cannot be directly applied to cement-based composites [3.39]. It was hence concluded that a single-parameter fracture toughness criterion cannot objectively describe the failure of quasi-brittle cementitious materials. The formation of a fracture process zone causes a significant amount of energy dissipation. Furthermore, because of stable crack growth, initial notch length cannot be used as the critical crack size. The size of the process zone is a function of the method of measurement, material, specimen size, and geometry of loading; and depending on the method employed, the zone has been measured from a few millimeters to roughly 50 centimeters [3.40]. For fiber-reinforced composites, the presence of fibers in the process zone and their contribution to the load-carrying capacity of the cracks increase the toughness of the composite by significantly affecting the characteristics of the process zone (see Figure 3-25). These mechanisms are not presently well understood, and the experimental approaches described in Section 3.7 are being used to provide further insight into such complex mechanisms.

Several approaches using two or more fracture parameters have been proposed recently for the fracture of plain concrete. These approaches can be categorized as cohesive crack models and equivalent-elastic fracture models (or effective crack models). In cohesive crack models, the fracture process zone is modeled by applying traction forces across the surfaces of newly formed cracks. These forces are due to processes such as aggregate interlock, grain bridging, or crack tortuosity; they tend to close the crack and decrease G, the strain energy release rate. The first application of cohesive crack models was by Hillerborg, Modeer, and Petersson, who introduced the fictitious crack model [3.41].

The form of uniaxial tensile stress-separation curve is assumed a priori. The fracture parameters are obtained by using f_t' for the tensile strength, w_c the critical separation, and G_c the fracture energy.

The effective crack models are based on representing the fracture process through an elastically equivalent effective crack [3.42, 3.43]. The fracture properties of the elastically equivalent crack can be obtained by using various approaches, as discussed later, among which are the size-effect law proposed by Bazant, Kim, and Pfeiffer and the two-parameter model proposed by Jenq and Shah [3.44, 3.45]. The Jenq and Shah model can be extended to the case of fiber-reinforced cementitious composites and will be described in detail.

3.10.1 Two-parameter fracture model of Jenq and Shah

To apply the principles of fracture mechanics, one has to take into account the stable crack growth region. The fracture model proposed by Jenq and Shah is based on the effective Griffith crack approach, which incorporates stable crack growth into an equivalent elastic crack [3.46]. It is assumed that an initial crack of length a_0 propagates steadily until it reaches an effective crack length of a_c. This critical crack length is computed so that the opening displacement (CTOD) of the initial crack tip calculated using LEFM at the peak load is equal to a critical value. Jenq and Shah as well as others have shown that based on this approach the material properties that are geometry- and size-independent can be represented by critical crack tip–opening displacement (CTOD_c) at the tip of the initial crack a_0, and the critical stress-intensity factor K_{IC}^s at the tip of the effective crack a_c. Figure 3-30 represents the definition of the terms. These two parameters can be obtained from a notched-beam test. The critical crack

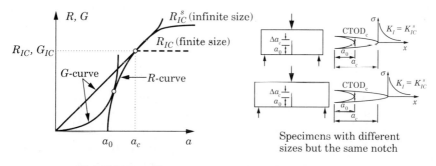

Material properties:
      critical stress-intensity factor—K_{IC}^s
      critical crack tip–opening displacement—CTOD_c

Figure 3-30 Definition of the R-curve and the parameters for the two-parameter fracture model.

length a_c is obtained by comparing the initial compliance with that obtained by unloading the specimen immediately after the peak load is reached in a crack mouth–opening displacement (CMOD)-controlled test. From these two compliances and the value of the maximum load, one can determine K_{IC}^s and CTOD$_c$ using the relevant LEFM formula. The specification for obtaining these two parameters are proposed as a standard by the RILEM Committee on Fracture Mechanics Testing [3.47].

The parameters K_{IC}^s and CTOD$_c$ can be used to obtain the parameters to an R-curve (fracture resistance curve). The R-curve, which is material-, geometry-, and size-dependent, is a convenient means of studying the stable crack growth and the toughening effects. It can furthermore be used to predict the load-displacement response and the toughness effects of various fiber types.

3.10.2   *R-curve formulations*

To predict the geometrical and size dependency of fracture parameters in a given material, an R-curve (fracture resistance curve) approach can be used. The prediction of the toughening mechanism owing to fiber reinforcement can also be incorporated in the same context. The objective is to model the possible toughening of a single, growing crack in the presence of fiber. No attempt is made to model possible toughening mechanisms such as multiple cracking and crack shielding. Based on experimental observation, a single-notch specimen subjected to uniaxial tension is considered.

Definition of the R-curve based on equations (3.13, 3.14) is shown in Figure 3-30. The R-curve is defined as the locus of fracture energy release rates G for differently sized specimens with the same initial crack length [3.48, 3.49]. Because of the existence of stable crack growth, it is assumed that the critical crack length a_c is proportional to the initial crack length a_0 (see Figure 3-30), that is,

$$a_c = a_0 + \Delta a_c = \alpha A_o \qquad (3.26)$$

where Δa_c is the precritical stable crack growth; α is a parameter representing the ratio of effective critical crack length to initial notch length for a given loading geometry. Based on the above formulation it can be shown that the R-curve can be approximately derived as

$$R = G_c = \beta(a - a_0)^d \qquad (3.27)$$

$$d = \frac{1}{2} + \frac{\alpha - 1}{\alpha} - \left[\frac{1}{4} + \frac{\alpha - 1}{\alpha} - \left(\frac{\alpha - 1}{\alpha}\right)^2\right]^{\frac{1}{2}} \qquad (3.28)$$

Equations (3.25, 3.26) are an extension to the R-curve formulation used by Broek [3.50], and it can be shown that they are geometry-

and size-dependent. Note that according to this approach the resistance to fracture is a function of the precritical crack growth, whereas the material properties and the geometrical effects are included in parameters α and β.

3.10.3   R-curve formulation for cement-based composites

For plain mortar or concrete the values of α and β can be determined based on matrix properties K_{IC}^s and CTOD_c introduced by Shah [3.47]. The critical crack length a_c is obtained by comparing the initial compliance with that obtained by unloading the specimen immediately after the peak load is reached in a crack mouth–opening displacement (CMOD)-controlled test. From these two compliances and the value of the maximum load, one can determine K_{IC}^s and CTOD_c using the relevant LEFM formula.

Because of the existence of fibers, stress intensity factor and crack tip–opening displacement consist of contributions from far-field stresses and closing pressure. The closing pressure reduces the stress intensity factor and crack opening displacement at the crack tip. Therefore, the following failure criteria are obtained

$$K_{IC}^s = K_I^m + K_I^f \qquad (3.29)$$

and

$$\text{CTOD}_c = \text{CTOD}_m + \text{CTOD}_f \qquad (3.30)$$

where K_I^m and CTOD_m are the stress intensity and crack tip–opening displacement at the tip of an effective traction-free crack in the matrix caused by far-field stress. The terms K_I^f and CTOD_f are the stress intensity and crack tip–opening displacement caused by the closing pressure of fibers and represent negative magnitudes. For a plain matrix, both K_I^f and CTOD_f are zero, and equations (3.27, 3.28) reduce to Jenq and Shah's fracture model. Since there is no closed-form solution for the calculation of the stress intensity factor caused by an arbitrary shape of closing pressure, the K_I value of a crack in a strip of unit thickness subjected to a unit point load was used as Green's function to calculate K_I^f. On the substitution of appropriate parameters into Equations (3.30, 3.31), the failure criterion can be defined as

$$K_{IC}^s = K_I^m\left(\sigma_c, \frac{a_c}{b}\right) - \int_0^{a_c} p\left(\frac{x}{a_c}\right) K_I^f\left(\frac{x}{a_c}\right) dx \qquad (3.31)$$

$$\text{CTOD}_c = \text{CTOD}_m\left(\sigma_c, \frac{a_c}{b}, \frac{a_0}{a_c}\right) - \int_0^{a_c} p\left(\frac{x}{a_c}\right) Q\left(\frac{a_c}{b}, \frac{a_0}{a_c}, \frac{x}{a_c}\right) dx \qquad (3.32)$$

where σ_c represents the critical far-field stress (i.e., the BOP or ultimate tensile strength of a matrix) and a_c is the critical crack length. Parameter Q represents Green's function for the closure of a crack of length a_c/b measured at the initial crack mouth a_0/a_c, caused by a unit force applied at an arbitrary point x/a_c along the crack face. The term $p(x/a)$ represents the distribution of closure forces along the crack face. In this formulation the experimental pull-out slip response of fibers is used as the closing pressure. The term $K_I^f(x/a_c)$ is the stress intensity caused by a unit load applied at an arbitrary point x. The terms K_I^m, CTOD_m, K_I^f, and Q can be obtained based on LEFM [3.51].

Equations (3.31, 3.32) present two coupled nonlinear integral equations and must be solved for the two unknowns a_c and σ_c. Once a_c is known, the values of α and β can be calculated from

$$a_c = \alpha a_0 \tag{3.33}$$

$$R_{IC}^s = \frac{(K_{IC}^m)^2}{E_m} = \beta\,(\alpha a_0 - a_0)^d \tag{3.34}$$

where E_m is Young's modulus of a matrix.

An R-curve for a toughened matrix is determined once the values of α and β are known. The load-displacement relation can thus be computed according to the energy balance condition, $R = G$, where G represents the energy release rate during a crack propagation. A typical specimen for experimental testing is shown in Figure 3-31.

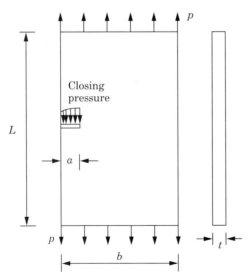

Single notch specimen subjected to tension

Figure 3-31   Uniaxial tensile specimen used and the fiber closing pressure applied on the crack surfaces.

The effect of fibers is modeled by a closing pressure on the crack surface. The fiber pull-out versus slip curve, obtained by Li, Mobasher, and Shah, was used as the closing pressure, which includes the debonding of the fiber [3.34]. Figure 3-32 represents the experimental results of the fiber pull-out for steel and glass fiber strands. Note that as the load increases the response deviates from linearity, suggesting the formation of a debonding zone. The slope of the ascending portion characterizes the elastic properties of the interface and was found to be independent of fiber length. The peak load is dependent on the type of fiber used and the interface conditions. The postpeak region of the loading history is characterized by a significant energy dissipation owing to the frictional forces of pull-out. A single fiber pull-out force was converted to a distributed pressure by dividing the force by the specific fiber spacing.

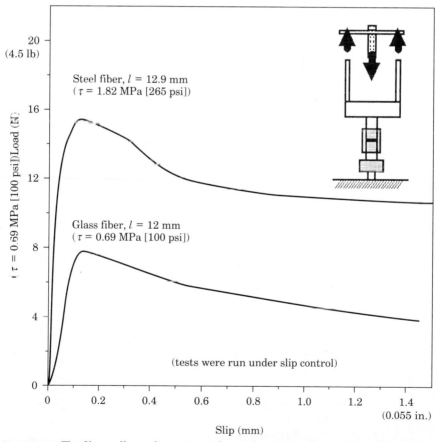

Figure 3-32 The fiber pull-out slip response for steel and glass fibers used as the closing pressure [3.34].

3.10.4   Comparison of theoretical and experimental results

Steel fiber composites.   The experimental results of matrix response in the presence of steel fiber were obtained from the work in Reference [3.52]. Specimens tested were $305 \times 25 \times 9.5$ mm, using a mortar mix of $1:2:0.5$ (cement : sand : water). Continuous steel wires of diameter 1.19 mm were used as the reinforcement. Specimens were tested in tension, and elongations were measured using a LVDT mounted across a gage length of 153 mm. Since required matrix properties for the mortar mix were not available, the average values of material properties (K_{IC}^s, CTOD_c, and E_m) from two mortars, which have the mix proportions of $1:2.6:0.65$ and $1:2.6:0.45$ (cement : sand : water), respectively, and obtained from a three-point bending specimen [3.46], were used as an input for theoretical solutions. For theoretical prediction, both σ_c and a_c were first solved from equations (3.31, 3.32). The R-curve can then be determined after the values of α and β are calculated according to equation (3.34). The theoretical bend-over point is defined as the ultimate strength of the matrix in the presence of the fibers. Figure 3-33 gives comparisons of the theoretically predicted strength at the BOP with the experimental data.

Glass fiber composites.   Figure 3-34 represents the theoretical (References [3.24, 3.53, and 3.54]) and experimental [3.46] effects of fiber volume fraction on the BOP for glass–fiber reinforced composites. The

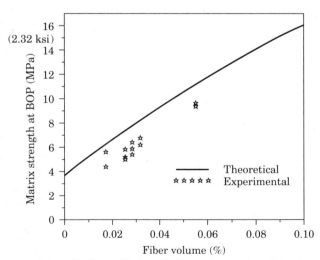

Figure 3-33   The theoretical prediction and experimental values for steel fiber composites.

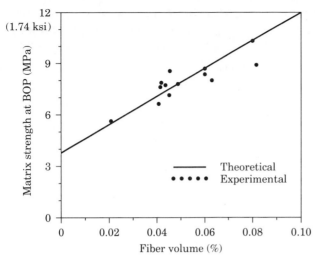

Figure 3-34 The theoretical prediction and experimental values for glass fiber composites [3.24].

theoretical results were obtained by using the material parameters (K_{IC}^s and $CTOD_c$) for mortar from the notched-beam tests. Note that although varying mix proportions were used by different investigators, the theoretical curve corresponding to only one mortar is shown.

3.11 Other Models for the Prediction of Failure Stress

Several theoretical models have been suggested for relating the properties of the fiber, matrix, and the interface with the strength of the composite. These models can be categorized according to the theory of multiple fracture, composite models, strain-relief models, interface mechanics models, micromechanics models, and fracture mechanics models. A review of the various models can be found in Reference [3.37].

Using an energy balance approach, Aveston, Cooper, and Kelly developed a model (ACK model) to predict the occupance of multiple cracking in fiber reinforced composites [3.7]. Using a constant frictional shear stress distribution at the interface, the model predicts the increase in the matrix fracture strain resulting from the presence of fibers. This increase can be obtained by

$$\epsilon_{mu} - \left[\frac{12\tau\gamma_m E_f V_f^2}{E_c E_m^2 r V_m} \right]^{1/3} \tag{3.35}$$

where E and V represent Young's modulus and the volume fraction, respectively; subscripts m, f, and c represent matrix, fiber, and composite, respectively, τ represents the interfacial shear strength, and

$$\gamma_m = 0.5 \frac{(K_{IC})^2}{E_m} \tag{3.36}$$

represents the surface fracture energy of the matrix. Note that based on the ACK model the strength is linearly proportional to the fracture strain, indicating a linear elastic response. Furthermore, a steady-state cracking is assumed at the BOP. Recently, multiple fracture formulations of fiber composites are proposed that provide solutions for the steady-state matrix-cracking stress using a fracture mechanics energy approach and various fiber-matrix interface conditions; [3.55, 3.56]. A stress intensity approach has been used, however, the stable crack-growth modeling is avoided by considering the two cases of short cracks and the long cracks approaching a steady-state condition [3.57]. The effect of fibers across a crack was modeled by means of a bridging force. Solutions for fully or partially bridged cracks in transversely isotropic materials have been provided [3.58]. The effect of interfacial frictional traction on the toughness properties of ceramics has been studied [3.59].

Fracture mechanics–based approaches in cementitious materials have been used [3.60]. And a fracture mechanics–approach based on the two-parameter fracture model has also been proposed [3.46]. The closing pressure of the fibers is also included in the process zone, and the R-curves for the composites are developed.

Strain-relief models, which are based on the approach developed by Irwin for unreinforced materials, have been extended to composite materials [3.61]. In these models, it is assumed that the strain energy released by formation of a transverse Griffith crack is contributed only by an elliptical zone surrounding the crack. For a crack of a given length, energy contributed by this elliptical zone is compared with the energy required to form new cracked surfaces. The elastic strain energy in the matrix and the fibers, and the energy absorbed in the debonding (assuming a constant shear stress at the interface and a frictional type of bond), are included in the analysis. The computation of elastic strain energy is achieved by breaking the elliptical zone into strips and integrating over the entire volume. The strain field in the ellipsoid is assumed a priori. For example, a piecewise linear strain field within the elliptical zone has been assumed [3.62]. Other formulations for the strain field approximation are also available [3.62]. The rate of energy released is computed by incrementing the crack by a small length and computing the change in the strain energy released. The mismatch between the fiber strain and matrix strain results in frictional energy dissipation. A condition of failure is defined when the far-field strain reaches magnitudes such that the energy released is higher than the energy absorbed by the friction.

The composite materials models have been developed for random two- and three-dimensional fiber orientations [3.63]. And the model has been extended to various interface conditions [3.64]. A probabilistic approach toward the strength of fiber composites has been used [3.65]. This model utilizes the weakest-link approach for a tensile specimen. After the matrix has cracked and the fiber pull-out is the operating mechanism, empirical parameters to the rule of mixtures as described in Section 3.3 are obtained. Statistical approaches to FRC fracture have been developed that simulate a series of noninteracting cracks in a tension specimen using a Weibull weakest-link theory [3.66]. Damage mechanics–based models present yet another approach in the formulation of the composite response [3.67]. Although most models were developed for use with steel fibers, studies have also been conducted for glass, polymeric, and natural fibers. The stress-strain behavior of these fibers is different from the stress-strain behavior of steel, and special formulations are therefore needed to predict the strength of a composite containing these fibers [3.68].

It should be noted that we are just beginning to understand the interaction between fibers and quasi-brittle matrices. All the models developed so far represent only part of reality. For example, the R-curve model described earlier does not include possible energy absorbed during debonding. Depending on the fiber parameters, debonding at the fiber-matrix interface may take place prior to reaching the BOP [3.69]. Much more research is needed before a general model can be developed.

3.12   References

3.1   Shah, S. P.; Stroeven, P.; Dalhuison, D.; and Van Steleelenburg, P. "Complete Stress-Strain Curves for Steel Fiber–Reinforced Concrete in Uniaxial Tension and Compression," Proceedings, RILEM Symposium on Testing and Test Methods of Fiber Cement Composites, Construction Press, U.K. 1978, pp. 399–408.

3.2   Baggot, R., "Polypropylene Fiber Reinforcement of Lightweight Cementitious Matrices," International Journal of Cement Composites and Lightweight Concrete, Vol. 5, No. 2, 1983, pp. 105–114.

3.3   Mobasher, B.; and Shah, S. P. "Test Parameters in Toughness Evaluation of Glass Fiber Reinforced Concrete Panels," ACI Materials Journal, Vol. 86, No. 5, 1989, pp. 448–458.

3.4   Shah, S. P.; and Rangan, V. B. "Fiber Reinforced Concrete Properties," Journal of American Concrete Institute, Vol. 68, No. 2, 1971 pp. 126–135.

3.5   Lawrence, P. "Some Theoretical Considerations of Fiber Pullout from Elastic Matrix," Journal of Materials Science, Vol. 7, No. 1, 1972, pp. 1–6.

3.6   Nemat-Nasser, S; and Hori, M. "Toughening by Partial and Full Bridging of Cracks in Ceramics and Fiber Reinforced Composites," Mechanics of Materials, Vol. 6, 1986, pp. 245–269.

3.7   Aveston, J; Cooper, G. A.; and Kelly, "Single and Multiple Fracture," *The Properties of Fiber Composites,* Conference Proceedings of the National Physical Laboratory, Guildford, U.K. IPC Science and Technology Press 1971, pp. 15–26.

3.8   McCartney, L. N. "Mechanics of Matrix Cracking in Brittle-Matrix Fiber Reinforced Composites," Proceedings, *Royal Society of London,* A409, 1987, pp. 329–350.

3.9   Budiansky, B.; Hutchinson, J. W.; and Evans, A. G. "Matrix Fracture in Fiber Reinforced Ceramics," *Journal of Mechanics, Physics, and Solids,* Vol. 343, No. 2, 1986, pp. 167–189.

3.10 Marshall, D. B.; Cox, B. N.; and Evans, A. G. "The Mechanics of Matrix Cracking in Brittle Matrix Fiber Composites," *Acta Metallurgy,* Vol. 33, No. 11, 1985, pp. 2013–2021.

3.11 Jenq, Y. S.; and Shah, S. P. "Crack Propagation in Fiber-Reinforced Concrete," ASCE, *Journal of Structural Engineering,* Vol. 112, No. 1, 1986, pp. 19–34.

3.12 Mobasher, B.; Ouyang, C.; and Shah, S. P. "Modelling of Fiber Toughening in Cementitious Materials Using an R-Curve Approach," *International Journal of Fracture,*. Vol. 50, 1991, pp. 199-219.

3.13 Gopalaratnam, V. S.; and Shah, S. P. "Tensile Failure of Steel Fiber-Reinforced Mortar," ASCE, *Journal of Engineering Mechanics Division,* Vol. 113, No. 5, May 1987, pp. 635–652.

3.14 Gopalaratnam, V. S.; and Shah, S. P. "Softening Response of Plain Concrete in Direct Tension," *Journal of American Concrete Institute,* Vol. 82, No. 3, 1985, pp. 310–323.

3.15 Shah, S. P. "Do Fibers Increase Tensile Strength of Cement-Based Matrices," *ACI Materials Journal,* Vol. 88, No. 6, 1991, pp. 595–602.

3.16 Krenchel, H., *Fibre Reinforcement,* Akademisk Forlag, Copenhagen, 1964, p. 158.

3.17 Laws, V., "The Efficiency of Fibrous Reinforment of Brittle Matrices, " *Journal of Physics D: Applied Physics,* Vol. 4, 1971, pp. 1737–1746.

3.18 Aveston, J.; Mercer, R. A.; and Sillwood, J. M. "Fiber Reinforced Cements—Scientific Foundations and Specifications." In Proc. NPL Conf. on Composites—Standards of Testing and Design, April 1974. IPC Science and Technology Press, Guildford, U.K., p. 93.

3.19 Laws, V.; Lawrence, P.; and Nurse, R. W. B. "Reinforcement of Brittle Material by Glass Fibers, " *Journal of Physics, Series D: Applied Physics,* Vol. 6, No. 19-B, 1973, pp. 523–37.

3.20 Brandt, A. M. "On the Optimization of the Fiber Orientation in Cement Based Composite Materials," Seminar Proceedings, Present State of Investigations and Applications of Fiber Reinforced Cement Based Materials, RILEM—Polish Academy of Sciences, Cracow Tech. University, September 1983, pp. 13–22.

3.21 Li, V.; Wang, Y.; and Becker, S. "Effect of Inclining Angle, Bundling, and Surface Treatment on Synthetic Fiber Pull-out from a Cement Matrix," *Composites,* Vol. 21, No. 2, 1990, pp. 132–140.

3.22 Hannant, D. J. *Fibre Cements and Fibre Concretes,* Wiley, Chicester, U.K., 1978, p. 219.

3.23 Stang, H.; Mobasher, B.; and Shah, S. P. "Quantitative Damage Characterization in Polypropylene Fiber Reinforced Concrete," *Cement and Concrete Research,* Vol. No. 20, 1990, pp. 540–558.

3.24 Mobasher, B.; Stang, H.; and Shah, S. P. "Microcracking in Fiber Reinforced Concrete," *Cement and Concrete Research,* Vol. 20, 1990, pp. 665–676.

3.25 Mobasher, B.; Castro-Montero, A.; and Shah, S. P. "A Study of Fracture in Fiber Reinforced Cement-Based Composites Using Laser Holographic Interferometry," *Experimental Mechanics,* Vol. 30, 1990, pp. 286–294.

3.26 Krenchel, H.; and Stang, H. "Stable Microcracking in Cementitious Materials," Brittle Matrix Composites, Proceedings, Second International Symposium on Brittle Matrix Composites (BMC2), Cedzyna, Poland, September 1988, pp. 20–33.

3.27 Mobasher, B. "Reinforcing Mechanism of Fibers in Cement Based Composites" Ph.D. Dissertation, Northwestern University, Evanston, Illinois, June 1990.

3.28 Mobasher, B.; and Shah, S. P. "Interaction between Fibers and the Cement Matrix in Glass Fiber Reinforced Concrete," *Thin-Section FRC and Ferrocement,* ACI Special Publication SP-124, 1990, pp. 137–148.

3.29 Maji, A.; Ouyang, C.; and Shah, S.P. "Fracture Mechanisms of Quasi-Brittle Materials Based on Acoustic Emission," *Journal of Materials Research,* Vol. 5, 1990, No. 1, pp. 206–217.

3.30 Griffith, A. A. "The Phenomena of Rupture and Flow in Solids," *Philosophical Transactions Royal Society of London,* A221, 1920, pp. 163–98.

3.31 Inglis, C. E. "Stresses in a Plate Due to the Presence of Cracks and Sharp Corners," Proceedings of Institute of Naval Architecture, 55, 1913, pp. 219–30.

3.32 Stang, H.; and Shah, S. P. "Failure of Fiber Reinforced Concrete by Pull-out Fracture," *Journal of Material Science,* 21 No. 3, 1986, pp. 953–957.

3.33 Stang, H.; Li, Z.; and Shah, S. P. "Pull-Out Problem: Stress versus Fracture Mechanical Approach," ASCE, *Journal of Engineering Mechanics,* Vol. 116, No. 10, pp. 2136–2150, 1990.

3.34 Li, Z.; Mobasher, B.; and Shah, S. P. "Characterization of Interfacial Properties of Fiber Reinforced Cementitious Composites," *Journal of the American Ceramic Society,* Vol. 74, No. 9, 1991, pp. 2156–64.

3.35 Shah, S. P.; and Naaman, A. E. "Mechanical Properties of Steel and Glass Fiber Reinforced Concrete," *Journal of American Concrete Institute,* 73, No. 1, 1976, pp. 50–53.

3.36 Jenq, Y. S.; and Shah, S. P. "Fracture Toughness Criteria for Concrete," *Engineering Fracture Mechanics,* Vol 21, No. 5, 1985, pp. 1055–69.

3.37 Gopalaratnam, V. S.; and Shah, S. P. "Failure Mechanisms and Fracture of Fiber Reinforced Concrete," *Fiber Reinforced Concrete Properties and Applications,* American Concrete Institute SP 105 1987 pp. 1–26.

3.38 Kaplan, M. F. "Crack Propagation and Fracture of Concrete," *Journal of the American Concrete Institute,* Vol. 58, No. 5, November 1961.

3.39 Shah, S. P.; and McGarry, F. J. "Griffith Fracture Criterion and Concrete," ASCE, Journal of Engineering Mechanics Division, December 1971, pp. 1663–1675.

3.40 Mindess, S. "The Fracture Process Zone in Concrete," in *Toughening Mechanisms in Quasi-Brittle Materials,* S. P. Shah (ed)., Kluwer, Academic, Norwell, Massuchusetts, 1991, pp. 271–286.

3.41 Hillerborg, A.; Modeer, M., and Petersson, P. E. "Analysis of Crack Formation and Crack Growth in Concrete by Means of Fracture Mechanics and Finite Elements," *Cement and Concrete Research,* Vol. 6, No. 6, 1976, pp. 773–782.

3.42 Karihaloo, B. L.; and Nallathambi, P. "Determination of Specimen-Size Independent Fracture Toughness of Plain Concrete," *Magazine of Concrete Research,* Vol. 38, 1986, pp. 67–76.

3.43 Planas, J.; and Elices, M. "Conceptual and Experimental Problems in the Determination of the Fracture Energy of Concrete," Proceedings, International Workshop on Fracture Toughness and Fracture Energy—Test Methods for Concrete and Rock, Tohoku University, Sendai, Japan, October 12–14, 1988.

3.44 Bazant Z. P.; Kim J. K.; and Pfeiffer, P. A. "Nonlinear Fracture Properties from Size Effect Tests," ASCE, *Journal of Structural Engineering,* Vol. 112, No. 2, 1986, pp. 289–307.

3.45 Bazant, Z. P.; Kazemi, M. T. "Determination of Fracture Energy, Process Zone Length and Brittleness Number from Size Effect, with Application to Rock and Concrete," *International Journal of Fracture,* Vol. 44, 1990, pp. 111–131.

3.46 Jenq, Y. S.; and Shah, S. P. "A Two-Parameter Fracture Model for Concrete," ASCE, *Journal of Engineering Mechanics,* Vol. 111, No. 10, 1985, pp. 1227–1241.

3.47 Shah, S. P. "Determination of Fracture Parameters (K_{IC}^s and $CTOD_c$) of Plain Concrete Using Three Point Tests," Proposed RILEM Recommendation, *Materials and Structures,* Vol. 23, 1990, pp. 457–60.

3.48 Ouyang, C.; Mobasher, B.; and Shah, S. P. "An R-Curve Approach for Fracture of Quasi-Brittle Materials," *Engineering Fracture Mechanics,* Vol. 37, No. 4, 1990, pp. 901–13.

3.49 Ouyang, C., and Shah, S. P.; "Geometry Dependent R-Curve for Quasi-Brittle Materials," *Journal of American Ceramic Society,* Vol. 74, No. 11, 1991, pp. 2831–36.

3.50 Broek, D. *Elementary Engineering Fracture Mechanics,* 4th Edition, Martinus Nijhoff, Hingham, Massachusetts, 1987, pp. 185–200.

3.51 Tada, H.; Paris, P. C.; and Irwin, G. R. *The Stress Analysis of Cracks Handbook,* 2nd Edition, Paris Production, St. Louis, Missouri 1985.

3.52 Somayaji, S.; and Shah, S. P. "Bond Stress versus Slip Relationship and Cracking Response of Tension Members," *ACI Journal,* Vol. 78, No. 3, 1981, pp. 217–225.

3.53 Oakley, D. R.; and Proctor, B. A. "Tensile Stress-Strain Behavior of Glass Fiber Reinforced Cement Composites," *RILEM Symposium on Fiber Reinforced Cement & Concrete,* A. Neville (ed.), 1975, pp. 347–359.

3.54 Ali, M. A.; Majumdar, A. J.; and Singh, B. "Properties of Glass-fiber Cement—The Effect of Fiber Length and Content," *Building Research Establishment Current Paper,* CP 94/75, October 1975.

3.55 McCartney, L. N. "Mechanics of Matrix Cracking in Brittle-Matrix Fiber Reinforced Composites," *Proceedings of Royal Society of London* 409A, 1987, pp. 329–350.

3.56 Budiansky, B.; Hutchinson, J. W.; and Evans, A. G. "Matrix Fracture in Fiber-Reinforced Ceramics," *Journal of Mechanics and Physics of Solids,* Vol. 34, No. 2, 1986, pp. 167–189.

3.57 Marshall, D. B.; Cox, B. N.; and Evans, A. G. "The Mechanics of Matrix Cracking in Brittle-Matrix Fiber Composites," *Acta Metallurgy,* Vol. 33, No. 11, 1985, pp. 2013–2021.

3.58 Nemat-Nasser, S.; and Hori, M. "Toughening by Partial or Full Bridging of Cracks in Ceramics and Fiber Reinforced Composites," *Mechanics of Materials,* Vol. 6, 1986, pp. 245–269.

3.59 Bennison, S. J.; and Lawn, B. R. "Role of Interfacial Grain-Bridging Sliding Friction in the Crack-Resistance and Strength Properties of Non-Transforming Ceramics," *Acta Metallurgy,* Vol. 37, No. 10, 1989, pp. 2659–2671.

3.60 Foote, R. M. L. "Crack Growth Resistance Curves in Fiber Cements," Ph.D. Dissertation, University of Sidney, 1986.

3.61 Hannant, D. J.; Hughes, D. C.; and Kelly, A. "Toughening of Cement and of Other Brittle Solids with Fibers," *Philosophical Transactions,* Royal Society of London, Vol. 310A, pp. 175–190, 1983.

3.62 Korczynskyj, Y.; Harris, S. J.; and Morley, J. G. "The Influence of Reinforcing Fibers on the Growth of Cracks in Brittle Matrix Composites," *Journal of Material Science,* Vol. 16, 1981, pp. 1533–1547.

3.63 Cox, H. L. "The Elasticity and Strength of Paper and Other Fibrous Materials," *British Journal of Applied Physics,* Vol. 3, March 1952, pp. 72–79.

3.64 Piggot, M. R. *Load Bearing Fiber Composites,* Pergamon, U.K. 1980.

3.65 Naaman, A. E.; Moavenzadeh, F.; and McGarry, F. J. "Probabilistic Analysis of Fiber Reinforced Concrete, ASCE, *Journal of Engineering Mechanics Division,* Vol. 100, No. EM2, April 1974, pp. 397–413.

3.66 Hu, X. Z.; Mai, Y. W.; and Cotterell, B. "A Statistical Theory of Time-Dependent Fracture for Brittle Materials, in *Fracture of Concrete and Rock,* S. P. Shah and S. E. Swartz (eds.), Springer, Berlin, pp. 37–46.

3.67 Stang, H. "Mathematical Modelling of Damage Evolution in Concrete and FRC Materials," SEM/RILEM International Conference on Fracture of Concrete and Rock, Houston, Texas, June 1987, S. P. Shah, and S. E. Swartz, (eds.), pp. 158–169.

3.68 Soroushian, P. and Lee, C. D. "Constitutive Modeling of Steel Fiber Reinforced Concrete under Direct Tension and Compression," *Fiber Cements and Concretes—Recent Developments,* Elsevier, 1989, pp. 363–377.

3.69 Yang, C. C.; Mura, T.; and Shah, S. P. "Micromechanical Theory and Uniaxial Tensile Tests of Fiber Reinforced Cement Composites," *Journal of Materials Research,* Vol. 6., No. 11, 1991, pp. 2463-2473.

Basic Concepts and Mechanical Properties: Bending

In many applications fiber-reinforced concrete is subjected to bending action. Hence, the behavior of composites in bending has been investigated extensively. This chapter deals with the basic mechanism by which fibers contribute to the composite under conditions of bending. The discussion applies to all types of fibers. The prediction of the bending behavior and test methods used to quantify the fiber contribution are also discussed.

4.1 Mechanism of Fiber Contribution to Bending

The basic mechanism of fiber contribution to bending resistance and to ductility can best be understood by observing the load-deflection behavior of the composite. Figure 4-1 shows typical load-deflection curves for fiber-reinforced composites with various types and volume fractions of fibers. These curves, which are similar to the load-elongation curves in tension, consist of a common initial linear portion but dissimilar postcrack branches. For composites reinforced with a high volume of fibers the precrack curve can have even steeper slopes. The difference in behavior in the postcrack region can be attributed to the fiber contribution, explained as follows.

Consider strain and stress distributions in a typical beam cross section, shown in Figure 4-2. In the initial elastic portion of load-deflection behavior, both strains and stresses can be assumed to be distributed linearly across the cross section. If the fiber volume fraction is low, the behavior of the beam can be described using classical bending theory.

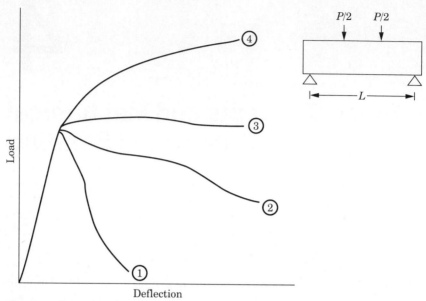

Figure 4-1 Typical load-deflection curve for fiber composites.

As the load increases, the maximum compression and tension stresses keep increasing. When the maximum tension stress reaches the tensile strength of the matrix, it cracks. Once the matrix cracks, the stress distribution across the cross section changes drastically even though the strain distribution can be assumed to be linear until the crack widens considerably. When the matrix cracks, the load carried by the matrix is transferred to the fibers, as in the case of

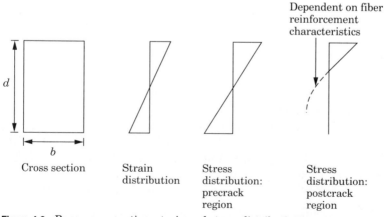

Figure 4-2 Beam cross section, strain and stress distributions.

tension specimens. The major difference between beam and tension specimens is the presence of strain gradient in beams. If the number of fibers bridging the crack is small and they can sustain only a small fraction of the force sustained by the matrix before cracking, the load capacity drops. This case is represented by curve 1 in Figure 4-1.

If there are a sufficient number of fibers to carry a good portion of the tension force, then the curve might look like curve 2 in Figure 4-1. In cases 1 and 2, the force that can be resisted by fibers is less than the force resisted by uncracked matrix. The behavior represented by these curves is called load-softening behavior. If the fibers are able to generate equal or greater force than the matrix in the precracked tension zone, then curves 3 or 4 will result. Since the fiber volume fractions are rarely greater than 10 percent, the fibers should have much higher strengths than the matrix in order to exhibit the response shown in curve 4. The resistance provided by the fibers also depends on the bond between the fibers and the matrix, as explained in Chapter 2. For bulk concrete construction, the most common type of response is similar to the curves 1 or 2 in Figure 4-1. For such construction, the contribution of fibers is best measured by the so-called flexural toughness.

4.2   Flexural Toughness

One of the primary reasons for adding fibers to concrete is to improve the energy-absorbing capacity of the matrix, which can be evaluated by determining the area under the stress-strain curve or by the load-deformation behavior. In the case of bending, the area under the load-deflection curve is used to estimate the energy-absorbing capacity or toughness of the material. Increased toughness also means improved performance under fatigue, impact, and impulse loading. The toughening mechanism provides ductility. The composite's ability to undergo larger deformations before failure is often measured using a toughness index.

There is no question about the contribution of fiber to toughness. However, how to measure this toughness and express it as an index useful for design purposes is still under debate [4.1–4.11]. The major factors that affect the load-deflection performance and hence the evaluation of toughness are the following: fiber type, fiber geometry, fiber volume fraction, matrix composition, specimen size, loading configuration, loading rate, deflection-measuring accuracy, the type of control (such as load or deformation control), and stiffness of the machine compared with the stiffness of the specimens. After the load-deflection curve is obtained, how to evaluate the factors that contribute to the improved performance is also under debate. Since improved toughness is an important attribute of

FRC, techniques and variabilities in its measurement are discussed in detail in the following sections.

4.2.1 Techniques for toughness measurement

The most common method to measure toughness is to use the load-deflection curve obtained using a simply supported beam loaded at mid-third points (four-point loading), as in Figure 4-3. There are at least three procedures that are suggested to quantify toughness using the load-deflection curve obtained in static bending. As reported by ACI Committee 544 on Fiber Reinforced Concrete [4.3], the ideal way to define toughness index I_t is by the following ratio:

$$I_t = \frac{\text{Area under the load-deflection curve until the load reaches zero for fiber composite}}{\text{Area under the load-deflection curve until the load reaches zero for plain matrix}} \qquad (4.1)$$

In equation (4.1), the energy absorbed by a fiber-reinforced composite beam is normalized by dividing it by the energy absorbed by an identical beam made of plain matrix. The computation of the ratio minimizes and in some cases eliminates the factors that influence the load-deflection response. As mentioned earlier, these factors include the following:

- specimen size and span-to-depth ratio
- load configuration
- type of test control
- loading rate

Typically, $4 \times 4 \times 14$ in. ($100 \times 100 \times 350$ mm) or $6 \times 6 \times 21$ in. ($150 \times 150 \times 530$ mm) beams are tested over a simply supported span of 12 or 18 in. (300 and 450 mm) respectively. The load configuration could be third-point loading, with loads located at 4 or 6 in. (100 or 150 mm) from the supports for shorter and longer beams, or center-point loading, with the load applied at the midspan. The test controls used include load control, crosshead displacement control, and load-point deflection control. The loading rate can be very slow to very rapid.

Even though equation (4.1) is conceptually sound, there are a number of practical difficulties. For example, it is not always possible to cast identical plain concrete beams. In some instances, such as glass fiber reinforced cement composites, the load may not reach zero for a very large deflection. Therefore, the equation was modified to simplify the testing procedure. In the procedure suggested in ASTM C1018,

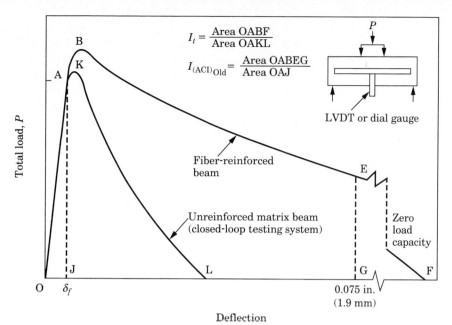

Figure 4-3 Presently available measures of toughness and toughness index definitions [4.5]: ACI.

the concept of nondimensionalization is preserved, but the areas are calculated using the single load-deflection curve as explained below [4.4].

In the ASTM C1018 procedure, the denominator in equation (4.1) is taken as the area under the load-deflection curve up to the first crack. The first crack is assumed to occur at the point where the load-deflection curve deviates from the initial linear portion. The numerator is taken as the area under the load-deflection curve up to a certain specified deflection. Three levels of deflection, namely, 3δ, 5.5δ, and 10.5δ, are suggested for the numerator. The term δ is the deflection up to first crack. Deflection values greater than 10.5δ can also be chosen for composites that can carry considerable loads at large deflections. The three suggested indices, called I_5, I_{10}, and I_{20}, are defined by the following equations. If δ is the deflection at first crack,

$$I_5 = \frac{\text{Area under the load-deflection curve up to } 3\delta}{\text{Area under the load-deflection up to } \delta} \qquad (4.2)$$

$$I_{10} = \frac{\text{Area up to } 5.5\delta}{\text{Area up to } \delta} \qquad (4.3)$$

$$I_{20} = \frac{\text{Area up to } 10.5\delta}{\text{Area up to } \delta} \qquad (4.4)$$

The procedure is schematically shown in Figure 4-4. In addition to these three indices, the ratios I_{10}/I_5 and I_{20}/I_{10} are also used to evaluate the toughness at larger deflections. Higher ratios represent better performance. The deflections for numerators in equations (4.2–4.4) were chosen using elastic–perfectly plastic behavior as the datum. For example, consider the case where the load-deflection curve is linearly elastic up to the first crack and perfectly plastic after cracking (Figure 4-5). The area under the curve up to 3δ would be 2.5 $P\delta$ and the area of up to δ would be $0.5P_c\delta$. Therefore, the I_5 value will be 5. Similarly for 5.5δ and 10.5δ, I_{10} and I_{20} values will be 10 and 20 respectively. If the composite being tested exhibits load-softening behavior after the first crack, then the toughness index values will be less than 5, 10, and 20 for I_5, I_{10}, and I_{20} respectively. For elastic–perfectly brittle material (i.e., if the beam collapses right after the first crack), the toughness values for all three would be 1.0. If plain concrete beams are tested under load control, the beam would break in two pieces right after the crack, thus indicating a toughness of 1.0. However, the beams were shown to have reproducible descending curves if servocontrol systems are used to reduce the load right after first crack [4.5]. In this case, even the plain concrete beam has a toughness value greater than 1. Since the load-carrying capacities decrease much faster (steeper descending branch) for plain concrete beams, the ratios I_{10}/I_5 and I_{20}/I_{10} tend to be 1.0.

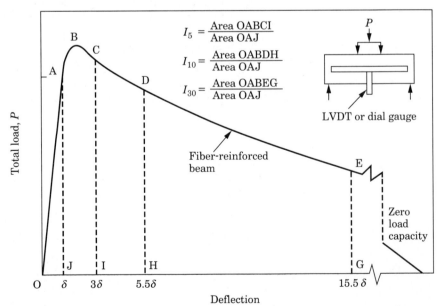

Figure 4-4 Presently available measures of toughness and toughness index definitions [4.5]: ASTM.

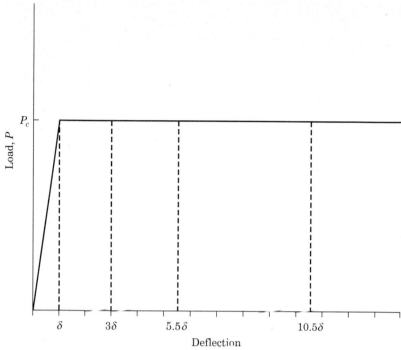

Figure 4-5 Load-deflection curve for elastic–perfectly plastic behavior.

These ratios are higher for composites that can withstand higher loads at larger deflections

ASTM procedure allows for testing different-size beams. The minimum specimen size is recommended to be greater than at least three times the fiber length. This minimum size is recommended in order to obtain a random orientation of fibers.

On the basis of data from a multi-university study the ASTM C1018 toughness indices (I_5, I_{10}, I_{30}) were found to be relatively insensitive to fiber type, fiber volume fraction, and specimen size [4.5]. Toughness as a measure of absolute energy, like the T_{JCI} explained below, can distinguish among composites with different fiber types, different fiber volume fractions, and different specimen sizes.

The method for evaluating toughness suggested by the Japan Concrete Institute [4.5] and the Japan Society of Civil Engineers [4.6] involves determining the energy required to deflect the beam a specified amount. The index T_{JCI} is defined as the area under the load-deflection curve up to a deflection of span length/150 (Figure 4-6). For fiber lengths shorter than 40 mm, a beam cross section of 100×100 mm (4×4 in.) is recommended. For longer fibers a beam cross section of 150×150 mm (6×6 in.) is recommended. The recommended span

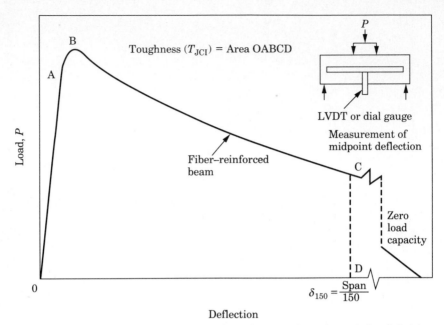

Figure 4-6 Presently available measures of toughness and toughness index definitions [4.5]: Japan Concrete Institute.

lengths for smaller and larger beams are 300 and 450 mm (12 and 18 in.) respectively. The factor T_{JCI} is designated as absolute toughness. Another factor, called flexural toughness factor σ_b, is defined by the equation

$$\sigma_b = \frac{T_{JEI}}{\delta_{150}} = \frac{S}{Wd^2} \tag{4.5}$$

where d = depth
          S = span
          W = width
          δ_{150} = $S/150$

The value of σ_b is also designated as equivalent flexural strength.

In all the aforementioned procedures for computing toughness index, the accurate measurement of center-point or load-point deflection is essential. The deflection values can vary greatly, depending on the test setup used for measurement. This aspect is discussed in detail in the following section, because the magnitude of reported toughness index can be in considerable error if proper precautions are not taken for accurate measurement of deflections.

Fracture toughness obtained using uniaxial tensile, third-point flexural, and instrumented impact tests are compared in Mobasher and

Shah [4.10]. Their study was conducted using specimens reinforced with glass fibers in the range of 4.45 to 5.25 percent. The observed response depended on the specimen geometry, loading geometry, gauge length, and strain rate. The gauge length and specimen geometry effects in turn depend on the inelastic deformation before and after the peak load. The toughness value measured using a bending specimen does not correspond to the value measured using a tension test.

4.2.2   Methods of deflection measurement and their influence on toughness index

The load-deflection curves recorded using three different techniques, shown in Figures 4-7 and 4-8, indicate the general variability that can be expected [4.5]. Figure 4-7 represents the load-deflection behavior of 4×4 in. (100×100 mm) beams reinforced with 0.1 volume percent of fibrillated polypropylene fibers, whereas Figure 4-8 presents the curves for fiber volume fraction of 0.5 percent. The three deflection measurements represent (*i*) ram (or crosshead) displacement, (*ii*) center point deflection measured from a fixed point and the bottom of the beam, and (*iii*) center point deflection measured from a frame attached to the beam at the midpoint of the beam over the supports. In the last system, the measurement represents the deflection at the

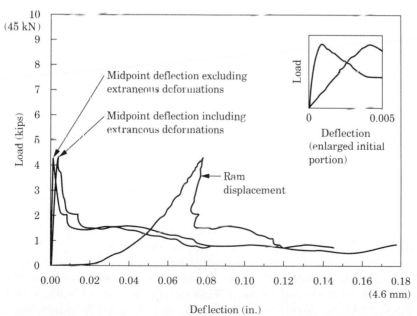

Figure 4-7   Load-deflection plots for a large unnotched polypropylene fiber reinforced beam (fiber volume = 0.1 percent) showing the influence of different methods of deflection measurement [4.5].

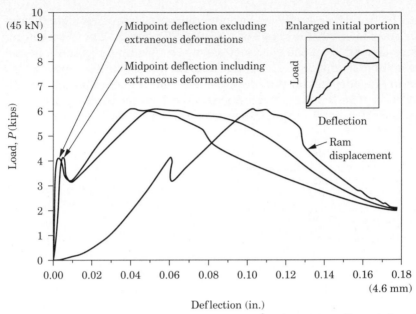

Figure 4-8   Load-deflection plots for a large unnotched polypropylene fiber reinforced beam (fiber volume = 0.5 percent) showing the influence of different methods of deflection measurement [4.5].

midspan between the undeflected and deflected positions of the beam. In this case the deformations that occur at the supports as a result of seating or other problems are not included in the measurement. Hence, this measurement provides the most accurate representation. The values of modulus of elasticity calculated using this deflection also match the values computed in a compression test, further assuring the accuracy.

Figures 4-7 and 4-8 clearly show that ram displacements should not be used for calculating toughness values. These displacements are highly machine-dependent and almost always greatly overestimate the deflections; the toughness values calculated using these curves will always be in error. The difference between the other two curves does not seem to be substantial for large deflections, but if one needs to use the first-crack deflection for toughness computation (ASTM procedure), then the difference can be substantial. First-crack deflection that excludes support deformations can be about one-fourth of the deflection that includes support deformations.

The implications of the inaccuracies in deflection measurements are twofold with regard to the ASTM C1018 toughness indices such as I_5, I_{10}, and I_{30}. First, since these indices are based on multiples of deflection at first-crack, the end points up to which they are computed are significantly influenced by the amount of extraneous deformations in the different test setups. Second, the area under the curve up to the first crack, which is the denominator in these indices, is also influenced. These two influences offset each other somewhat and inconsistently, depending upon the magnitude of errors in the deflection measurement and the shape of the load deflection diagram in the postcracking regime. For composites exhibiting an essentially elastic-plastic load-deflection response up to the range of interest, the errors in the numerator and denominator in such indices will approximately offset each other. For other types of composite load-deflection responses (postcracking softening, postcracking strengthening, and responses with a postcracking second peak), the actual toughness index may be higher or lower than those computed using erroneous deflection at firstcrack values. Absolute toughness values are less sensitive to errors in deflection measurement if deflection limits used for such toughness determination are not related to the deflection at first-crack (e.g., of the type with fixed limits or limits as a function of the span or other geometric parameters). Absolute toughness values computed from erroneous deflections are consistently conservative (i.e., less than those computed from actual deflections).

The best way to check the accuracy of deflection measurements is to compute the modulus of elasticity using the load-deflection curve and compare it with the modulus calculated using compression or other tests.

4.2.3 Effect of fiber type and volume fraction on toughness

For a given fiber type a higher volume fraction provides more energy absorbing capacity or toughness as long as the fibers can be properly mixed and the composite can be cast and compacted properly. This result should be expected because more fibers provide more resistance, especially in the tension zone.

For a given fiber geometry, longer fibers typically provide greater toughness. The effect is predominant in the case of straight smooth fibers. For deformed fibers the mechanical deformations provide more anchorage, and therefore the length/diameter ratio is less influential compared with straight fibers. For a given fiber volume and length, deformed fibers provide better toughness. Fibers with end-anchorages provide the best performance.

Polymeric fibers typically have a lower modulus of elasticity than steel fibers. Hence, beams reinforced with a small volume of these fibers undergo larger deformations before the fibers become effective. There is always a drop in load after the first crack. The load may increase at larger deflections for high fiber volume fractions, but the deformations (deflections) are typically higher for these fibers because the crack must widen in order to induce sufficient strains in the fibers. Note that for most polymeric fibers, the modulus of elasticity is much smaller than that of concrete.

The difference between beams reinforced with steel and polymeric fibers with respect to energy absorption for a given deflection is shown in Figures 4-9 and 4-10 [4.5]. In these figures energy absorption is plotted against midspan deflection for 4×4 and 6×6 in. (100×100 and 150×150 mm) beams. It can be clearly seen that for a given deflection, steel fibers can provide much higher absorption than polypropylene fibers.

4.2.4   Effect of specimen size and notch

FRC beams were found to be insensitive to notch up to a notch/depth ratio of 0.125 [4.5]. Apparent elastic modulus, sensitivity to errors in deflection measurement, toughness at first crack for equivalent depth, and toughness values for equivalent depth were all found to be the same for both notched and unnotched beams.

Studies conducted using beams with 4×4 in. (100×100 mm) and 6×6 in. (150×150 mm) cross sections show that first-crack strength slightly decreases with increase in beam size [4.5]. ASTM toughness I_5 was found to be insensitive to beam size. For a given curvature the

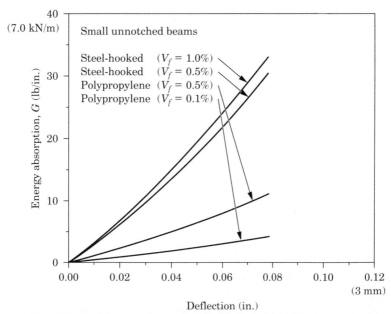

Figure 4-9 Energy-absorption capacity at various deflection limits showing actual sensitivity to fiber type and volume fraction: small unnotched beams [4.5].

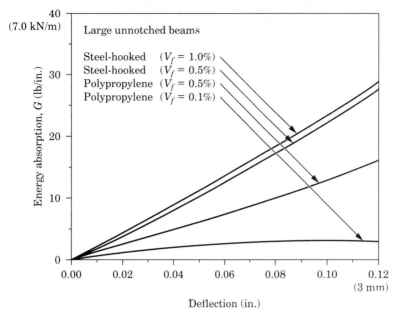

Figure 4-10 Energy-absorption capacity at various deflection limits showing actual sensitivity to fiber type and volume fraction: large unnotched beams [4.5].

larger beams have to undergo higher tensile strain at the tension face, and hence the energy absorption is lower for these beams than for smaller beams. Toughness measures that use the area under the load-deflection curve as an absolute measure of energy are also sensitive to specimen size [4.5].

4.2.5 Possible new techniques for toughness measurement

Two new techniques, involving the use of nonlinear fracture mechanics and crack mouth opening displacement (CMOD), were suggested recently for computing toughness [4.5]. The technique using CMOD is very similar to the techniques used with deflection measurements and is briefly discussed in this section.

As mentioned earlier, bending tests can be conducted in load, crosshead movement, or deflection control. In addition, the load can be applied using crack mouth–opening displacement (CMOD) control. In this procedure a notch is cut at the midspan of the beam (Figure 4-11). Since the section with the notch is now the weakest section, crack initiation and growth occur at this location. Opening of the mouth of this crack is used as feedback to apply the load. This procedure provides the most stable postcrack response. Right after the crack

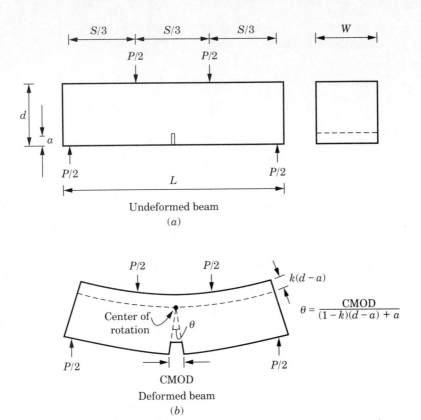

Figure 4-11 Schematic illustrating the load-CMOD approach for toughness measurements: (a) undeformed beam; (b) deformed beam.

forms, the crack tends to open very fast. Since crack opening is used as a control, the load is reduced by the servocontrol mechanism of the testing machine, avoiding sudden failure. Complete load-deflection curves, including reproducible descending branches, have been obtained using this method even for plain cement matrix specimens.

The increase in width of the crack represents the fracture energy of the beam and, hence, can be used as a measure of toughness of the material. The crack mouth opening can be related to the rotation θ that occurs at the cracked section. If we assume small rotations, rotation θ and crack mouth–opening displacement CMOD can be related using the equation

$$\theta = \frac{\text{CMOD}}{(1 - k)(d - a) + a} \tag{4.6}$$

where a = depth of the notch
d = depth of the beam
kd = depth of neutral axis

Since the maximum bending moment M is $PS/6$ for beams loaded at midthird points with load P over a span of S, the energy absorbed $M\theta$ can be expressed as:

$$\int_0^\theta M\,d\theta = \int_0^{\text{CMOD}} \frac{M\,d(\text{CMOD})}{(1-k)(d-a)+a} \tag{4.7}$$

$$= \int_0^{\text{CMOD}} \frac{PS\,d(\text{CMOD})}{6[(1-k)(d-a)+a]} \tag{4.8}$$

$$= \frac{S}{6[(1-k)(d-a)+a]} \int_0^{\text{CMOD}} P\,d(\text{CMOD}) \tag{4.9}$$

Equation (4.9) can be used to compute the energy absorbed up to a chosen CMOD opening limit. This limit is similar to the limits used for deflection in toughness computations, I_5, I_{10}, and I_{20}. In this case crack width is used as a limiting parameter, as opposed to the deflection. The limits on crack opening can be more useful in applications where crack widths rather than deflections should be controlled. For example, durability and corrosion of reinforcing materials are more directly related to crack width rather than to deformation. Experimental measurements of CMOD are less prone to errors than measurements of deflection.

Load-CMOD plots can be used to identify the CMOD at first-crack. Toughness values can be nondimensionalized using the energy required to create the first crack as in the case of toughness indices computed using load-deflection curves. For example, the toughness at five times the first crack CMOD can be obtained using the equation:

$$I_{\text{CMOD5}} = \frac{\text{Energy required for opening CMOD by five times the CMOD at first crack}}{\text{Energy required for the initiation of crack}} \tag{4.10}$$

The energy required can also be expressed in absolute values for any chosen crack mouth opening or residual load capacity. It has been found that toughness values calculated using CMOD values are generally less than those using deflections [4.5]. If the values are expressed as ratios, the toughness indices I_5, I_{10}, and I_{30} computed using the CMOD are found to be the same as the values obtained using load-deflection responses. The scatter between samples (experimental variation) was found to be less for values computed using CMOD.

4.3   Prediction of Load-Deflection Response

A few attempts have been made to predict the load deflection response of fiber-reinforced concrete beams [4.8–4.11]. The currently available prediction models are based on fracture mechanics. The crack mouth opening displacements and the corresponding deflections are computed using the energy needed to propagate the crack. The predicted results compare well with the experimental results obtained using notched beams.

It should also be possible to predict the load-deflection response using the stress-strain behavior of the composite in axial tension and axial compression. Unfortunately, the number of stress-strain curves available in the published literature is limited especially for FRC containing coarse aggregates. In addition, strain localization and the consequent nonlinear rotation that occur at the location of the first-crack also present problems [4.9]. Micromechanical finite element models are being developed to predict the response of fiber-reinforced concrete.

More details regarding the various models can be found in Reference [4.11].

4.4　References

4.1　Johnston, C.D. "Steel Fiber Reinforced Mortar and Concrete—A Review of Mechanical Properties," *Fiber Reinforced Concrete,* American Concrete Institute, Detroit, Michigan, SP44, 1974, pp. 127–142.

4.2　RILEM Committee on Fracture Mechanics of Concrete, "Determination of the Fracture Energy of Mortar and Concrete by Means of Three-Point Bend Tests on Notched Beams," *Materials and Structure,* Vol. 18, No. 106, 1985, pp. 285–290.

4.3　ACI Committee 544; "Measurement of Properties of Fiber Reinforced Concrete," *ACI Materials Journal,* Vol. 85, No. 8, 1988, pp. 583–593.

4.4　ASTM; "Standard Method of Test for Flexural Toughness of Fiber Reinforced Concrete," ASTM Standards for Concrete and Mineral Aggregates, Vol. 04.02, C1018, 1989.

4.5　Gopalaratnam, V. S.; Shah, S. P.; Batson, S. P.; Criswell, M.; Ramakrishnan, V.; and Wecharatna, H. "Fracture Toughness of Fiber Reinforced Concrete," *ACI Materials Journal,* Vol. 88, No. 4, 1991, pp. 339–353. (Also Report, Task Group on CMRC NSF Research/ACI Committee 544.)

4.6　Japan Concrete Institute; "Method of Test for Flexural Strength and Flexural Toughness of Fiber Reinforced Concrete," Standard SF4, *JCI Standards for Test Methods of Fiber Reinforced Concrete,* 1983, pp. 45–51.

4.7　Japan Society of Civil Engineers; "Method of Tests for Steel Fiber Reinforced Concrete," Standard JSCE-SF4 for Flexural Strength and Fexural Toughness of SFRC, *Concrete Library of JSCE,* No. 3, June 1984, pp. 58–66.

4.8　Wecharatana, M.; and Shah, S. P. "A Model for Predicting Fracture Resistance of Fiber Reinforced Concrete," *Cement and Concrete Research,* Vol. 13, No. 11, November 1983, pp. 819–830.

4.9　Jenq, Y. S.; and Shah, S. P. "Crack Propagation in Fiber-Reinforced Concrete," ASCE, *Journal of Structural Engineering,* Vol. 112, No. 1, January 1986, pp. 19–34.

4.10　Mobasher, B.; and Shah, S. P. "Test Parameters for Evaluating Toughness of Glass-Fiber Reinforced Concrete Panels," *ACI Materials Journal,* Vol. 86, No. 5 1989, pp. 448–458.

4.11　Shah, S. P.; and Ouyang, C. "Mechanical Behavior of Fiber-Reinforced Cement-Based Composites," *Journal of American Ceramic Society,* Vol. 74, No. 11, 1991, pp. 2727–2738.

Properties of Constituent Materials

The two major components of fiber-reinforced cement composites are the matrix and the fiber. The matrix generally consists of portland cement, aggregates, and admixtures. In some cases the cementing material can be made of nonportland cement materials. The fibers can be metallic, mineral, polymeric, or organic. This chapter deals with the basic properties of cement, mortar, concrete, and fibers. The materials presented on cement and concrete are very brief. The reader can refer to a number of excellent books that deal with properties of cement and concrete exclusively.

5.1 Cement

The most commonly used cement is called normal portland cement, designated as Type I by the American Society of Testing and Materials, ASTM [5.1]. Concrete made with this cement is tested for standard strength after a curing (and hydration) period of 28 days. Other cement types commercially available include high early strength cement, low-heat cement, and sulfate-resistant cement. The chemical compositions of these cements are listed in Table 5.1. All these cement types can be used to produce fiber-reinforced concrete. The type of structure, strength development needed, and exposure conditions control the type of cement selected. For example, if high early strength is needed, Type III cement can be used instead of Type I.

TABLE 5.1 Typical Chemical Composition of Portland Cements Used in Construction

Cement Type: ASTM Designation	Chemical components (%)								Remarks
	C_3S	C_2S	C_3A	C_4AF	$CaSO_4$	CaO	MgO		
Type I (Normal)	49	25	12	8	2.9	0.8	2.4	General Purpose
Type II (Modified)	45	29	6	12	2.8	0.6	3.0	Relatively low heat generation
Type III (High early strength)	56	15	12	8	3.9	1.4	2.6	Faster strength gain compared to type I
Type IV (Low heat)	30	46	5	13	2.9	0.3	2.7	Low heat generation during hydration; used for mass concrete
Type V (Sulfate-resisting)	43	36	4	12	2.7	0.4	1.6	Used for structures exposed to aggressive environments

5.2 Aggregates

Again, the aggregates suitable for plain concrete are suitable for FRC. The aggregates are normally divided into two categories, namely, fine and coarse.

Fine aggregate normally consists of natural, crushed, or manufactured sand. Natural sand is the usual component for normal-weight concrete. In some instances, manufactured lightweight particles are used for lightweight concrete or mortar. Heavyweight particles made of metallic components are sometimes used to produce heavyweight concrete for nuclear shielding purposes.

Fine aggregate is needed for both fiber-reinforced concrete and fiber-reinforced mortar. Fiber-reinforced mortar is primarily used for manufacturing thin-sheet items such as glass fiber–reinforced cement products and for fiber-reinforced boards (using either polymeric or naturally occurring fibers). The maximum grain size and size distribution depend on the type of product being manufactured. For example, fine sand is normally used for manufacturing thin sheets and relatively small-diameter pipes, whereas sand containing coarse particles is used for shotcreting applications and for large-diameter pipes with wall thicknesses exceeding 1 in. (25 mm). Stringent quality control over size distribution and cleanliness of the aggregates is needed for sheet products. The grading requirements for fine aggregates used in concrete are listed in ASTM specification C-33 [5.1]. The maximum size is limited to 0.25 in. (6mm) and at least 80 percent of the particles should be smaller than 0.125 in. (3 mm).

Coarse aggregates can be normal-weight, lightweight, or heavyweight in nature, even though heavyweight aggregate usage is very limited. Normal-weight coarse aggregates can be made of natural gravel or crushed stone. Lightweight coarse aggregates are normally made of expanded clay (such as shale or pumice) or blast furnace slag. Concrete made with normal-weight coarse aggregate weighs about 140 lb/ft^3 (2240 kg/m^3), whereas the structural lightweight aggregate concrete weighs in the range of 90–110 lb/ft^3 (1440–1760 kg/m^3). Nonstructural lightweight components such as boards or noise barriers can weigh as little as 20 lb/ft^3 (320 kg/m^3). Fibers are sometimes used in these very light products to improve their strengths. Fiber-reinforced concretes containing normal-weight or lightweight aggregates have been successfully used in a number of field applications, such as pavements and overlay.

In the area of grading for coarse aggregates, guidelines prescribed in ASTM standards are normally applicable for FRC, even though a more restrictive grading requirement is shown to reduce mixing problems [5.2]. The grading requirements recommended by the ACI Committee

TABLE 5.2 Recommended Grading Requirement for Aggregate Used in Steel Fiber–Reinforced Concrete

U.S. sieve	Standard size (mm)	Percent passing for maximum size of				
		3/8″ (10 mm)	1/2″ (13 mm)	3/4″ (19 mm)	1″ (25 mm)	1½″ (38 mm)
2	51	100	100	100	100	100
1 ½	38	100	100	100	100	85–100
1	25	100	100	100	94–100	65–85
¾	19	100	100	94–100	76–82	58–77
½	13	100	93–100	70–88	65–76	50–68
⅜	10	96–100	85–96	61–73	56–66	46–58
# 4	5	72–84	58–78	48–56	45–53	38–50
# 8	2.4	46–57	41–53	40–47	36–44	29–43
# 16	1.1	34–44	32–42	32–40	29–38	21–34
# 30	0.6	22–33	19–30	20–32	19–28	13–27
# 50	0.3	10–18	8–15	10–20	8–20	7–19
# 100	0.15	2–7	1–5	3–9	2–8	2–8
# 200	0.08	0–2	0–2	0–2	0–2	0–2

Note: Aggregates should be well graded from the largest to the smallest size. Aggregate should not vary from near the maximum allowable percent passing one sieve to near the minimum allowable percent passing the next sieve size.

544 for steel fiber-reinforced concrete are shown in Table 5.2. FRC made with aggregates complying with the guidelines shown in Table 5.2 also provide better workability [5.2].

5.3 Water and Water-Reducing Admixtures

Water is needed for hydration of cement and molding of concrete to the desired shape. The relationship between compressive strength and water-cement ratio is well established. An increase in water-cement ratio leads to a reduction in compressive strength. A water-cement ratio of about 0.28 provides sufficient water for hydration. However, a water-cement ratio of about 0.6 is needed to obtain a plastic workable mixture that can be transported, placed, properly compacted, and finished to the final form. Therefore, chemical admixtures have been developed to improve the workability at lower water-cement ratios. These are called water-reducing admixtures.

Water-reducing admixtures have become (almost) an integral part of fiber-reinforced composites. The addition of fibers to a cement matrix (either mortar or concrete) normally reduces the workability. But the advent of water-reducing admixtures made it possible to maintain the workability of a fiber-reinforced matrix without adding more water. Since the addition of extra water almost always reduces strength, increases shrinkage, and enhances the tendency to crack, resulting in durability problems, it is always advisable to use the minimum amount of water possible.

There are two types of water-reducing admixtures available in the market. The first type can reduce the water demand by 10% to 20%. The second class of admixtures, known as high-range water-reducing admixtures or superplasticizers, can be used to obtain flowable mixtures even at a water-cement ratio of 0.28. The high-range water-reducing admixtures have been successfully used for both cast-in-place concrete and shotcrete applications. Details regarding these admixtures can be found in books dealing with concrete.

5.4   Mineral Admixtures

The most widely used mineral admixtures are fly ash and silica fume. Fly ash is used to improve the workability of fresh concrete, to reduce heat of hydration, to improve economy, and to enhance permeability characteristics. Silica fume is added mainly to obtain high strength. Use of mineral admixtures, especially silica fume, became more widespread after the introduction of high-range water-reducing admixtures. In the case of fiber-reinforced composites, these admixtures produce a denser matrix, resulting in better mechanical properties of the composite. For shotcrete applications, such as tunnel linings, the addition of silica fume was found to reduce rebound (i.e., the material that falls out during the shotcreting process). The addition of silica fume was also found to improve the bond between fibers and matrix. Silica fume also improves the durability of certain fibers added to concrete. In carbon fiber–reinforced composites, silica fume acts as a dispersing agent. The contribution and use of silica fume is discussed in detail in the following chapters.

5.5   Other Chemical Admixtures

Air entraining as well as accelerating and retarding admixtures have also been used in FRC. Air entrainment is the most commonly used admixture for exposed structures. Studies have shown that air entrainment is needed for exposed FRC structures such as pavements and linings, since FRC is as susceptible as plain concrete to freeze-thaw cycling [5.3]. Accelerating admixtures are normally used for shotcreting applications to hasten the setting process. Retarding admixtures are used when a reduction in the heat of hydration is needed.

5.6 Special Cements

Cementing materials other than portland cement can also be used for fiber composites. There are primarily two classes of cementing materials in this category. The first consists of cementing materials developed for repairs. These are either blended portland cements, such as rapid-set cement, or they are chemicals that can act as cementing agents themselves, such as magnesium phosphate, which can develop compressive strengths up to 6000 psi (40 MPa) within an hour. Addition of fibers to these cementing materials was found to improve the shrinkage characteristics and ductility of the matrix [5.4].

The second category of cementing materials consists of low-cost materials such as lime or clay. These matrices are normally used with naturally occurring fibers in developing countries. One exception is the use of gypsum in industrialized nations for manufacturing a variety of building products.

5.7 Metallic Fibers

Metallic fibers are made of either carbon steel or stainless steel. The tensile strength ranges from 50 to 200 ksi (345 to 1380 MPa). The minimum strength specified in ASTM is 50 ksi (345 MPa). The modulus of elasticity for metallic fibers is about 29,000 ksi (200 GPa).

The fiber cross section may be circular, square, crescent-shaped, or irregular. The length of the fibers is normally less than 3 in. (75 mm) even though longer fibers have been used. The length-diameter ratio typically ranges from 30 to 100 or more.

5.7.1 Fiber geometry and manufacturing methods

Some of the fiber shapes used in the field are shown in Figure 5-1. Essentially, the fibers may be categorized as straight, deformed, rippled, with special ends (e.g., enlarged or hooked ends), and with irregular cross sections.

Round, straight steel fibers are produced by cutting into pieces thin wires having a diameter in the range of 0.01 to 0.04 in. (0.25 to 1 mm), Figure 5-1a. Flat, straight steel fibers are produced either by shearing thin sheets that are about 0.006 to 0.016 in. (0.15 to 0.41 mm) thick or by flattening wires. These fibers have a width in the range of 0.01 to 0.04 in. (0.25 to 1 mm), Figure 5-1a. Crimped or deformed fibers are produced by crimping the full length, Figure 5-1b, or bent or enlarged at the ends only, Figure 5-1c and Figure 5-1d. Deformations are also induced by bending or flattening of wires to

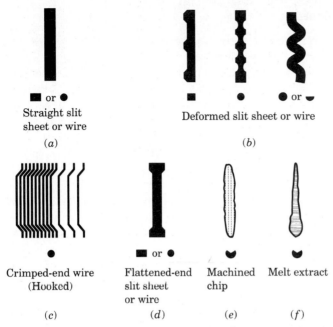

Figure 5-1 Various shapes of steel fibers used in fiber-reinforced concrete. (*a*) Straight slit sheet or wire (*b*) Deformed slit sheet or wire (*c*) Crimped-end wire (*d*) Flattened end slit sheet or wire (*e*) Machined chip (*f*) Melt extract.

increase bonding. Fibers with crimped (hooked) ends, Figure 5-1*c,* are also available in collated form. Collation is done by gluing the fibers together along their sides with water-soluble glue. The glue dissolves during the mixing process, facilitating the distribution of individual fibers. Hence, larger fiber volume fractions as well as fibers with higher length-diameter (aspect) ratios can be incorporated into the concrete without balling of the fibers.

Fibers are also produced from wires that have been shaved down in the steel-wool-making process. These wires, which have a crescent-shaped cross section, are chopped and crimped to produce deformed fibers.

Fibers produced by melt extraction processes have an irregular surface and are crescent-shaped in cross section, Figure 5-1*f.* In the melt extraction process, a rotating wheel contacts a molten metal surface and lifts off liquid metal, which rapidly solidifies into fibers.

Elongated chips produced by chatter machining, Figure 5-1*e,* are also being used as fibers. These fibers have a rough, irregular surface.

5.8   Polymeric Fibers

Synthetic polymeric fibers have been produced as a result of research and development in the petrochemical and textile industries. Fiber types that have been tried with cement matrices include acrylic, aramid, nylon, polyester, polyethylene, and polypropylene. They all have a very high tensile strength, but most of these fibers (except for aramids) have a relatively low modulus of elasticity. The quality of polymeric fibers that makes them useful in FRC is their very high length-to-diameter ratios; their diameters are on the order of micrometers. Table 5.3 presents a summary of physical properties of various polymeric fibers.

Polymeric fibers are available in single filament or fibrillated form. The lengths used in FRC range from 0.5 to 2 in. (12 to 50 mm). Some types of fibers are available in very short lengths (pulp form) of only a few millimeters. On the other end of the spectrum, very long fibers are available for applications that require continuous fiber reinforcement. The following sections provide a brief description of the commercially available polymeric fibers.

5.8.1   Acrylic

Fibers that contain at least 85 percent by weight of acrylonitrile are classified as acrylic fibers. These fibers are denser than water and have

TABLE 5.3   Physical Properties of Polymeric Fibers

Fiber Type	Eff. dia. $\times 10^{-3}$ in. (10^{-3} mm)	Specific gravity	Tensile strength, ksi (MPa)	Elastic modulus, ksi (GPa)	Ultimate elongation(%)
Acrylic	0.5–4.1 (13–104)	1.17	30–145 (207–1000)	2000–2800 (14.6–19.6)	7.5–50.0
Aramid I	0.47 (12)	1.44	525 (3620)	9,000 (62)	4.4
Aramid II (high modulus)	0.40 (10)	1.44	525 (3620)	17,000 (117)	2.5
Nylon		1.16	140 (965)	750 (5.17)	20.0
Polyester		1.34–1.39	130–160 (896–1100)	2500 (17.5)	
Polyethylene	1.0–40.0 (25–1020)	0.96	29–35 (200–300)	725 (5.0)	3.0
Polypropylene		0.90–0.91	45–110 (310–760)	500–700 (3.5–4.9)	15.0

a slightly higher modulus of elasticity than other polymeric fibers except for aramid fibers.

5.8.2   Aramid

Because of their high modulus of elasticity, aramid fibers can enhance the mechanical properties of FRC, including tensile and bending strength. The primary limitation to the use of these fibers in concrete is their high cost compared with other fibers. These fibers are also available in strand form.

5.8.3   Nylon

Commercially available nylon fibers are made of Nylon 6. They are available in various lengths in single-filament form. Since these fibers are very thin, the number of fibers per pound (fiber count) is in the range of 35 million per pound (0.45 kg) for a fiber length of .75 in. (19mm).

5.8.4   Polyester

Polyester fibers are made of ethyl acetate monomers. Their physical and chemical properties can be changed substantially by altering manufacturing techniques. The higher modulus of elasticity and better bonding to concrete that is important for FRC applications can be achieved by some of these modifications.

5.8.5   Polyethylene

Polyethylene fibers are available both in standard lengths (0.5 to 2 in., 12 to 50 mm) and in pulp form. The longer fibers available in the market have wart-like surface deformations, enabling better bond to concrete. The fibers that are available in pulp form have been promoted as a replacement for asbestos fibers in concrete. These short fibers can also be used in cement matrix to improve ductility, impact resistance, and fatigue strength.

5.8.6   Polypropylene

Polypropylene fibers are available both in single-filament and fibrillated form in lengths ranging from 0.25 to 2 in. (6 to 50 mm). Short fibers in the form of pulp are also available. Polypropylene pulp seem to have lower strength than polyethylene pulp made with oriented molecules.

5.9   Carbon Fibers

Until the mid-1980s the high cost of carbon fibers limited their use in portland cement composites. More recently, low cost carbon fibers have been manufactured with petroleum and coal pitch. Even though their cost is still higher than polymeric fibers, carbon fibers have potential for special applications that require high tensile and flexural strength.

Carbon fibers have elastic moduli as high as steel and are two to three times stronger than steel, yet they are very light, with a specific gravity of about 1.9. They are inert to most chemicals.

Carbon fibers are typically produced in strands (tows) that can contain up to 12,000 individual filaments. These strands are normally spread before incorporation into cement matrices.

5.10   Glass Fibers

Glass fibers are primarily used for glass fiber-reinforced cement (GFRC) sheets. Regular E-glass fibers were found to deteriorate in concrete. This observation led to the development of alkali-resistant, AR-glass fibers. Properties and uses of these fibers are discussed in detail in Chapter 13.

5.11   Naturally Occurring Fibers

The oldest forms of fiber-reinforced composites were made with naturally occurring fibers such as straw and horse hair. Modern technology has made it possible to extract fibers economically from various plants, such as jute and bamboo, to be used in cement composites. A unique aspect of these fibers is the low amount of energy required to extract these fibers. The primary problem with the use of these fibers in concrete is their tendency to disintegrate in an alkaline environment. Efforts are being made to improve the durability of these fibers in concrete by using admixtures to make the concrete less alkaline and by subjecting the fibers to special treatment.

Natural fibers used in portland cement composite include akwara, bamboo, coconut, flax, jute, sisal, sugarcane bagasse, wood, and others. Mechanical properties of some of these fibers are presented in Table 5.4. The following sections provide a brief description of these fibers.

5.11.1   Akwara fibers

Akwara is a natural fiber derived from a plant stem grown in large quantities in Nigeria. They are made of a cellular core covered with a smooth sheath. Akwara fibers were found to be durable in alkaline

TABLE 5.4 Properties of Naturally Occurring Fibers [5.5]

Fiber type	Coconut	Sisal	Sugarcane Bagasse	Bamboo	Jute	Flax	Elephant grass	Musamba	Wood fiber (kraft pulp)
Fiber length (mm)	50–350	NA*	NA	NA	180–200	500	NA	NA	2.5–5
Fiber diameter (mm)	0.1–0.4	NA	0.2–0.4	0.05–0.4	0.1–0.2	NA	NA	NA	0.015–0.08
Specific gravity	1.12–1.15	NA	1.2–1.3	1.5	1.02–1.04	NA	NA	NA	1.5
Modulus of elasticity (GP$_e$)	19–26	13–26	15–19	33–40	26–32	100	4.9	0.9	NA
Ultimate tensile strength (MPa)	120–200	280–568	170–290	350–500	250–350	1000	178	83	700
Elongation at break (%)	10–25	3–5	NA	NA	1.5–1.9	1.8–2.2	3.6	9.7	NA
Water absorption (%)	130–180	60–70	70–75	40–45	NA	NA	NA	NA	50–75

*Properties are not readily available or not applicable.
1 mm = 0.04 in.
1 MPa = 0.145 ksi

environment of cement matrix, and they are also dimensionally stable under wetting and drying conditions. The disadvantages are their low elastic modulus and brittleness [5.5].

5.11.2   Bamboo fibers

Bamboo, which is a member of the grass family, grows in tropical and subtropical regions. Plants can grow up to a height of 15 m. Their hollow stalks have intermediate joints; the diameters of these stalks range from 0.4 to 4.0 in. (1 to 10 cm). Special techniques are needed to extract the fibers from bamboo. Bamboo fibers are strong in tension, but have a relatively low modulus of elasticity (Table 5.4). Their tendency to absorb water adversely affects the bonding between the fibers and the matrix during the curing process.

5.11.3   Coconut fibers

A mature coconut has an outer fibrous husk. Coconut fibers, called coir, can be extracted simply by soaking the husk in water or, alternatively, by using mechanical processes. These short (only a few inches) stiff fibers have been used for making rope for centuries. Coir has a low elastic modulus and is also sensitive to moisture changes.

5.11.4   Flax and vegetable fibers

Flax is grown mainly for its fiber. Flax fibers are strong under tension and also possess a high modulus of elasticity (Table 5.4).

Fibers extracted from other plants such as elephant grass, water reed, plantain, and musamba have also been tried as reinforcements for concrete. Most of these fibers are removed from the stems of the plants manually.

5.11.5   Jute fibers

Jute, grown solely for its fiber content, is cultivated mainly in Bangladesh, China, India, and Thailand. Jute plants can grow to a height of 8 ft (2.4 m) with stalk diameters normally less than 1 in. (25 mm). The bark of these plants contains the fibers, which are extracted from the stalk simply by soaking them in water for about four weeks, thereby loosening the fibers. The fibers are then removed manually and dried. Recently mechanical equipment has been developed for removing the fibers.

Jute fibers are relatively strong in tension (Table 5.4). Traditionally these fibers had been used for making ropes and for weaving into gunny cloth used in bags to transport grains.

5.11.6 Sisal fibers

A number of investigators in Australia and Sweden have studied the properties of cement composites made with sisal fibers [5.5]. These fibers, extracted from the leaves of *Agave sisalana,* are primarily made of hemicellulose, lignin, and pectin. Sisal fibers are relatively strong (Table 5.4) but are not durable in alkaline environments.

5.11.7 Sugarcane bagasse fibers

Sugarcane bagasse is the fibrous material left after the extraction of juice from sugarcane. In most instances, bagasse has about 50 percent fiber content. The physical properties of the fibers depend on the variety of the sugarcane, its maturity, and the efficiency of the milling plant.

5.11.8 Wood fibers (cellulose fibers)

Wood fibers constitute the major portion of the natural fibers used in concrete worldwide. Their use in portland cement composite is gaining popularity as a replacement for asbestos fibers. The advantages are their availability, high tensile strength, high modulus of elasticity, and the well-developed technology to extract the fibers.

The primary disadvantages are their vulnerability to decomposition in the alkaline environment present in concrete and other portland cement-based matrices. However, recent research efforts have identified methods and processes to minimize the disintegration of fibers in an alkaline environment [5.6].

The process of extracting fibers from wood is called pulping. The process may be mechanical, chemical, or semichemical. The properties of the resultant fibers depend very much on the pulping process. The primary components of woods are cellulose, hemicellulose, and lignin. Lignin has an adverse effect on the strength of the fibers; hence, the pulping process that removes the most lignin provides the best fibers. Tensile strengths for delignified cellulose fibers have been recorded as high as 290 ksi (2000 MPa), whereas cellulose fibers from which lignin has not been removed have tensile strengths in the vicinity of 70 ksi (500 MPa).

Chemical pulping processes reduce the lignin content in wood fibers more effectively than mechanical or semichemical methods. The two common chemical methods are the kraft or sulfate process and the sulfite process. As mentioned earlier, removal of more and more lignin provides better fibers. However, lignin removal can limit the amount of fibers that can be extracted from certain woods, hence, lignin-free fibers are normally more expensive.

5.11.9 Other fibers

Other natural fibers that have been tried as reinforcing material include palm and elephant grass fibers. Fibers have also been extracted from basalt rock.

Elephant grass fibers are extracted from elephant grass stems, which grow up to 10 ft (3 m) tall and are packed with tough, sharp fibers bonded together by lignin. Extraction of these fibers is difficult, but the fibers are dimensionally stable under different moisture conditions and they are also alkali resistant.

5.12 References

5.1 ASTM; *Annual Book of ASTM Standards,* Vol. 04.02, Concrete and Mineral Aggregates, 1991, 824 pp.

5.2 ACI Committee 544; "A State-of-the-Art Report on Fiber Reinforced Concrete," (ACI 544. IR-82), American Concrete Institute, Detroit, Michigan, 16 pp.

5.3 Balaguru, P. N.; and Ramakrishnan, V. "Freeze-Thaw Durability of Fiber Reinforced Concrete," *ACI Journal,* Vol. 83, No. 3, 1986, pp. 374–382.

5.4 Balaguru, P. "Fiber-Reinforced Rapid-Setting Concrete," *Concrete International,* American Concrete Institute, Detroit, Michigan, Vol. 14, No. 2, 1992, pp. 64–67.

5.5 Balaguru, P.; and Shah, S. P. "Alternative Reinforcing Materials for Developing Countries," *International Journal for Development Technology,* Vol. 3, 1985, pp. 87–105.

5.6 Soroushian, P.; and Marikunte, S. "Reinforcement of Cement-Based Materials with Cellulose Fibers," *Thin-Section Fiber Reinforced Concrete and Ferrocement,* SP-24, American Concrete Institute, Detroit, Michigan, 1990, pp. 99–124.

6

Mixture Proportions, Mixing and Casting Procedures

As mentioned in Chapter 1, FRC is a general term used to describe portland cement composites reinforced with relatively strong fibers. Based on the matrix composition, fiber volume fraction, type of fibers, and the manufacturing process, the following specialized products can be identified:

FRC with Coarse Aggregates. This composite contains fine and coarse aggregates and discontinuous fibers. The matrix is usually proportioned following the procedures used for plain concrete. The volume fraction of fibers ranges from 0.4% to 2% ($\sim$ 50 to 250 lb/yd^3, 30 to 150 kg/m^3) for steel fibers and 0.06% to 0.5% (1 to 8 lb/yd^3, 0.6 to 4.8 kg/m^3) for polymeric fibers. The mix proportions obtained for plain concrete are slightly modified to maintain workability, easy fiber mixing, and good fiber distribution.

Fiber Reinforced Cement Mortar. This term applies to a wide variety of manufactured products such as glass fiber–reinforced cement sheets (GFRC) as well as panels and tiles made using other fibers such as naturally occurring and polymeric fibers. The manufacturing process for this composite (which contains cement and fine aggregate) is quite different from the procedures used for FRC with coarse aggregates. The fiber volume fraction ranges from 1% to 5%.

Fiber Reinforced Cement Products. These are similar to fiber-reinforced cement mortar but contain little or no fine aggregate. The widely known asbestos cement sheets fall into this category. These products are usually manufactured using the Hatschek process, in which a mat of fibers is dipped into a cement slurry and then dewatered to form the fiber-cement sheets. These can be formed into a variety of products such as corrugated sheets and pipes. Since the use of asbestos fibers is restricted in most countries, a large number of replacement fibers are being investigated. These fibers are relatively short (a few millimeters long) and have very high length-to-diameter ratios. These fibers also help to retain cement during the manufacturing process. Fibers that have been marketed to replace asbestos include polyethylene, polypropylene, polyvinyl alcohol, aramid, cellulose, and carbon. The fiber volume fraction ranges from 3% to 6%. In the case of asbestos cement sheets, up to 20% of fibers has been used.

Recently, cement composites containing a high volume of steel fibers have been developed. These are categorized as slurry-infiltrated fiber concrete (SIFCON). This composite is cast by infiltrating a bed of fibers with cement or mortar slurry. The fiber volume fraction used ranges from 4% to 22%.

Special Cement Products.   These composites are made using cements other than portland cements, such as magnesium phosphate or gypsum, or modified cements like rapid-set cements. Products falling into this category include gypsum boards and composites used for rapid repairs (e.g., overlaying a bridge deck that has to be opened to traffic in a few hours).

This chapter deals only with FRC with coarse aggregates that is cast in place. Procedures used for other composites are presented in separate chapters (Chapters 12 to 16).

6.1   Mixture Proportions for FRC Containing Coarse Aggregates

As mentioned in Chapter 5, the constituent materials used for FRC are cement, fine aggregates, coarse aggregates, water, admixtures, and fibers. The water-cement ratio is the primary controlling variable for compressive strength. The other major variables that control strength and workability are cement content, maximum aggregate size and gradation (size distribution), and the presence of entrained air. In FRC the major variables controlling workability are the fiber content and the fiber aspect (length-diameter) ratio. The objective is to obtain a mix that produces the required (compressive) strength, is workable, and contains a minimum amount of cement. Since cement is the most expensive component in plain concrete, reduction of cement usually results in better economy.

Procedures for obtaining the mix proportion of plain concrete are well established. These procedures can be found in textbooks on plain, reinforced or prestressed concrete. The mix design procedure recommended by ACI Committee 211 is one of the common methods used [6.1]. It involves selection of the various constituent materials using a set of tables. (All the tables can be found in Reference [6.1]). The major steps involved are as follows:

Step 1. If slump is not specified, choose a slump suitable for the type of construction. Recommended slumps [6.1] vary from 1 to 2 in. (25 to 50 mm) for mass concrete, and up to 4 in. (100 mm) for beams and columns. Slump (or workability) can be increased by using water-reducing admixtures. Nevertheless, it is advisable to choose a base-slump value that is independent of admixture for mix design.

Step 2. Choose the maximum size of aggregate. The maximum size should be smaller than

- one-fifth of the narrowest dimension between forms,
- one-third of the depth of slab, and
- three-fourths of the clear spacing between reinforcing bars.

The maximum size is limited to 1.5 in. (38 mm) except for thick sections or mass concrete.

Step 3. Decide on the amount of water and entrained air. A table is provided in Reference [6.1] for choosing these quantities. The amount of water required is presented as a function of slump, maximum aggregate size, and amount of entrained air. Higher slump values, smaller aggregate sizes, and lower air contents lead to more water demand. The amount of air required depends on the type of exposure. For structures subjected to severe conditions of freezing and thawing, wetting and drying, more entrained air is recommended. More entrained air provides better workability but may reduce the compressive strength.

Addition of fibers affects both workability and entrained air. Hence, these quantities must be readjusted based on trial mixes.

Step 4. Choose the amount of cement needed to obtain the specified compressive strength. Typically a water-cement ratio [6.1] is chosen to obtain the required strength. Since the water content is already estimated in Step 3, the water-cement ratio can be used to obtain the cement content.

Step 5. Choose the volume of coarse and fine aggregates. The ratio of coarse-to-fine aggregate is again decided based on workability requirements. A slightly higher sand content than that used for plain concrete seems to provide better results for fiber-reinforced concrete.

Step 6. Make adjustments in the amount of water to be added based on the moisture present in the coarse and fine aggregates.

For a given project, trial mixes based on the mix design have to be made to ascertain the workability and the strength requirements. If the ready-mix plant supplying the concrete has already supplied concrete in the strength range required, their proportions can be used for trials.

The mix has to be designed to obtain an average compressive strength f'_{cr} that is higher than the specified compressive strength f'_c. The term f'_{cr} is defined as the average required compressive strength in the ACI Code. Guidelines to estimate f'_{cr} are provided in the ACI Code 318 [6.2]. The overdesign (difference between f'_{cr} and f'_c) depends mainly on the level of quality control maintained at the ready mix plant.

6.1.1   Special requirements for FRC with steel fibers (SFRC)

The slump test is not a reliable test to obtain the workability of SFRC. Since SFRC should be vibrated in place for proper compaction, either the inverted slump cone test or V-B test should be used for measuring workability (these test methods are described in Chapter 7). Consequently, after the trial proportion is established, the water or water-reducing admixture has to be adjusted to obtain the required inverted slump cone (or V-B) time.

Typically, SFRC mixtures require higher cement and higher fine aggregate content for maintaining the strength and workability. Table 5.2 provides guidelines for combined aggregate gradations for good workability of SFRC. Fiber characteristics such as length, length-diameter ratio, and shape play an important role. Trial batches are needed for any fiber types to be used. Most fiber manufacturers maintain records and provide trial mix proportions for their fibers. Table 6.1 presents the general range for the mix proportion with steel fibers. Table 6.2 presents mix proportions for two typical mixes using high-range water-reducing admixtures [6.3].

6.1.2   Mixes with lightweight aggregate

Mix design procedures for lightweight aggregate concrete are different compared with normal-weight concrete. ACI Committee 211 has presented a set of recommendations for selecting proportions for structural lightweight concrete [6.4]. A few researchers have investigated the performance of fiber-reinforced lightweight concrete and have reported details regarding mix composition, workability, special precautions, and strength development [6.5, 6.6].

TABLE 6.1   General Range of Proportions for Normal-Weight Steel Fiber–Reinforced Concrete

	$^3/_8$ in. max sized aggregate	$^3/_4$ in. max sized aggregate	$1^1/_2$ in. max sized aggregate
Cement (lb/yd³)	600–1000	500–900	470–700
W/C ratio	0.35–0.45	0.35–0.50	0.35–0.55
Percentage of fine to coarse aggregate	45–60	45–55	40–55
Entrained air content (%)	4–8	4–6	4–5
Fiber content (volume percent)			
deformed fiber	0.4–1.0	0.3–0.8	0.2–0.7
smooth fiber	0.8–2.0	0.6–1.6	0.4–1.4

1 lb/yd³ = 0.59 kg/m³
1 in. = 25.4 mm
1 steel fiber volume percent = 132.3 lb/yd³ (78.5 kg/m³)

6.1.3   Special requirements for concrete reinforced with polymeric fibers

The volume fraction of polymeric fibers currently used in the field is very low. In most cases it is limited to 0.1%. This translates to about 1.6 lb/yd³ (1 kg/m³) of concrete. At this fiber-loading level, the change in workability is minimal. Still, slight modifications may be needed to maintain workability and air content level. Fibers can be added at the ready-mix plant or at the site. Mixing for at least 10 minutes is recommended after the addition of fibers.

Researchers have used up to 2 percent volume fraction of fibers (lengths 0.5 to 2 in., 12 to 50 mm) with conventional mixing. For

TABLE 6.2   Mixture Proportions for Two Typical Mixes Using Fibers with Hooked Ends (Adapted from Reference [6.3])

	Mix 1	Mix 2
Cement content (lb/yd³)	611	799
Ratio of fine to coarse aggregate	50/50	50/50
Max. coarse aggregate size (in.)	1.5	1.5
Water-cement ratio	0.40	0.30
High-range water-reducing admixture percent by weight of cement	1.0	1.2
Air-entraining agent percent by weight of cement	0.15	0.20
Fiber content (lb/yd³)	75	75
28 day compressive strength (ksi)	6.4	7.5

Note: Both the mixes were highly workable.
1 lb/yd³ = 0.59 kg/m³
1 in. = 25.4 mm
1 ksi = 6.89 MPa

volume fractions above 0.2%, special precautions are necessary because the workability is reduced considerably and the amount of
entrapped air increases. The latter can cause severe problems with
respect to consolidation and strength reduction. Workability problems
may be solved by using higher dosages of high-range water-reducing
admixtures. Aggregate gradation and mixing sequences have to be
adjusted to control the amount of entrapped air.

Single-filament fibers tend to reduce the workability more than fibrillated fibers. Smaller fiber diameters that result in higher fiber count
also reduce workability more than larger diameters. Fiber length also
plays an important role. Fibers in pulp form with lengths limited to a
few millimeters can be mixed up to a volume fraction of 5%. In general, fiber lengths used in the field vary from 0.5 to 2 in. (12 to 50
mm).

6.2   Mixing and Casting Procedures

Only fiber-reinforced concrete containing steel and polymeric fibers in
the volume fraction range of 0.1% to 2% are presented in this chapter.
Procedures used for other matrices such as mortar or cement slurry
and for other fiber types such as glass or carbon are presented in separate chapters. Procedures dealing with composites placed in position
by shotcreting are also presented in a separate chapter (Chapter 12).
Note that in most concrete applications the fibers used predominantly
are steel and synthetic polymers.

6.2.1   Concrete reinforced with steel fibers

6.2.1.1   Mixing.   The primary concern in mixing is the uniform distribution of fibers throughout the matrix. A collection of long thin steel
fibers, usually with aspect (length/diameter) ratios higher than 100,
will interlock to form a mat, or a ball, during mixing. Once these balls
have formed, separating the fibers is extremely difficult. Clumping is
one of the reasons why straight smooth fibers cannot be successfully
used in the field. Higher aspect ratios are needed to develop sufficient
bond strength between fiber and matrix. Even with length-to-diameter
ratios of 100 or more, about 2% volume fraction of fibers are needed
to develop sufficient ductility. The combination of higher aspect ratio
and higher volume fraction needed for straight fibers makes mixing
an impossible task.

Besides the aspect ratio, balling of fibers is typically a function of
fiber volume fraction, gradation of aggregates used in the mix, fiber geometry, and the procedure used for the addition of fibers in the mixer.
Larger aspect ratios and larger maximum-size aggregates reduce the

maximum volume fraction of fibers that can be incorporated without balling. For a given fiber type, mixing becomes more difficult for fibers with higher aspect ratios. For a given aspect ratio, strong stiff fibers allow better mixing because they do not clump so easily.

In the 1970s the development of deformed fibers with better anchoring characteristics gave a big boost to the use of fibers in concrete. The deformations provided better bonding. As a result, shorter-length fibers could be used. Since anchoring was more efficient, a smaller volume fraction of fibers could generate sufficient ductility.

One should avoid feeding clumped fibers into the mixer. The possibility for clumping of fibers exists whenever (a) fibers drop from one conveyor belt onto another, (b) the conveyor belts carrying the fibers bounce over their rollers, (c) an overload of fibers reaches the sides of the mixing drum, (d) fibers are tossed against the side of the charge opening, and (e) they are transferred from high-speed to slower-speed conveyor belts.

Several mixing sequences have been successfully used both in the laboratory and in the field. The following mixing sequences have been found to work efficiently for most of the mixture proportions [6.7]:

1. Add the clump-free fibers directly to the mixer once the other ingredients have been uniformly mixed. The recommended rate of addition is 100 lb (45 kg) per minute. The fibers can be added manually, by emptying the containers into the truck hopper, or via a conveyor belt or blower either at the batch plant or at the job site. The mixer should be rotating at full speed as the fibers are being added. After the fiber addition is complete, the contents should be mixed for at least another 50 revolutions at normal recommended mixing speed.

2. Or add fibers to the aggregates before charging into the mixer. The common procedure is to add the fibers to the aggregates as they are moving on the charging belt. They can either be placed directly on top of the aggregates or be carried on a separate belt that empties onto the charging belt. Fibers should be spread out as much as possible to avoid heavy concentrations.

3. Or place the fibers on top of the aggregates, weighed and ready to be charged into the mixer. The flow of aggregates from the weight batchers to the mixer will distribute the fibers within the aggregate. Fibers can be added manually or using a conveyor belt.

4. Or mix the fibers in by feeding them simultaneously with aggregates, cement, admixtures, and about 90 percent of water. This procedure is achieved by slowing down the aggregate feed and adding the other ingredients.

It is essential that the method chosen for the addition of fibers be tested in the field using the actual working crew members. The mixed concrete should be inspected to ensure uniform distribution of clump-free fibers.

6.2.1.2 Transporting and placing. Transporting and placing of FRC with steel fibers can be done with conventional equipment. This section deals with some of the special precautions needed in handling concrete containing steel fibers [6.7].

Trucks carrying concrete with higher fiber contents should be loaded to less than their full capacity. The maximum should be limited to about 85 percent. Typically, fiber concrete is more cohesive than unreinforced concrete, and more power is needed to rotate the drum. Hence, the decreased load will help not only to reduce the total weight but also to maintain proper rotation of the drum. The same is true for pan mixers used in plants making precast concrete.

A well-proportioned FRC mix barely slides down the chute when discharged from the mixer. Slope of the chute should be increased slightly for easy discharge. If the mix is stiff, the concrete may have to be pulled down manually. The addition of high-range water-reducing admixture was found to eliminate this problem.

If concrete buckets are used, they should have steep hopper slopes and large gate openings. If fibers bridge the opening, FRC may not fall freely when the buckets are opened. The solution is to attach a vibrator to the side of the bucket that is activated when the bucket opens, facilitating the discharge of the concrete.

When FRC is transported through long vertical access shafts, concrete cannot be just dumped on the hopper. Fiber bridging may totally block the pipe. Vibrating the concrete in the hopper with an immersion vibrator will make the concrete fluid enough to facilitate flow. This method has been successfully used in the field.

FRC has also been pumped as far as 950 ft (285 m) through lines involving several bends and elbows and flexible hose at the end. The transport line included a 100 ft (30 m) vertical drop. A properly proportioned FRC mix can be pumped without problems. Even though the mix may look harsh, it flows through smoothly. In general it is advisable to (a) use pumps with a slightly larger capacity than needed for plain concrete, (b) use larger-diameter (>6 in., 150 mm) lines, (c) minimize the use of flexible hoses, (d) provide screens to avoid fiber clumps entering the pump, and (e) avoid mixtures with excessive slump. Pumping of mixtures with excessive slump was found to result in the separation of fibers from the paste, leading to formation of plugs.

6.2.1.3 Finishing. Only minor adjustments are needed in finishing FRC compared with plain concrete. Open slab surfaces should be struck off with a vibrating screed, preferably a metal screed, with slightly rounded edges. A "jitterbug" can be used in areas inaccessible to vibrating screeds. Chamfers or rounds should be provided at edges and corners to avoid protrusion of fiber ends.

Magnesium floats can be used to close up tears or open areas caused by the screeds. Wood floats normally leave rough surfaces with some fibers on the surface. For certain applications such as pavements further finishing may be necessary. If a texture is required for skid resistance, a broom or roller can be used before initial set. Burlap drags are not recommended because they can get caught on the fibers. Even though they are more difficult to use, larger floats provide flatter and better finishes. Floats should not be moved on edges when finishing, or they will pick up and move the fibers.

Loose fibers on the finished surface should be removed because they are a potential hazard, especially on airport runways used by high-speed jets. The fibers may become airborne missiles that result in injuries. In some airport pavement applications, "roller bugs" are used to press down fibers near the surface.

With careful workmanship, FRC can be finished to any desired smoothness and flatness. Extensive checks on a curved spillway invert and a flat stilling-basin floor with straight edges and templates showed maximum variations of only 0.04 to 0.08 in. (1 to 2 mm) over a 10 ft (3 m) length.

6.2.2 Concrete reinforced with polymeric fibers

The most common mixing method for polymeric fiber–reinforced concrete is batch mixing. The fibers can simply be added to the wet mix directly from bags, boxes, or feeders. It is recommended that the concrete be mixed for at least 10 minutes after the addition of fibers. For ready-mix concrete, fibers are usually added at the batch plant. If the fiber volume fraction exceeds 0.2%, special precautions may be needed in mixing. As mentioned earlier, the volume fraction currently is limited to about 0.1% for concrete (containing coarse aggregates) applications.

Transporting, placing, and finishing techniques for polymeric fiber–reinforced concrete are the same as those used for plain concrete. For some fiber types the slump values may be slightly less, but if vibration is used for compaction there should be no workability problems. The fiber concrete can be pumped using conventional equipment used for plain concrete. Excess water should be avoided because fibers that are lighter than water may tend to float. Occasionally, certain fibers tend to produce a hairy finish. This can be avoided with proper finishing techniques.

6.3 REFERENCES

6.1 ACI Committee 211, Standard Practice for Selecting Proportions for Normal, Heavy-weight, and Mass Concrete, ACI 211.1-89, American Concrete Institute, Detroit, Michigan, 1989, 38 pp.

6.2 ACI Committee 318, *Building Code Requirements for Reinforced Concrete,* ACI 318–89 *and Commentary,* ACI 318R–89, American Concrete Institute, Detroit, Michigan, 1989, 353 pp.

6.3 Ramakrishnan, V.; and Coyle, W. V. "Steel Fiber Reinforced Superplasticized Concretes for Rehabilitation of Bridge Decks and Highway Pavements," No. DOT/RSPA/DMA-50/84-2, available from NTIS, Springfield, Virginia, 1983, 408 pp.

6.4 ACI Committee 211, Standard Practice for Selecting Proportions for Structural Lightweight Concrete, ACI 211-2.91, American Concrete Institute, Detroit, Michigan, 1991, 18 pp.

6.5 Balaguru, P.; and Ramakrishnan, V. "Properties of Lightweight Fiber Reinforced Concrete," *Fiber Reinforced Concrete Properties and Applications,* American Concrete Institute, Detroit, Michigan, SP105, 1987, pp. 305–322.

6.6 Dipsia, M. "Lightweight High Strength Concrete," M.S. Thesis, submitted to Rutgers–The State University of New Jersey, New Brunswick, 1987.

6.7 ACI Committee 544, "Guide for Specifying Mixing, Placing, and Finishing Steel Fiber Reinforced Concrete," ACI SP81, 1984, pp. 441–447.

Properties of Freshly Mixed FRC Containing Coarse Aggregates

This chapter deals with the properties of freshly mixed FRC containing coarse aggregates used for cast-in-place applications. Typical fiber loadings range from 50 to 250 lb/yd^3 (30 to 150 kg/m^3) of steel fibers or 1 to 8 lb/yd^3 (0.6 to 4.8 kg/m^3) of polymeric fibers. Fiber lengths generally range from 0.5 in. to 2.5 in. (12 to 64 mm). Properties of cement composites containing higher volume fractions of fibers and manufactured using special techniques such as shotcreting are covered in Chapters 12 to 15.

The quality control parameters most often used for fresh concrete are workability and air content. The other parameters measured include unit weight, concrete temperature, air temperature, and relative humidity. Workability can be measured using a standard slump cone test, an inverted slump cone test, or a V-B test. For FRC, especially for stiff mixes, the standard slump cone test does not provide an accurate indication of workability. Hence, either the inverted slump cone or the V-B test is recommended. However, for fluid mixes with 2 in. (50 mm) or more slump, the standard slump cone tests can still be used as a quality control measure.

In general, addition of fibers makes the mix more stiff. Water-reducing admixtures may be added to improve the workability. Typically, a higher dosage of air-entraining admixture is needed for steel fiber–reinforced concrete to maintain the air content. Polymeric fibers have no effect on air entrainment if volume fractions are less than 0.5%. They are known to entrap more air when volume fractions exceed 0.5%.

7.1   Workability Tests

Adequate workability is needed for proper placement, consolidation, and finishing. Only the minimum amount of water that is needed should be in the mix because excess water results in segregation and bleeding as well as inferior hardened concrete in terms of both strength and durability. The test methods described in the following sections are used for estimating the workability and for controlling the amount of water in the mix. These test methods are applicable for both steel and polymeric fiber–reinforced concrete as long as the fiber volume fractions are limited to 2% in the case of steel fibers and 0.5% in the case of polymeric fibers.

7.1.1   Slump cone test

The slump test, described in ASTM C143, is the most commonly used workability test and is described in almost all books on concrete. As mentioned earlier, this test can be used for FRC only if slump values exceed 2 in. (50 mm). It can also be used to monitor the FRC consistency from batch to batch.

7.1.2   Inverted slump cone test

This test, specifically developed for FRC, is described in ASTM C995 [7.1]. In this test, the time, in seconds, needed for concrete to flow through a standard slump cone kept in the inverted position is determined. The flow of concrete is aided by an immersion vibrator to simulate the workability of concrete compacted by vibration. This test is not suitable for concrete with more than a 4 in. (100mm) slump made using water-reducing admixtures because the concrete will flow through the cone too quickly. For such fluid mixtures, the standard slump test is recommended.

Figure 7-1 shows the relationship between standard and inverted slump cone test results for plain and fiber concrete [7.2]. It can be seen that when standard slump is less than 3 in. (75 mm), FRC flows better under vibration than plain concrete. For mixes with slump of more than 3 in. (75 mm), the difference between FRC and plain concrete is negligible. The variations are similar for polymeric fiber concrete except that the standard slump value is reduced drastically if the fiber volume fraction exceeds 0.2 percent. Typical results are presented in Section 7.4.

7.1.3   V-B test

This test is described in British Standards Institute Standard BS1881, "Methods of Testing Concrete, Part 2" [7.3]. In this test, the concrete is subjected to external vibration. The consistency of the mix is

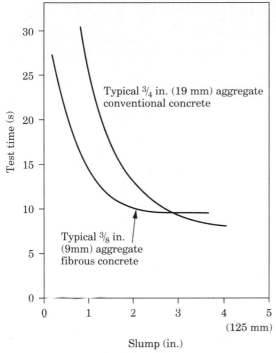

Figure 7-1 Slump, as determined in the slump cone test, versus time, as determined in the inverted slump cone test [7.2].

determined by the time, in seconds, needed for a certain amount of concrete to flow. The V-B consistometer is not suitable for field use because of its size and weight.

Figure 7-2 shows the relations between standard slump, inverted slump, and V-B time [7.2, 7.4]. From this figure it can be seen that inverted cone and V-B tests provide comparable results. Their relation with respect to standard slump is parabolic even though a concrete with a slump of 2 in. (50 mm) has an inverted cone time of about 7 seconds compared with 3 seconds in V-B time. Inverted slump cone times are typically higher than V-B times for all levels of consistency.

7.2   Tests for Air Content, Yield, and Unit Weight

ASTM tests described in ASTM C138, C173, and C231 for plain concrete can also be used for FRC [7.1]. Air content can be measured using gravimetric (C138), volumetric (C173), or pressure (C231) methods. When testing FRC, consolidation using internal or external vibration is recommended rather than rodding. Some vibration is essential when

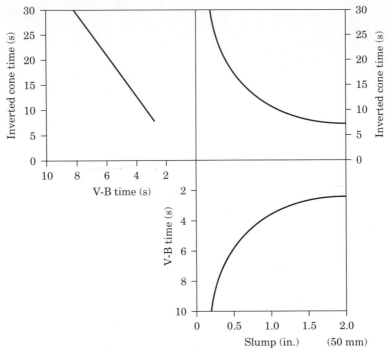

Figure 7-2 Relationship between slump, V-B time, and inverted cone time [7.2, 7.4].

the mix is stiff and cannot be compacted properly using rodding. Use of vibration is permitted in ASTM Standard C138. The determination of unit weight and yield are covered in ASTM C138. These are simple tests common to both plain and fiber-reinforced concrete.

7.3   Steel Fiber–Reinforced Concrete

Factors that influence the behavior of freshly mixed steel fiber-reinforced concrete include matrix composition, fiber type, fiber geometry, fiber volume fraction, and fiber-matrix interfacial bond characteristics. As mentioned earlier, addition of fibers will make the composite look stiffer. However, the mix can be highly workable if vibrators are used for placement and compaction.

All fiber geometries mentioned in Chapter 5 have been used in concrete. Typically, longer fibers and fibers with higher aspect (length/diameter) ratios tend to reduce workability. Balling (discussed in Chapter 6) during mixing and placing may be eliminated for certain fiber configurations by using smaller maximum–size aggregates, lower aspect ratios, and lower volume fractions.

Balaguru and Ramakrishnan have reported the results of an extensive investigation using fibers with hooked ends [7.5]. Variations in slump, air content, and V-B time were studied for various cement and fiber contents. The water-cement ratio was varied from 0.28 to 0.50. A large number of mixtures with and without fibers were evaluated. Tables 7.1 and 7.2 present the mix proportions for concretes with and without fibers, respectively. Slump, V-B time and air content values for these mixtures are shown in Tables 7.3 and 7.4, as well as air temperatures since workability is sensitive to air temperatures. These results reflected the influences of water-cement ratio, cement content, fiber content, and high-range water-reducing admixture dosage. Extensive details are presented because these are field usable, realistic mixtures. The details can also be used for proportioning mixtures.

The analysis of variance showed that the factors that influence slump (in descending order) are the following [7.6]:

- Water-cement ratio
- Combined effect of water-cement ratio and cement content
- Cement content
- High-range water-reducing admixture content
- Fiber content
- A combination of water-cement ratio, cement content, and fiber content

Within the range of the test variables, especially cement and fiber contents, the influence of fiber content was found to be 1000 times less than the water-cement ratio and 300 times less influential than the combination of cement content and water-cement ratio. This result leads to the observation that fiber contents in the range of 55 to 100 lb/yd^3 (32 to 60 kg/m^3) have very little influence on slump if high-range water-reducing admixtures are used.

Variance analysis for V-B time showed the following variables to be influential. Again the variables are listed in descending order of importance.

- High-range water-reducing admixture content
- Water-cement ratio
- Cement content
- Fiber content
- A combination of water-cement ratio, cement content, and fiber content

The accuracy of analysis, measured using the coefficient of determination, R^2, was 0.725 for slump and 0.629 for V-B time.

TABLE 7.1 Mixture Designation and Proportions: Steel Fiber Concrete [7.5, 7.6]

Mixture designation	W/C Ratio (by weight)	Cement (lb/yd³)*	Fibers (lb/yd³)*	High-range water-reducer (wt %)‡	Air-entraining agent (wt %)‡
F1	0.36	611	55	1.0	0.20
F2,F2R	0.32	564	65	0.8	0.50
F3,F3R	0.32	658	65	0.8	0.20
F4,F4R	0.40	564	65	0.8	0.10
F5,F5R	0.40	658	65	0.8	0.13
F6,F6R	0.32	564	65	1.2	0.20
F7,F7R	0.32	658	65	1.2	0.13
F8,F8R	0.40	564	65	1.2	0.10
F9,F9R	0.40	658	65	1.2	0.10
F10	0.36	611	75	0.6	0.20
F11	0.28	611	75	1.0	0.40
F12	0.36	517	75	1.0	0.20
F13	0.36	705	75	1.0	0.10
F14	0.44	611	75	1.0	0.10
F15	0.36	611	75	1.4	0.20
F16,F16R	0.32	564	85	0.8	0.30
F17,F17R	0.32	658	85	0.8	0.20
F18,F18R	0.40	564	85	0.8	0.13
F19,F19R	0.40	658	85	0.8	0.16
F20,F20R	0.32	564	85	1.2	0.40
F21,F21R	0.32	658	85	1.2	0.20
F22,F22R	0.40	564	85	1.2	0.13
F23,F23R	0.40	658	85	1.2	0.13
F24	0.36	611	95	1.0	0.20
F25	0.36	611	75	1.0	0.30
F26	0.36	611	75	1.0	0.25
F27	0.36	611	75	1.0	0.40
F28	0.36	611	75	1.0	0.25
F29	0.36	611	75	1.0	0.30
F30	0.36	611	75	1.0	0.13
F31	0.36	611	75	1.0	0.20
F32,F32R	0.40	611	75	1.0	0.15
F33,F33R	0.30	799	75	1.2	0.20
F34	0.40	611	75	1.0	0.15
F35	0.30	799	75	1.2	0.20
F36	0.40	611	75	1.0	0.15
F37	0.30	799	75	1.2	0.20
F38	0.40	611	75	1.0	0.15
F39	0.40	611	75	1.0	0.15
F40	0.40	611	75	1.0	0.15
F41	0.43	690	100	1.3	−†
F42	0.50	690	100	−†	−†
F43	0.43	690	100	0.86	0.101
F44	0.43	690	100	1.09	0.185
F45	0.43	690	100	1.00	0.254
F46,F46R	0.30	799	75	1.2	0.20
F47,F47R	0.40	611	75	1.0	0.15
F48	0.40	611	75	1.0	0.15
F49	0.30	799	75	1.2	0.10
F50	0.30	799	75	1.2	0.20

Note: Replicate mixture proportions (designated by R) are the same as those of the original.

*1 lb/yd³ = 0.59 kg/m³   †Data discarded   ‡Percent by weight of cement

TABLE 7.2 Mixture Designation and Proportions: Plain Concrete [7.5, 7.6]

Mixture designation	W/C ratio (by weight)	Cement (lb/yd³)*	Air-entraining agent (wt %)‡	High-range water-reducer (wt %)‡
		Mixture proportions		
S1	0.38	611	0.1	1.0
S2,S2R	0.33	654	0.095	0.8
S3,S3R	0.33	658	0.095	0.8
S4,S4R	0.43	564	0.070	0.8
S5,S5R	0.43	658	0.095	0.8
S6,S6R	0.33	564	0.130	1.2
S7,S7R	0.33	658	0.120	1.2
S8,S8R	0.43	564	0.095	1.2
S9,S9R	0.43	658	0.085	1.2
S10	0.38	611	0.095	0.6
S11	0.28	611	0.120	1.0
S12	0.38	517	0.200	1.0
S13	0.38	611	0.050	1.0
S14	0.38	705	0.095	1.0
S15	0.48	611	0.050	1.0
S16	0.38	611	0.095	1.4
S17,S17R	0.33	564	0.130	0.8
S18,S18R	0.33	658	0.120	0.8
S19,S19R	0.43	564	0.130	0.8
S20,S20R	0.43	658	0.095	0.8
S21,S21R	0.33	564	0.150	1.2
S22,S22R	0.33	658	0.095	1.2
S23,S23R	0.43	564	0.095	1.2
S24,S24R	0.43	658	0.095	1.2
S25	0.38	611	0.085	1.0
S26	0.28	705	0.095	1.2
S27	0.28	705	0.095	1.6
S28	0.30	705	0.095	1.2
S29	0.30	705	0.120	1.6
S30	0.28	752	0.120	1.2
S31	0.28	752	0.130	1.4
S32	0.28	752	0.130	1.4
S33	0.28	799	0.130	1.2
S34	0.28	799	0.130	1.0
S35	0.28	799	0.160	1.0
S36	0.28	846	0.130	1.0
S37	0.28	846	0.130	1.2
S38	0.28	846	0.130	0.8
S39	0.30	799	0.130	0.8
S40	0.30	799	0.130	1.0
S41	0.43	564	0.130	0.8
S42	0.43	564	0.095	0.8
S43	0.33	564	0.095	1.2
S44	0.33	658	0.095	1.2
S45	0.28	611	0.12	1.0
S46	0.38	517	0.12	1.0
S47	0.38	611	0.095	1.0
S48	0.38	611	0.070	1.0

(continued)

TABLE 7.2—cont'd

Mixture designation	W/C ratio (by weight)	Cement (lb/yd³)*	Air-entraining agent (wt %)‡	High-range water reducer (wt %)‡
			Mixture Proportions	
S49	0.38	611	0.095	1.4
S50	0.33	564	0.095	0.8
S51	0.43	564	0.095	0.8
S52	0.33	564	0.11	1.2
S53,S53R	0.38	611	0.08	1.0
S54,S54	0.28	799	0.18	1.2
S55	0.38	611	0.08	1.0
S56	0.28	799	0.18	1.2
S57	0.38	611	0.08	1.0
S58	0.28	799	0.18	1.2
S59	0.38	611	0.08	1.0
S60	0.28	799	0.18	1.2
S61	0.38	611	0.08	1.0
S62	0.28	799	0.18	1.2
S63	0.38	611	0.08	1.0
S64	0.28	799	0.18	1.2

Note: Replicate mixture proportions (designated by R) are the same as those of the original.
*1 lb/yd³ = 0.59 kg/m³
‡Percent by weight of cement

TABLE 7.3   Properties of Plastic Fiber Concrete [7.5, 7.6]

Mixture designation	Air Temperature (°F)*	Air content (vol %)	Slump (in.)†	V-B time (s)
F1	88	5.7	2.25	8.0
F2	80	4.4	0.0	15.0
F2R	72	3.4	0.0	9.0
F3	80	9.6	6.25	1.5
F3R	76	5.2	0.625	6.0
F4	79	4.2	1.75	8.6
F4R	73	7.2	2.0	8.0
F5	80	6.8	4.5	1.5
F5R	70	9.8	7.625	0.0
F6	80	4.2	0.0	16.0
F6R	74	5.4	1.0	13.5
F7	80	6.8	7.875	0.0
F7R	78	6.0	2.75	5.5
F8	80	8.0	8.0	0.0
F8R	78	8.2	7.875	0.0
F9	84	10.2	7.5	0.0
F9R	70	10.0	9.25	0.0
F10	84	3.5	0.0	27.0
F11	80	3.8	0.0	16.0
F12	84	3.2	0.0	19.5
F13	80	9.2	9.125	0.0
F14	77	7.2	8.0	0.0

(continued)

TABLE 7.3—cont'd

Mixture designation	Air temperature (°F)*	Air content (vol %)	Slump (in.)†	V-B time (s)
F15	85	7.5	5.25	3.0
F16	80	4.8	0.0	19.0
F16R	72	3.0	0.0	15.0
F17	83	3.6	0.0	28.0
F17R	74	5.8	0.0	14.5
F18	79	3.8	0.375	16.0
F18R	75	5.0	0.5	10.8
F19	80	9.6	7.3	0.0
F19R	78	8.0	6.0	3.0
F20	83	5.2	0.0	16.0
F20R	76	7.0	1.375	6.0
F21	86	4.2	0.5	7.8
F21R	74	5.4	1.0	11.0
F22	86	5.0	1.625	11.0
F22R	75	5.8	0.0	8.0
F23	80	8.8	8.5	0.0
F23R	74	11.2	8.25	0.0
F24	88	4.6	0.25	11.7
F25	81	9.2	4.5	1.4
F26	83	8.6	4.5	4.6
F27	82	7.4	1.5	6.2
F28	83	6.8	2.5	6.0
F29	90	6.0	1.3	6.4
F30	80	4.0	1.75	11.7
F31	81	4.5	1.5	9.4
F32	68	9.0	7.00	0
F32R	72	5.2	3.125	4.0
F33	68	3.3	0.75	11.0
F33R	70	3.8	1.50	6.0
F34	76	8.0	4.25	3.0
F35	76	4.4	2.5	8.0
F36	74	10.8	7.75	0
F37	78	6.6	4.0	3.0
F38	68	4.2	2.25	8.0
F39	66	3.8	2.0	7.5
F40	75	4.2	1.0	9.5
F41	72	2.2	1.0	14.0
F42	68	1.2	6.0	1.8
F43	73	2.6	3.25	3.0
F44	71	3.0	4.0	4.8
F45	71	4.5	4.0	4.2
F46	71	7.4	5.0	5.0
F46R	69	8.4	3.75	8.0
F47	71	8.2	6.375	0
F47R	70	10.8	7.50	0
F48	71	9.10	7.0	0
F49	70	3.6	0	11.0
F50	74	4.8	4.0	3.0

Note: Replicate mixture proportions (designated by R) are the same as those of the original.
*°C = (°F − 32) × $\frac{5}{9}$
†1 in. = 25.4 mm

TABLE 7.4   Properties of Plastic Plain Concrete [7.5, 7.6]

Mixture designation	Air temperature (°F)*	Air content (vol %)	Slump (in)†	V-B time (s)
S1	76	9.3	7.5	0
S2	68	3.6	0	6.5
S2R	68	3.9	0.5	4.0
S3	69	4.1	1.5	4.0
S3R	70	3.0	0	11.0
S4	70	7.2	7.625	0
S4R	68	4.6	2.75	2.5
S5	67	8.6	9.0	0
S5R	67	8.0	9.625	0
S6	66	4.9	1.125	5.0
S6R	70	3.2	0	9.0
S7	71	9.2	8.125	0
S7R	71	11.0	8.5	0
S8	76	7.2	7.625	0
S8R	71	9.0	4.0	2.5
S9	67	9.2	10.5	0
S9R	69	6.6	9.750	0
S10	72	7.4	3.625	1.5
S11	76	4.4	0	7.5
S12	73	6.8	1.625	4.5
S13	74	8.0	8.625	0
S14	72	9.0	8.25	0
S15	68	4.0	10.25	0
S16	70	10.2	9.5	0
S17	68	6.4	0	7.5
S17R	68	6.6	0	7.5
S18	71	7.0	3.125	
S18R	69	6.8	3.5	2.5
S19	71	8.8	3.25	3.0
S19R	66	11.8	7.375	0
S20	69	9.6	8.125	0
S20R	74	9.2	7.5	0
S21	68	5.6	1.125	5.5
S21R	68	9.2	4.375	2.5
S22	67	7.6	5.5	1.5
S22R	68	5.4	2.0	4.0
S23	68	11.2	9.25	0
S23R	72	9.2	7.0	0
S24	75	9.0	9.5	0
S24R	72	9.8	9.75	0
S25	68	10.4	8.125	0
S26	75	4.8	0	7.5
S27	71	4.3	3.5	6.0‡
S28	68	4.6	2.25	5.0
S29	68	4.2	7.5	0
S30	70	3.0	0	7.5
S31	65	9.6	9.0	0
S32	68	7.0	9.0	0
S33	70	6.9	4.5	1.5

(continued)

TABLE 7.4—cont'd

Mixture designation	Air temperature (°F)*	Air content (vol %)	Slump (in)†	V-B time (s)
S34	70	3.8	2.0	3.5‡
S35	70	4.3	5.75	1.0
S36	70	4.0	4.0	3.5
S37	67	4.6	7.125	0
S38	69	3.4	0	7.0
S39	70	3.4	1.375	4.7
S40	70	4.0	2.125	3.5
S41	73	11.4	7.625	0
S42	68	10.6	8.0	0
S43	67	5.3	2.0	5.0
S44	72	6.1	4.125	—
S45	74	5.6	0	7.5
S46	73	4.2	1.125	6.0
S47	68	11.8	9.0	0
S48	74	8.0	8.625	0
S49	70	10.2	9.5	0
S50	68	5.0	0	8.0
S51	70	4.5	1.5	5.0
S52	68	6.0	1.75	4.5
S53	69	9.2	7.0	0
S53R	71	8.6	6.0	1.0
S54	70	8.6	3.0	2.0
S54R	72	8.6	4.0	2.0
S55	83	3.2	2.5	3.8
S56	84	4.0	0	7.6
S57	72	8.1	3.375	2.0
S58	74	9.8	5.375	2.0
S59	70	4.2	2.25	3.0
S60	70	6.2	2.0	3.0
S61	71	9.8	8.5	0
S62	71	5.0	3.125	3.0
S63	70	10.2	8.0	0
S64	72	6.8	6.125	1.0

Note: Replicate mixture proportions (designed by R) are the same as those of the original.

*°C = (°F − 32) × $\frac{5}{9}$

†1 in. = 25.4 mm

‡Not used in the computation; slump is not consistent with flow table spread.

Note that the amount of water reducer plays a more important role in influencing V-B time than slump. Apparently, the water reducer is much more effective under vibration. Water reducer was almost twice as influential as water-cement ratio. Fiber content was about 20 times less influential than water reducer and 10 times less influential than water-cement ratio. Considering the variables tested, the authors found V-B time is more sensitive than the slump to fiber presence.

The factors that influence air content were found to be the following:

- Water-cement ratio
- Combination of water-cement ratio and cement content
- Cement content
- High-range water-reducer dosage
- Fiber content
- Combination of water-cement ratio, cement content, and fiber content

Note that air-entraining agent dosage was not used as a variable. It had already been found that a larger dosage of air-entraining admixture is needed to produce the same amount of air content in the presence of fibers [7.6].

The loss of workability with time is a problem in the construction field. The problem is more acute when a high-range water-reducing admixture is used, because these admixtures lose their effectiveness with time [7.7]. Loss of slump and reduction in air content with time, measured for comparable plain and fiber concretes, are presented in Figures 7-3 through 7-6 [7.7]. The details of the proportions for the four mixes are shown in Table 7.5. The fibers used, which had hooked ends, were 50 mm (2 in.) long, 0.5 mm (0.02 in.) in diameter and were made of low-carbon steel. From Figures 7-3 and 7-4 it can be seen that the

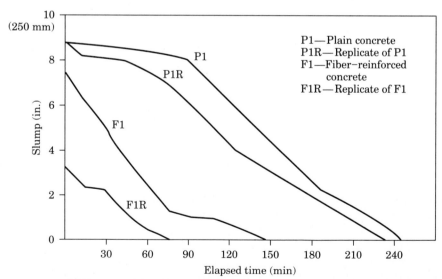

Figure 7-3 Comparison of slump loss (F1 and P1) in replicated mixtures of plain and fiber-reinforced concrete [7.7].

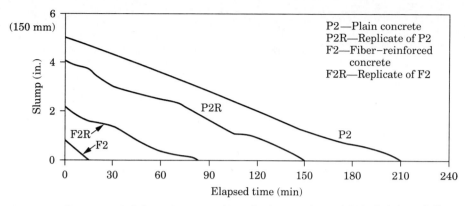

Figure 7-4 Comparison of slump loss in replicated mixtures (F2 and P2) of plain and fiber-reinforced concrete [7.7].

slump values fall considerably after about 30–60 minutes. The rate of drop for fiber and plain concrete is about the same, but the absolute values decrease much faster for fiber concrete.

The reduction in air content seems to be less dramatic (Figures 7-5 and 7-6). While the values drop considerably in the first 30 minutes, they tend to stabilize after 60 minutes. The variation with time can be considered about the same for fiber and plain concretes.

It has been shown that concrete can be retempered using a high-range water-reducing admixture without adversely affecting its mechanical properties [7.6]. Retempering is a process in which either water or admixtures are added to improve workability. Retempering more than once is not recommended even though experimental results show that retempering can be done twice.

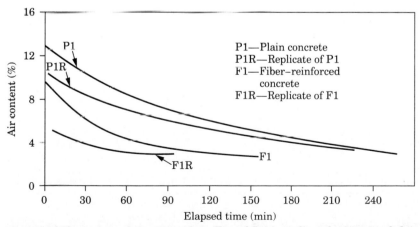

Figure 7-5 Comparison of air content loss (F1 and P1) in replicated mixtures of plain and fiber-reinforced concrete [7.7].

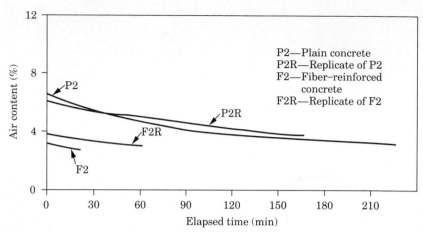

Figure 7-6 Comparison of air content loss in replicated mixtures (F2 and P2) of plain and fiber-reinforced concrete [7.7].

Properties of lightweight fiber-reinforced concrete resembles that of normal-weight concrete except for air entrainment [7.8]. Controlling air content is the primary problem in lightweight fiber concrete. By incorporating high-range water-reducing admixtures, one can formulate lightweight fiber concrete (unit weight of about 110 lb/ft^3, 65 kg/m^3) that is highly workable.

Investigations conducted using deformed fibers indicate that workability can be controlled by adjusting the dosage of high-range water-reducing admixtures [7.9]. The tests were conducted using 2 in. (50 mm) long fibers at 100 lb/yd^3 (60 kg/m^3). Fibers with both circular and rectangular cross sections were used. A slump of 0.75 in. (19 mm) could be obtained even at a water-cement ratio of 0.4 using high-range water-reducing admixtures.

Straight steel fibers are not commonly used anymore. If straight fibers have to be used for some reason, an aspect ratio of less than 200 and a volume fraction of less than 2 percent are recommended to

TABLE 7.5   Mixture Proportions Used for Slump Loss Study [7.7]

Mixture designation	F1*	F2*	P1	P2
Cement content (lb/yd^3)†	611	799	611	799
Water-cement ratio (by weight)	0.4	0.3	0.38	0.28
High-range water-reducing admixture‡	1.0	1.2	1.0	1.2
Air-entraining agent‡	0.15	0.20	0.1	0.13

*Fiber-reinforced concrete mixtures F1 and F2 had a fiber content of 75 lb/yd^3 (44 kg/m^3).
†1 lb/yd^3 = 0.59 kg/m^3
‡Percent by weight of cement

avoid balling of fibers. The maximum coarse aggregate size should be limited to 3/8 in. (9 mm).

Fibers are also being extensively used for refractory concretes. In this application stainless steel fibers are used most because of the corrosive environment in refractory concrete. Volume fractions of 0, 0.5, 1.0, and 1.5 percent produced V-B times of 6.0, 6.0, 10.0, and 11.0 seconds respectively [7.10]. All the mixes were found to be highly workable and finishable. The slump was reported to be almost zero for mixes containing 1.0 and 1.5 percent of stainless steel fibers manufactured using melt-extract process. These fibers were 1 in. (25 mm) long and 0.018 in. (0.46 mm) in diameter.

7.4   Polymeric Fiber–Reinforced Concrete

Polypropylene, Nylon 6, and polyester fibers are some of the polymeric fibers currently being used to reduce the plastic shrinkage cracking of concrete. Typical fiber volume fraction is 0.1%, although percentages as low as 0.05 have been tried. Polypropylene fibers are available both in fibrillated form and as single filaments.

Investigations have been carried out on the workability of polymeric fiber–reinforced concrete for a number of variables, including fiber type, fiber length, fiber content, and concrete strength [7.11–7.14].

Wu et al. [7.10] and Balaguru [7.11] investigated the effects of various polymer fiber types at volume fractions ranging from 0.075% to 0.5%. Slump, inverted slump, and air content of concretes containing various volume fractions of polypropylene, Nylon 6, and polyester fibers are shown in Table 7.6. where N6 refers to Nylon 6 fibers, whereas MF and NU indicate fibrillated polypropylene and polyester fibers respectively. All these fibers were 0.75 in. (19 mm) long. N6M refers to mixes containing equal volumes of Nylon 6 fibers that are 0.75, 1.0, and 1.5 in. (19, 25, and 38 mm) long. Designations 075, 100, and 150 refer to fiber volume fractions of 0.75, 1.00, and 1.50 lb/yd^3 (0.45, 0.6, and 0.9 kg/m^3). The mix designated as CON was the control mix that had no fibers.

The concrete targeted for a 28 day compressive strength of 3000 psi (20 MPa) had cement, natural sand, and crushed stone contents of 517, 1370, and 1800 lb/yd^3 (307, 813, and 1068 kg/m^3) respectively. The water-cement ratio was 0.57. Both water-reducing and air-entraining admixtures were used. The targeted air content was 6%.

Figure 7-7 shows the variation of slump values for the concretes having three concentrations of Nylon 6 fibers. Even at low volume fractions, the fibers make the mix cohesive, as one can see by the consistent reduction of slump values. The workability is affected very little, as shown by the very low flow times in inverted slump cone tests (Table 7.6).

TABLE 7.6   Properties of Polymeric Fiber–Reinforced Fresh
Concrete: Volume Fraction Less Than 0.1 percent [7.12]

Mix designation	Slump (in.)	Inverted slump (s)	Air content (%)	Unit weight (lb/ft^3)
CON*	7.00	—	5.50	145.7
N6 075	5.50	3	6.00	148.2
N6 100	5.25	4	6.00	143.1
N6 150	4.00	4	5.00	147.4
N6M 100	4.00	4	5.25	146.5
MF 100	5.25	4	5.25	148.2
MF 150	6.50	3	6.00	144.8
NU 100	5.25	3	5.75	148.2

*Plain concrete
1 lb/ft^3 = 16 kg/m^3
1 in. = 25.4 mm

Figures 7-8 and 7-9 show comparative slump values for Nylon 6, polyester, and polypropylene fibers. As mentioned earlier, Nylon 6 and polyester fibers incorporated in the concrete were made of single filaments, whereas polypropylene fibers were fibrillated. At a fiber volume fraction of 1 lb/yd^3 (0.6 kg/m^3), there is very little difference between fiber types. When the volume fraction is increased to 1.5 lb/yd^3 (0.9 kg/m^3), fibrillated polypropylene fibers show less reduction in slump values than single filament Nylon 6 fibers (Figure 7-9). The difference may be attributed both to fibrillation and to fiber count. Nylon 6 fibers had a much smaller fiber diameter and much larger fiber count. Longer fibers, as indicated by mixture N6M, result in greater reduction of slump (Figure 7-8).

The variations in air content within fiber types and fiber volume fractions are not significant (Table 7.6). In each case the air content was within the permissible limits of the targeted value of 6%.

The aforementioned fiber types were also evaluated at a fiber content of 8 lb/yd^3 (4.8 kg/m^3). The slump values decreased from 6 in. (150 mm) for plain concrete to about zero slump for fiber concretes. However, the inverted slump cone time showed much less variation. The polypropylene fibers had almost the same inverted cone time as the plain concrete. Nylon 6 and polyester fibers had 20% and 100% increases in inverted slump cone time respectively. However, the slump cone time was still less than 11 seconds for all mixes, and they were all highly workable.

Results for fiber volume fractions of 0.1%, 0.2%, and 0.3% were reported by Ramakrishnan et al. [7.13]. Fibrillated 0.75 in. (19 mm) long polypropylene fibers were used in a concrete mix containing 658 lb/yd^3 (390 kg/m^3) cement. The 28 day compressive strength was about 6000 psi (40 MPa). The water-cement ratio was 0.4, and a high-range water-reducing admixture was used to improve the workability. The

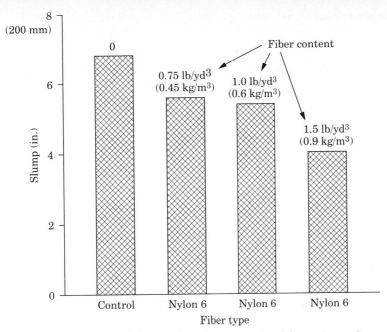

Figure 7-7 Comparison of slump values for a control and for various volume fractions of Nylon 6 [7.12].

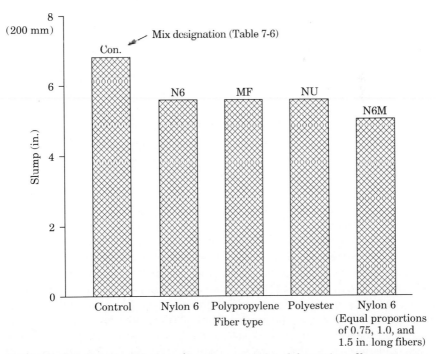

Figure 7-8 Comparison of slump values for a control and for various fiber types at a constant fiber content of 1.0 lb/yd³ (0.6 kg/m³) [7.12].

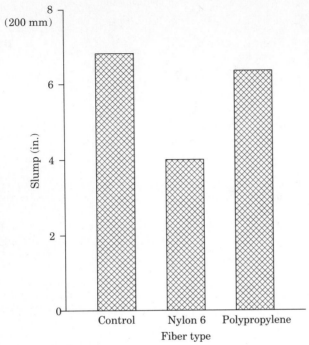

Figure 7-9 Comparison of slump values for a control and for various fiber types at a constant fiber content of 1.5 lb/yd³ (0.9 kg/m³) [7.12].

slump values for plain concrete and for concrete containing 0.1, 0.2, and 0.3 volume percent of fibers were 8.25, 6.00, 7.00, and 5.75 in. (210, 152, 178, and 146 mm) respectively. The plain concrete mix had a higher dosage of water-reducing admixture. Hence, if the fiber volume fraction is less than 0.5%, fiber addition can be assumed to influence workability only to a limited extent [7.13].

A volume fraction of 1% was found to increase the inverted slump cone time considerably [7.14]. In this investigation, fibrillated, 0.75 in. (19 mm) long polypropylene fibers were used at volume fractions of 0.1%, 0.5%, and 1%. The mix proportion details and the plastic concrete properties are shown in Tables 7.7 and 7.8 respectively.

For the first group of specimens, cement content and water-cement ratio were kept constant for all volume fractions. The effect of added fibers can be seen on both standard slump values and inverted slump cone times. The slump decreased from 235 mm for the control to almost zero for a fiber volume fraction of 1%. Inverted slump cone times increased from about zero to 90 seconds. The V-B time showed less change than inverted slump cone time.

For the second group of specimens, high-range water-reducing admixture dosages were adjusted to obtain comparable workability for plain and fiber concretes. The results show that workability can be controlled well by adjusting the quantity of admixture. The range of inverted slump cone times was narrower for the second group than for the first group.

The change in air content for the fiber volume range of 0.1% to 1% (Table 7.8) is within the experimental variations. Higher air content typically produces higher slump for both plain and fiber concretes.

In summary, workability of polymeric fiber-reinforced concrete can be assumed to be the same as that of the corresponding plain concrete for low-fiber volume fractions ($\leq 0.1\%$). If the volume fraction exceeds 0.5%, some modifications may be needed to keep the workability the same. The most efficient way to improve the workability is to use high-range water-reducing admixtures.

Air content does not seem to be influenced by the presence of polymeric fibers (Nylon 6, polyester, and fibrillated polypropylene) for volume fractions up to 0.5%. These observations may not be applicable to other types of fibers.

7.5   Other Fibers

Various types of other polymeric, glass, carbon, and naturally occurring fibers have been used in cement composites, mostly in thin sheet applications involving cement or cement mortar matrix. Limited investigations have been carried out using these fibers in concrete-containing coarse aggregates. Documented results on workability are rare in the published literature.

7.6   Relations between Slump, V-B Time, and Inverted Slump Cone Time and Their Significance

As mentioned earlier, the slump test does not provide a good indication of the workability of fiber-reinforced concrete. However, the test still has to be used for highly workable mixes because the concrete flows through the inverted slump cone too quickly. The V-B test can be used for most mixes, but the apparatus is too cumbersome for field use. Hence, a combination of all three tests has to be used in the field and the laboratory. This section provides a brief discussion of the interrelationships among the three test measures, for one measure (say, V-B time) can be used as a rough estimate of the other (slump). The field engineer might need this conversion, for example, to compare V-B time obtained in the laboratory with slump value obtained in the field.

Typically, inverted slump cone times are much higher than V-B times because it takes more time for the mix to flow through the cone.

TABLE 7.7 Mix Quantities for Polymeric Fiber Concrete for a 0.09 m³ (=0.12 yd³) Batch [7.14]

Mix series designation	Fiber* (kg)	Coarse aggregate (kg)	Sand (kg)	Cement (kg)	W/C ratio	Superplasticizer dosage (mL)	Air-entraining agent dosage (mL)
CON I†	None	85.19	85.19	35.73	0.40	180	25
I 0.1	.080	85.19	85.19	35.73	0.40	240	25
I 0.1	.080	85.19	85.19	35.73	0.40	240	25
I 0.5	.401	85.19	85.19	35.73	0.40	330	25
I 0.5	.401	85.19	85.19	35.73	0.40	330	25
I 1.0	.801	85.19	85.19	35.73	0.40	380	25
I 1.0	.801	85.19	85.19	35.73	0.40	550	30
CON II†	None	97.89	77.02	29.48	0.50	78	18
II 0.1	.080	97.89	77.02	29.48	0.50	78	18
II 0.5	.401	82.46	64.64	39.42	0.42	156	22
II 1.0	.801	82.46	64.64	39.42	0.50	208	23

*Fibrillated polypropylene fibers
†CON−control
1 kg = 2.202 lb
1 mL = 0.034 oz

TABLE 7.8   Properties of Polymeric Fiber-Reinforced Fresh Concrete: Fiber Volume Fraction Less Than or Equal to 1 Percent [7.14]

Mix series designation	Slump (mm)	Concrete temperature (°C)	Air content (vol %)	V-B time (s)	Inverted cone time (s)
CON I*	235	21.9	5.2	0.7	†0
I 0.1	209	21.2	9.0	2.0	†0
I 0.1	148	26.6	4.4	2.0	7.8
I 0.5	133	21.6	5.4	3.5	23.0
I 0.5	95	25.4	3.2	3.7	23.7
I 1.0	3	21.1	4.4	10.0	90.0
I 1.0	3	25.1	3.2	9.5	62.0
CON II*	95	26.7	5.2	2.2	9.5
II 0.1	31	26.3	3.5	5.5	15.3
II 0.5	165	27.1	7.6	1.5	6.7
II 1.0	79	26.8	6.9	3.7	56.0

*CON—control
†Concrete goes through fast; therefore the time cannot be recorded.
1 mm = 0.04 in.
°F = (°C × $\frac{9}{5}$ + 32)

Figure 7-10 shows that a linear relationship exists between these two times and that the simpler inverted slump cone test provides a dependable indication of energy needed for compaction [7.15]. The linear relationship was found to exist for both straight and hooked-end steel fibers.

The relationships between slump and V-B time for plain concrete, plain concrete with a high-range water-reducing admixture, and

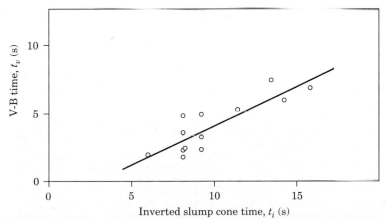

Figure 7-10 Correlation between V-B time and inverted slump cone time [7.15].

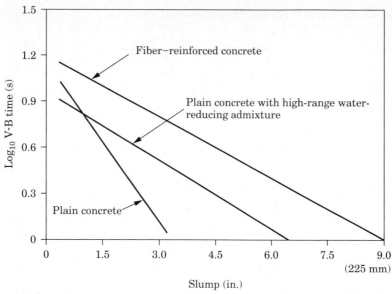

Figure 7-11 Comparison of slump and V-B time for three types of concrete [7.5].

fiber-reinforced concrete are shown in Figure 7-11 [7.5]. Decrease in slump results in logarithmic increase of V-B time. Figure 7-11 also shows that fiber-reinforced concrete is more cohesive under vibration than plain concrete. For example, plain concrete with no water-reducing admixture collapsed under vibration (V-B time ≈ 0) when the slump is more than 3 in. (75 mm), whereas the V-B times could be measured for slumps as high as 6 in. (150 mm) in concrete with admixture. For the stiff mixes with slump less than 1.5 in. (38 mm), the difference in V-B times is not significant. The lower slope of the regression line for fiber concrete indicates that the flow of concrete under vibration is more gradual for fiber concrete than plain concrete.

7.7   References

7.1 ASTM, *Annual Book of ASTM Standards,* Vol. 04.02, Concrete and Aggregates, 1991, 824 pp.

7.2 ACI Committee 544, "Measurements of Properties of Fiber Reinforced Concrete," American Concrete Institute, Detroit, Michigan, ACI544-2R-89, 1988, 11 pp.

7.3 BS 1881: Part 2, *Methods of Testing Concrete,* Standard III-1, Specification of Steel Fibers for Concrete, Concrete Library No. 50, 1983.

7.4 Johnston, C. D.; "Measures of the Workability of Steel Fiber Reinforced Concrete and Their Precision," *Cement, Concrete and Aggregates,* Vol. 4, No. 2, Winter 1982, pp. 61–67.

7.5 Balaguru, P.; and Ramakrishnan, V. "Comparison of Slump Cone and V-B Tests as Measures of Workability for Fiber Reinforced and Plain Concrete," *Cement, Concrete and Aggregates,* Vol. 9, No. 1, Summer 1987, pp. 3–11.

7.6 Ramakrishnan, V.; and Coyle, W. V. "Steel Fiber Reinforced Super-plasticized Concretes for Rehabilitation of Bridge Decks and Highway Pavements," Final Report to U.S. Department of Transportation, DOT/RSPA/DMA-50/84-2, 1983, 408 pp.

7.7 Balaguru, P.; and Ramakrishnan, V. "Properties of Fiber Reinforced Concrete: Workability, Behavior under Long-Term Loading, and Air-Void Characteristics," *ACI Materials Journal*, Vol. 85, May-June 1988, pp. 189–196.

7.8 Balaguru, P.; and Ramakrishnan, V. "Properties of Lightweight Fiber Reinforced Concrete," *Fiber Reinforced Concrete—Properties and Applications,* SP105, American Concrete Institute, Detroit, Michigan, 1987, pp. 305–322.

7.9 Ramakrishnan, V.; and Josifek, C. "Performance Characteristics and Flexural Fatigue Strength of Concrete Steel Fiber Composites," Proceedings of the International Symposium on Fiber Reinforced Concrete, Oxford and IBH Publishing Co. Pvt. Ltd., New Delhi, India, 1987, pp. 273–284.

7.10 Wu, G. Y.; Shivaroj, S. K.; and Ramakrishnan, V. "Flexural Fatigue Strength, Endurance Limit, and Impact Strength of Fiber Reinforced Refractory Concretes," *Fiber Reinforced Cements and Concretes: Recent Developments*, Elsevier, New York, 1989, pp. 261–273.

7.11 Balaguru, P. "Evaluation of New Synthetic Fibers for Use in Concrete," CE Report No. 88-10, Rutgers, New Brunswick, New Jersey, November 1988, 93 pp.

7.12 Khajuria, A.; and Balaguru, P. "Properties of Polymeric Fiber Reinforced Concrete," CE Report No. 89-13, Rutgers, New Brunswick, New Jersey, October 1989, 89 pp.

7.13 Ramakrishnan, V.; Gollapudi, S.; and Zellers, R. "Performance Characteristics and Fatigue Strength of Polypropylene Fiber Reinforced Concrete," *Fiber Reinforced Concrete—Properties and Applications,* SP105, American Concrete Institute, Detroit, Michigan, 1987, pp. 159–177.

7.14 Vondran, G. L.; Nagabhushanam, M.; and Ramakrishnan, V. "Fatigue Strength of Polypropylene Fiber Reinforced Concretes," *Fiber Reinforced Cements and Concretes: Recent Developments*, Elsevier, New York, 1989, pp. 533–543.

7.15 Ramakrishnan, V.; Brandchaug, T.; Coyle, W. V.; and Schrader, E. K. "A Comparative Evaluation of Concrete Reinforced with Straight Steel Fibers and Fibers with Deformed Ends Glued Together into Bundles," *ACI Journal*, Vol. 77, No. 3, 1980, pp. 135–143.

Properties of Hardened FRC

The behavior of hardened fiber-reinforced concrete under compression, tension, flexure, and freeze-thaw durability is discussed in this chapter. In addition, other physical properties such as permeability are also discussed briefly. As in Chapters 5 through 7, only concrete that contains coarse aggregate and fiber volume fractions less than 2% is considered in this chapter. The fibers under consideration are either metallic or polymeric. The long-term performance parameters, including creep and shrinkage, the behavior under fatigue and impact loading, and the plastic and drying shrinkage—are presented in Chapters 9 through 11.

8.1 Behavior under Compression: Steel Fibers

The increase in strength provided by steel fibers very rarely exceeds 25%. With increased use of deformed fibers, the fiber quantity is often limited to 100 lb/yd^3 (60 kg/m^3), or less than 0.75%. At this volume fraction the strength increase can be considered negligible for all design purposes. In special cases where the fiber content is more than 200 lb/yd^3 (120 kg/m^3), an increase in strength, though not significant in high-strength concrete, may be expected.

The other factors to be considered in design are modulus of elasticity, strain at peak load, and postpeak behavior. The change in modulus of elasticity can be considered negligible. On the other hand, fibers make a considerable contribution to ductility. Fiber addition increases the strain at peak load and results in a less steep and more reproducible descending branch (Figure 8-1). Overall, FRC can absorb much more energy before failure compared with its plain concrete counterpart.

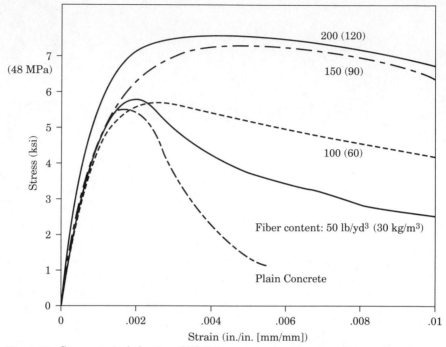

Figure 8-1 Stress-strain behavior of FRC in compression for normal-strength concrete containing 50mm hooked-end steel fibers.

The increase in ductility provided by the fibers depends on a number of factors, including (*i*) fiber volume fraction, (*ii*) fiber geometry, and (*iii*) matrix composition. An increase in fiber content results in an increase in energy absorption capacity. However, the relative magnitude of energy increase in the 0 to 0.7% fiber volume fraction range is much greater than any further energy increase at higher fiber contents.

With respect to fiber geometry, the length/diameter, or aspect, ratio is important for the performance of smooth straight fibers. As the aspect ratio increases, ductility increases as long as fibers can be properly mixed with the concrete. In the case of deformed fibers, hooked ends provide good energy absorption. Aspect ratio plays a role even with deformed fibers. However, the influence is not so significant as in FRC with straight fibers.

The matrix composition contributes in at least two ways to strength and energy absorption. The first is its bonding characteristics with the fiber. For example, a matrix containing silica fume tends to provide a better bond with the fibers and hence makes them more effective. Second is that the brittleness of the matrix itself plays an important role in the behavior of FRC. Normal-strength concrete tends to be less brittle than high-strength concrete, and the addition of fibers makes the composite even more ductile. Brittleness is even more pronounced for concrete containing fly ash and silica fume. Hence, a higher fiber volume fraction is needed for high-strength concrete to produce ductile failure. The contrast can be seen in stress-strain curves in Figures 8-1 and 8-2. For normal-strength concrete, about 100 lb/yd^3 (60 kg/m^3) of steel fibers with hooked ends is sufficient to produce a reasonably flat descending part, whereas for high-strength concrete a fiber content of 200 lb/yd^3 (120 kg/m^3) is needed to obtain a ductile behavior.

Compressive strength seems to govern the brittleness of both plain and fiber-reinforced concrete. Higher compressive strength always results in brittle mode of failure (steeper descending branches) for both normal-weight and lightweight concrete [8.1–8.3].

Cement mortars are more brittle than concrete. Fibers are also more effective in improving the ductility of mortar. Typically, mortars provide a better bond and, hence, fibers are better utilized in mortar than concrete. Fibers were also found to improve the ductility of a matrix made with cementitious materials other than portland cement [8.4]. Figure 8-3 shows typical stress-strain curves obtained for rapid hardening concrete made with magnesium phosphate cement. From Figure 8-3 one can see that the behavior of the composite is quite similar to portland cement concrete [8.4].

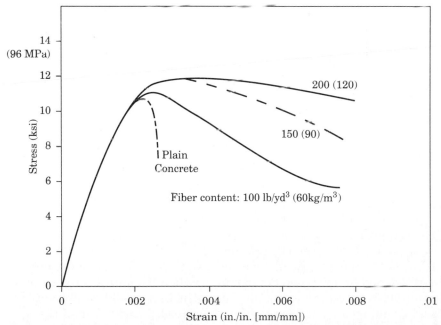

Figure 8-2 Stress-strain behavior of FRC in compression for high-strength concrete containing 30mm hooked-end steel fibers.

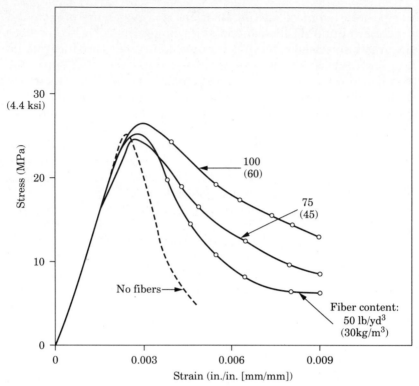

Figure 8-3 Stress-strain behavior of fiber-reinforced rapid-set materials in compression [8.4].

8.2 Behavior under Compression: Polymeric Fibers

As mentioned in earlier chapters, the most common fiber volume fraction used for polymeric fibers is 0.1%, although in certain cases slightly higher dosages are used. At this low fiber volume the fibers do not affect any property of hardened concrete, including the modulus of elasticity. Even at a fiber volume fraction of 0.5%, the change in modulus of elasticity is negligible [8.5].

The addition of fibers up to a volume fraction of 0.1% does not affect the compressive strength. Figure 8-4 shows the variation of compressive strength for concrete containing four different fiber contents of 0.75 in. (19 mm) single filament fibers. In some instances if more water is added to fiber concrete to improve its workability, a reduction in compressive strength can occur. This reduction should be attributed to additional water, not fiber addition. Figures 8-5 and 8-6 show the comparison of different types of polymeric fibers at about 0.1%. From these figures one can observe that the difference between the fiber types is negligible. At 0.5% fiber volume fraction, a 5% to 10% reduction in

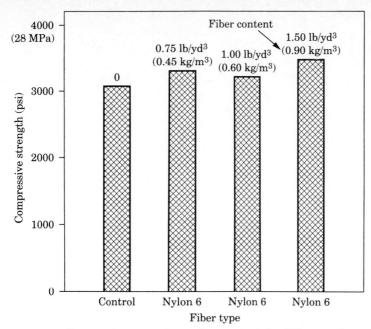

Figure 8-4 Compressive strength vs. fiber content for different volume fractions [8.6].

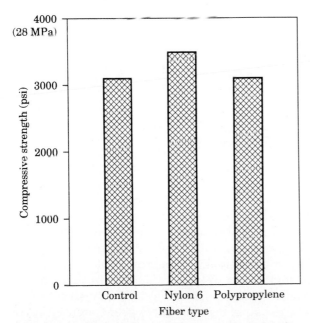

Figure 8-5 Comparison of compressive strength of two polymeric fiber-reinforced concretes and plain concrete; fiber content 1.5 lb/yd³ (0.9 kg/m³) and compressive strength in the range of 3 ksi (20 MPa) [8.6].

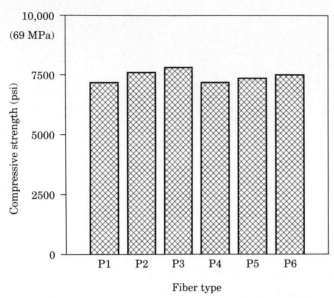

Figure 8-6 Comparison of compressive strength of six different polymeric fibers; fiber content 0.1% ($\approx$ 1.5 lb/yd³, 0.9 kg/m³) and compressive strength in the range of 7 ksi (46 MPa) [8.5].

compressive strength can occur. The strength reduction is mainly due to an increase in entrapped air [8.6]. The reduction does not occur in all cases.

The behavior of lightweight and normal weight concretes is essentially the same [8.6, 8.7]. The improvement in ductility is negligible at a 0.1% volume fraction. Fiber contribution seems to increase in the presence of continuous bars (as in reinforced concrete with fibers), especially under impact loading (Chapter 16).

8.3   Behavior under Tension: Steel Fibers

There are two types of tension tests used for concrete: direct tension and splitting tension. In the former, specimens such as dog-bone shapes are subjected to axial tension. Such tests are rarely used in practice for concrete containing coarse aggregates. In the splitting tension test, which is more popular, a cylindrical specimen is subjected to a splitting tension along its axis. Cubes can also be used for this test.

In most cases, fiber volume fractions of less than 2% do not improve the splitting tensile strength. The exception is concrete made using silica fume. High-strength concrete containing silica fume tends to bond better with fibers. The better bond produces fiber fracture rather than fiber pull-out, resulting in higher splitting tensile strength. An increase of up to 200% was reported for lightweight concrete containing silica fume [8.3].

Deformations are not commonly measured in splitting-tensile tests; hence, ductility under this type of loading is difficult to measure. However, attempts have been made to instrument the splitting tension test with some success [8.8]. Fibers were shown to contribute to ductility especially when failure occurs as a result of fiber pullout rather than fiber fracture.

8.4   Behavior under Tension: Polymeric Fibers

The effect of polymeric fiber addition is less pronounced than steel fibers for concrete containing coarse aggregates and a fiber volume less than 0.5%. Cement and mortar specimens containing 2% to 7% polymeric fibers behave quite differently than concrete containing less than 0.5% fibers. These composites are discussed in Chapters 13 and 14.

At a volume fraction less than 0.2%, the addition of fibers to concrete can be assumed to produce no effect on splitting tensile strength. At volume fractions of 0.5% or higher, there can be some strength reduction because of higher concentrations of entrapped air [8.5, 8.6].

8.5   Behavior under Flexure: Steel Fibers

Behavior under flexure is the most important aspect for FRC because in most practical applications the composite is subjected to some kind of bending load. Moreover, the addition of fiber improves the flexural toughness of plain concrete. Tests are usually done using $4 \times 4 \times 14$ in. ($100 \times 100 \times 350$ mm) beams under third-point loading (Chapter 4). The strength increases recorded for fiber contents less than 2% are not substantial except for concrete containing silica fume. In the latter case, fibers tend to provide more increase in strength because of better bond between fibers and concrete. In all cases, the increases in flexural strength are normally higher than increases in either compressive or splitting tensile strength. Both fiber volume fraction and aspect ratio play an important role. Longer fibers tend to provide preferred orientations along the length of the specimen, resulting in higher strength increase. For a given fiber geometry, higher aspect ratios result in greater strength increase if the composite is compacted properly. Overall, if the fiber content is less than 150 lb/yd^3 (90 kg/m^3), a level that is rarely exceeded in field application, the increase in flexural strength can be neglected for design purposes.

As mentioned numerous times throughout this book, the increase in flexural toughness provides the primary motive for using fibers in concrete. There are a number of ways to measure flexural-toughness (Chapter 4). Although a considerable amount of data is available on the topic of flexural toughness, only the information from References

[8.9 and 8.10] is reviewed here, because the deflection measurements reported in these references exclude extraneous deformations (for more details see Chapter 4).

Factors that influence flexural toughness include fiber type, fiber geometry, fiber volume fraction, matrix composition, and specimen size. The discussion here is limited to fibers with some kind of deformation, fiber contents less than or equal to 200 lb/yd^3 (120 kg/m^3), and concrete containing coarse aggregate. In most cases the specimens are $4 \times 4 \times 14$ in. ($100 \times 100 \times 350$ mm) prisms with no notches. A few samples with notches and a few $6 \times 6 \times 24$ in. ($150 \times 150 \times 610$ mm) prisms were also tested. ASTM C1018 procedures were used for toughness evaluation.

8.5.1 Influence of fiber volume fraction

The influence of fiber volume fraction is shown in Figures 8-7 and 8-8. Figure 8-7 illustrates the load-deflection responses obtained using $4 \times 4 \times 14$ in. ($100 \times 100 \times 350$ mm) beams tested over a simply supported

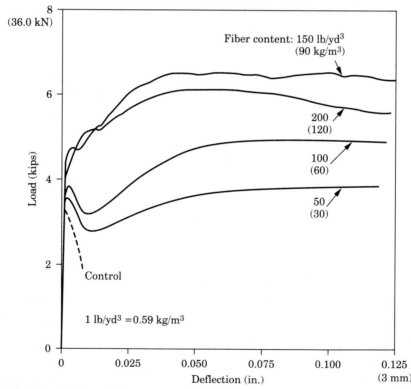

Figure 8-7 Influence of fiber content on load-deflection curves: 50 mm (1.97 in.) long hooked-end fibers.

span of 12 in. (300 mm). Third-point loads were applied at 4 in. (100 mm) from each support. The compressive strength of the matrix was 4000 psi (27 MPa). The fibers, which had hooked-ends, were 50 mm long and had a diameter of 0.5 mm. The following observations are based on Figure 8-7.

The load-deflection response is essentially linear up to 90% of first-crack load.

The improvement in energy absorption capacity (area under the load-deflection curve) in the 0–50 lb/yd^3 (0–30 kg/m^3) range is much higher than further increases for higher fiber content.

For 50 and 100 lb/yd^3 (30 and 60 kg/m^3) fiber content, there is a drop soon after the development of first-crack load. The amount of drop decreases with increasing fiber content. For other fiber geometries this drop is much higher, as discussed later.

For 150 and 200 lb/yd^3 (90 and 120 kg/m^3) fiber content, the postcrack increase in load is significant. This increase essentially provides the increase in flexural strength and a stable postcrack behavior. The curves also crisscross. At these fiber contents, the method of casting becomes very critical. As mentioned earlier, these fiber contents are relatively high for applications such as pavements. These volume fractions may be suitable for structural applications such as column-beam connections, corbels, and deep beams.

The toughness index values calculated using ASTM C1018 (see Chapter 4) are shown in Figure 8-8. These values represent the behavior of FRC beams reinforced with 50 mm long, 0.5 mm diameter hooked steel fibers. The values I_5 and I_{10} are insensitive to the fiber content. This observation agrees with interlaboratory study reported in Reference [8.9].

An increase in fiber content results in a consistent increase in toughness index for indices ranging from I_{20} to I_{100}. The differences between fiber contents become more pronounced for larger indices. The variations between fiber contents were similar for 30 and 60 mm long hooked-end fibers and fibers with other geometries such as deformed-end fibers and corrugated fibers.

8.5.2 Influence of fiber length

The influence of fiber length is very significant for straight fibers. Since these fibers are not currently used for practical applications they are not discussed further here. However, it is an established fact that longer fibers with higher aspect (length/diameter) ratios provide better performance in both strength increase and energy absorption as long as they can be mixed, placed, compacted, and finished properly. The following discussion is primarily based on results obtained using hooked-end fibers in concrete that had a compressive

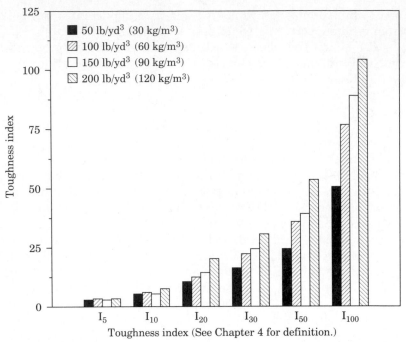

Figure 8-8 Influence of fiber content on toughness indices: 50 mm (1.97 in.) long hooked-end fibers.

strength of 4000 psi (27 MPa). The results are based on $4 \times 4 \times 14$ in. ($100 \times 100 \times 350$ mm) beams tested using third-point loading. The three fiber lengths were 30, 50, and 60 mm, and the corresponding diameters were 0.5, 0.5, and 0.8 mm. The load-deflection responses for two fiber contents and three fiber lengths are shown in Figure 8-9.

Energy absorption increases slightly with increase in apparent aspect ratio. Note that since the fibers have hooked-ends, length/diameter does not truly represent aspect ratio. Since hooked-ends provide good anchorage, an increase in aspect ratio of hooked-end fibers has less influence compared with straight steel fibers. The difference between fiber lengths becomes even less significant at higher volume fractions.

The toughness index values presented in Figure 8-10 confirm that the influence of fiber length is not significant. This figure compares toughness values I_{30}, I_{50}, and I_{100} for three fiber lengths and four volume fractions. The 60 mm long fiber performs better than the 30 or 50 mm fibers in most cases. Even though 50 mm fibers had a higher (apparent) aspect ratio, their performance was not better than that of 60 mm fibers. This result may be due to differences in the mixing and casting process. The 60 mm fiber tends to mix well, and the plastic concrete containing these fibers is easier to work with than concrete made with 50 mm fibers.

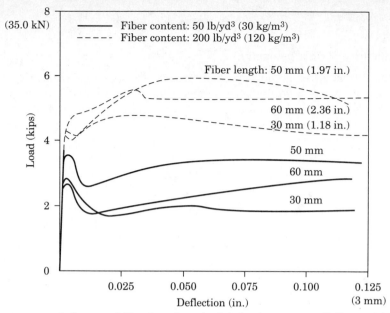

Figure 8-9 Influence of fiber length on load-deflection curves, all fibers with hooked ends.

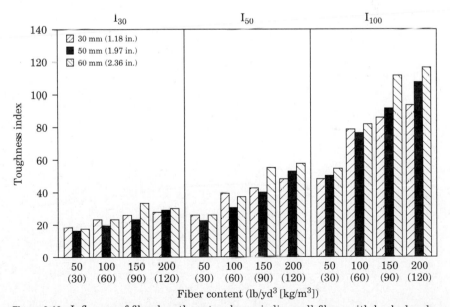

Figure 8-10 Influence of fiber length on toughness indices, all fibers with hooked ends.

8.5.3   Influence of fiber geometry

The influence of fiber geometry is shown in Figures 8-11 and 8-12. Figure 8-11 presents load-deflection curves for three fiber geometries, namely, hooked-end fibers, corrugated fibers, and deformed-end fibers. The hooked-end fibers were 30 mm long, whereas corrugated and deformed-end fibers were 25 and 30 mm, respectively. Concrete with hooked-end fibers had a higher tensile strength and a better postcrack response than the other two types. The drop after the first peak is much more pronounced for corrugated and deformed-end fibers.

The differences in behavior among the three fiber types can be seen even more clearly in Figure 8-12, which shows the toughness indices of I_{30}, I_{50}, and I_{100}. Hooked-end fibers perform better in almost all cases. The differences are more significant at larger deflections. For example, I_{100}, which is calculated using the area under the load-deflection curve up to 50.5 times the first-crack deflection, is much larger for hooked-end fibers. The results reported in Reference [8.9] dealing with hooked-end and corrugated fibers show similar trends.

8.5.4   Influence of specimen size and notch on toughness indices

Results of beams reinforced with hooked-end fibers are discussed here, but corrugated fibers exhibit a similar trend [8.9]. The toughness at

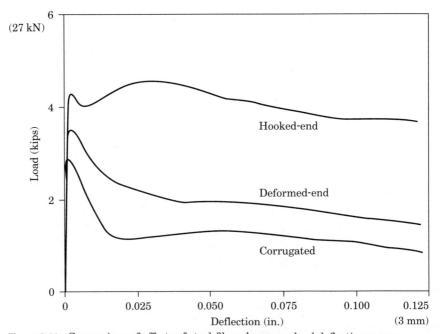

Figure 8-11   Comparison of effects of steel fiber shapes on load-deflection curves.

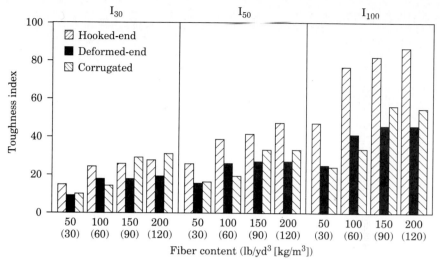

Figure 8-12 Comparison of effects of steel fiber shapes on toughness indices.

first crack is somewhat lower for notched beams. The toughness index I_5 was found to be insensitive to both specimen size and presence of notch. The difference between the two beam cross sections of 4×4 in. (100×100 mm) and 6×6 in. (150×150 mm) was not significant, even though larger beams had slightly higher toughness.

8.5.5   Influence of matrix composition

Within concrete containing coarse aggregates, the major variables are compressive strength, aggregate size, presence of admixtures such as fly ash and silica fume, and type of aggregate. Typically, as the matrix strength increases, concrete becomes more brittle and, hence, more fibers are needed to achieve the same amount of ductility. As mentioned earlier, the addition of silica fume makes the matrix more brittle and, hence, for the same fiber type and volume fraction, silica fume concrete beams have lower toughness. Figure 8-13 clearly shows a larger drop in the postcrack load-deflection curve for silica fume concrete than for concrete without silica fume. The total energy absorbed by the high-strength silica fume concrete could be higher than that of normal strength concrete but the toughness index values calculated using ASTM procedures are lower.

Concretes containing lightweight aggregates were found to behave similarly to normal weight concrete [8.2]. Here again, high-strength concrete containing silica fume shows a greater drop in load capacity after the first crack than concrete without silica fume [8.2, 8.3]. The use of longer fibers is recommended for concretes made with large-size aggregates.

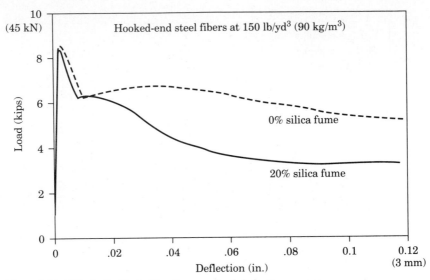

Figure 8-13 Effect of added silica fume on load-deflection curves.

A matrix consisting of nonportland cement containing fibers behaves quite similarly to portland cement concrete with fibers. The toughness indices of magnesium phosphate quick-set concrete reinforced with hooked-end, crimped, and deformed-end fibers are compared in Figure 8-14. From this figure one can see that hooked-end fibers provide the best performance. As for portland cement concrete, the differences in behavior become more significant at large deflections reflected in I_{30} and I_{50}. The results were based on $2 \times 2 \times 13$ in. ($50 \times 50 \times 330$ mm) beams tested using third-point loading [8.10].

8.6   Behavior under Flexure: Polymeric Fibers

The number of studies of FRC containing polymeric fibers is very limited [8.7, 8.9–8.12]. The variables investigated include fiber volume fraction (in the range of 0.075% to 0.5%), fiber type, and matrix composition. The fiber types evaluated include single-filament and fibrillated fibers; in both cases the fiber length was 0.75 in. (19 mm). The matrix compositions studied include normal weight concrete, lightweight concrete, and rapid-setting materials including nonportland cement matrices.

A typical load-deflection curve for beams reinforced with 0.5% fibrillated polypropylene fibers is shown in Figure 8-15. From this figure one can see that the load capacity drops considerably after the first crack. Both a low volume fraction and lower modulus of elasticity of fibers contribute to this behavior. The drop is even more marked for the

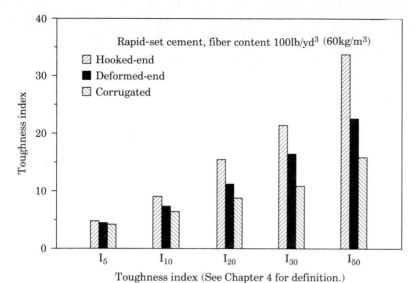

Figure 8-14 Toughness indices for rapid-set materials reinforced with steel fibers.

volume fraction of 0.1% [8.5, 8.6]. Fibers made with other polymeric materials show similar trends.

The effect of fiber volume fraction in the range of 2 to 8 lb/yd^3 (1.2 to 4.8 kg/m^3) on toughness indices is shown in Figures 8-16 and 8-17 for fibrillated and single filament fibers, respectively. By comparing these figures with Figures 8-8, 8-10, and 8-12, one can see

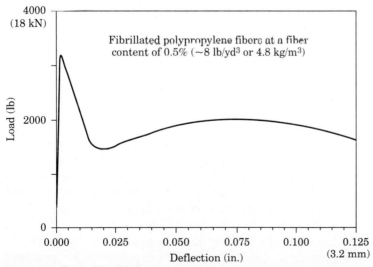

Figure 8-15 Typical load-deflection curve for polymeric fiber-reinforced concrete.

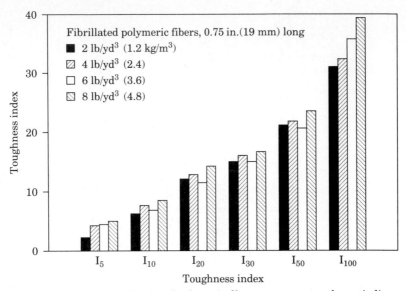

Figure 8-16   Influence of fibrillated polymeric fiber content on toughness indices.

that polymeric fibers are not so effective as steel fibers in improving toughness. The indices I_5 and I_{10} are very insensitive to changes in volume fractions and to fiber type as well. For example, there is very little difference in toughness indices between polymeric and steel fibers if one uses I_5 and I_{10} for comparison. The differences between steel and polymeric fibers do become significant for indices I_{50} and I_{100}, which are calculated using larger deflections. Between Nylon 6 and polypropylene fibers the differences between indices are not significant for the whole range, as shown in Figure 8-18.

As with steel fibers, there is not much difference in behavior between lightweight and normal-weight concrete. Changes in specimen sizes and the presence of notch also do not alter the properties significantly.

Polymeric fibers seem to contribute more to toughness in the case of rapid-set materials. Results obtained using 10 lb/yd^3 (6 kg/m^3) of either Nylon 6 or polypropylene fibers show that they can be quite effective (Figure 8-19). The fibers are as effective as some of the steel fibers with low tensile strength [8.10]. Note that the matrix has a high mortar content and can bond well to the polymeric fibers, resulting in a greater contribution to the tension force. A fiber content of 10 lb/yd^3 (6 kg/m^3) is higher than currently used fiber contents for portland cement concrete.

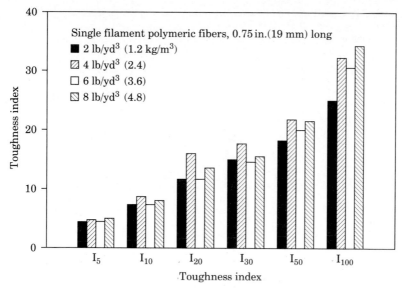

Figure 8-17 Influence of single filament polymeric fiber content on toughness indices.

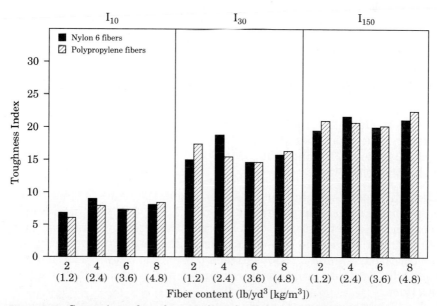

Figure 8-18 Comparison of toughness indices for Nylon 6 and polypropylene.

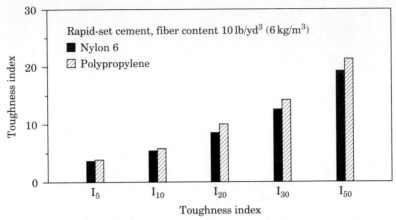

Figure 8-19 Toughness indices of rapid-set materials reinforced with Nylon 6 and polypropylene fibers.

8.7   Behavior under Shear, Torsion, and Bending: Steel and Polymeric Fibers

Knowledge of behavior under torsion, shear, bending, and combinations of bending, shear, and torsion is needed in order to design structural components. The bond between FRC and bar reinforcement plays an important role in the behavior of reinforced concrete members. Torsion, shear, and bending behaviors have been evaluated by a number of researchers, primarily using steel fibers (see Chapter 16). Limited studies have also been carried out using polymeric fibers. Chapter 16 provides a detailed discussion of these properties and their use for structural design. In general, the addition of fibers leads to a modest increase in shear strength and a significant improvement in ductility in all three modes of loading.

8.8   Unit Weight, Abrasion Resistance, Friction and Skid Resistance, Thermal and Electrical Conductivity: Steel and Polymeric Fibers

The unit weight of hardened FRC is about the same as plain concrete for both steel and polymeric fibers [8.1, 8.6, 8.7]. The unit weight tends to increase for steel fiber-reinforced concrete at volume fractions higher than 3%. In the case of polymeric fibers the unit weight tends to decrease if volume fractions exceed 0.5%. The weight reduction can be attributed to an increase in entrapped air at higher fiber volume fractions.

Standard abrasion tests conducted using ASTM C 779, Procedure C [8.13] show that steel fibers have almost no effect on the abrasion resistance of concrete [8.14]. Field observations of pavements and floor slabs under wheel loads also indicate that steel fibers do not affect abrasion resistance. Steel fibers were found to provide significant improvement in resistance to disintegration when abrasion is caused by particulate debris under high velocity [8.15–8.18]. These conditions typically occur at the downstream side of dams.

A simulated skid test was conducted to compare steel fiber-reinforced concrete with identical plain concrete for static friction, skid, and rolling resistance [8.19]. The static friction of FRC and plain concrete was found to be the same if there was no surface deterioration. If the surface has been subjected to erosion, then FRC had better skid and rolling resistances. The improvement was about 15 percent under dry, wet, or frozen conditions.

Thermal conductivity of concrete increases with the addition of steel fibers [8.20]. A fiber volume fraction of about 1% is needed to see a notable increase in thermal conductivity. The effect of fibers on electrical conductivity has not been well established.

0.9   Freeze-Thaw Durability

An extensive study conducted using ASTM Procedure C666 [8.21] indicates that the behavior of FRC is similar to that of plain concrete

TABLE 8.1   Mixture Proportions used for Durability Study [8.22]

Mixture designation	Cement (lb/yd^3)	Water-cement ratio	HRWR*	Fiber content (lb/yd^3)
A1	611	0.40	1.00	75
A2	611	0.40	1.00	75
A3	611	0.40	1.00	75
A4	611	0.40	1.00	75
B1	799	0.30	1.20	75
B2	799	0.30	1.20	75
C1	690	0.50	0	100
C2	690	0.43	1.30	100
C3	690	0.43	0.86	100
C4	690	0.43	1.09	100
C5	690	0.43	1.00	100
D1	611	0.38	1.00	0
D2	799	0.28	1.20	0
D3	690	0.50	0	0
D4	690	0.43	1.04	0
D5	690	0.43	0.42	0
D6	690	0.43	1.00	0
D7	690	0.43	1.00	0

*HRWR: High-range water reducer, percentage by weight of cement

[8.22]. The variables investigated in this study, listed in Table 8.1, include

Cement content: 611 to 799 lb/yd^3 (362 to 474 kg/m^3)

Water-cement ratio: 0.28 to 0.5

Fiber content: 0, 75, 100 lb/yd^3 (0, 45, 60 kg/m^3)

Air content: 1.2 to 10.8%

The plain concrete was made using ASTM Type I cement; natural sand and 1 in. (25 mm) maximum-size coarse aggregate; and high-range water-reducing and air-entraining admixtures. The hooked-end fibers used were 50 mm (2 in.) long and 0.5 mm (0.02 in.) in diameter.

The test specimens were $4 \times 4 \times 14$ in. ($100 \times 100 \times 350$ mm) prisms cured in a lime-saturated water bath for 14 days. The durability test was conducted according to ASTM C666, Procedure A [8.21]. In this test, specimens are subjected to a maximum of 300 freeze-thaw cycles. Typically, each freeze-thaw cycle takes about 4.5 hours. The freezing temperature is about 0°F (-17.8°C), and the thawing temperature is about 40°F (4.4°C).

After every 30 freeze-thaw cycles, the fundamental transverse resonance frequencies of the specimens were measured. The frequency at various stages is used to evaluate the soundness of the specimen. Typically, two factors are used for qualitative evaluations, namely, the relative dynamic modulus P_c and the durability factor DF. These two factors are defined as follows:

$$P_c = \frac{n_1^2}{n^2} \times 100 \qquad (8.1)$$

and

$$\text{DF} = \frac{PN}{M} \qquad (8.2)$$

where P = relative dynamic modulus at N cycles, expressed as a percentage,

P_c = relative dynamic modulus at C cycles, expressed as a percentage,

N = number of cycles at which P_c reaches the specified minimum value for discontinuing the test, or the specified number of cycles at which the exposure is to be terminated, whichever is less,

M = specified number of cycles at which the exposure is to be terminated (usually, 300 cycles)

n = fundamental transverse frequency at the zero cycle of freezing and thawing,

n_1 = fundamental transverse frequency after c freeze-thaw cycles

As the specimen deteriorates in response to exposure, the relative dynamic modulus decreases. Figures 8-20 to 8-22 show variations of relative dynamic modulus with respect to the number of freeze-thaw cycles. Figure 8-20 presents results for two concretes that had a fiber content of 75 lb/yd^3 (45 kg/m^3), and Figure 8-21 presents results for a concrete with a fiber content of 100 lb/yd^3 (60 kg/m^3). In both figures a principal variable is air content, which varied from 1.2% to 10.8%. Mix designations and air contents are marked on the figures. The following observations can be made.

As in the case of plain concrete, the addition of entrained air improves the freeze-thaw durability [8.22]. This trend is visible in almost all specimens.

For a water-cement ratio of more than 0.4 and a cement content of less than 700 lb/yd^3 (420 kg/m^3), a minimum of 6% (preferably 8%) air should be used to achieve a satisfactory freeze-thaw durability. This

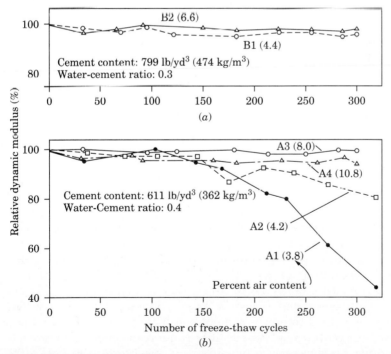

Figure 8-20 Effect of air content on relative dynamic modulus (fiber content 75 lb/yd^3, 45 kg/m^3) [8.22].

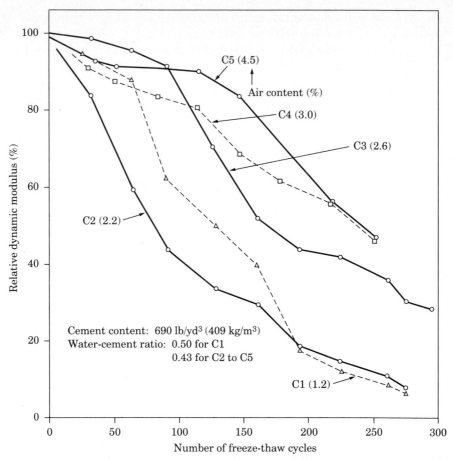

Figure 8-21   Effect of air content on relative dynamic modulus (fiber content 100 lb/yd³, 60 kg/m³) [8.22].

observation is particularly important because concrete used in the field usually has a water-cement ratio higher than 0.4 and contains less than 700 lb/yd³ (415 kg/m³) of cement.

For the same amount of entrained air, the durability is improved by increasing the cement and reducing the water-cement ratio. Figure 8-22 and Table 8.2 compare plain concrete and fiber-reinforced concrete. In concrete with either a high content of entrained air or a high content of cement (Figure 8-22a) the fiber-reinforced concrete exhibited slightly better durability than the plain concrete. However, both plain and fiber-reinforced concrete showed excellent durability, and the differences are not significant.

In concretes containing less air, fiber-reinforced concrete shows better durability for three out of five mixtures, as shown in Figure 8-22b through 8-22e. Here again, the differences in behavior do not appear

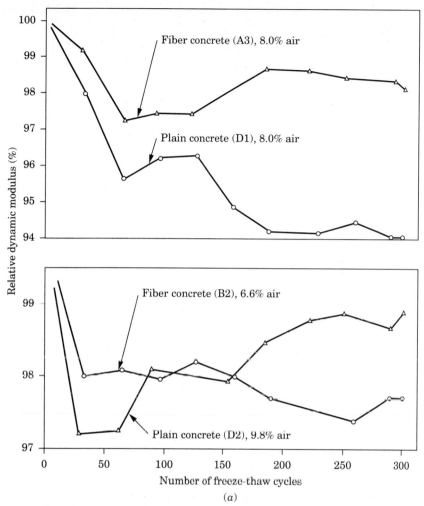

Figure 8-22a Comparison of freeze-thaw durability for plain and fiber-reinforced concretes: mixtures D1 and A3, and B2 and D2 [8.22].

to be significant, considering the variation to be expected in the experimental results for concrete.

Table 8.2 compares plain and fiber-reinforced concrete in terms of cycle number at failure (if there are fewer than 300 cycles) and durability factor. The durability factors reflect the same behavior pattern shown in Figure 8-22, as one would expect since durability factors are computed using relative dynamic modulus.

Since the modulus of rupture and the toughness index are important properties of fiber concrete, the effect of freeze-thaw cycling on them was also studied (Table 8.2) [8.22]. Flexural strength testing was done using third-point loading and a span of 12 in. (300 mm). The modulus of rupture generally decreased for specimens subjected to freeze-thaw

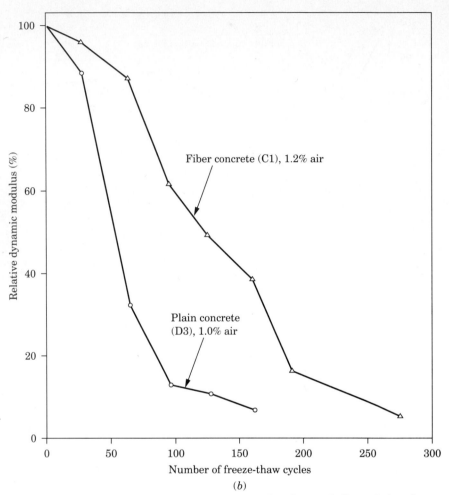

Figure 8-22b Comparison of freeze-thaw durability for plain and fiber-reinforced concretes: mixtures D3 and C1 [8.22].

cycles for both plain and fiber concretes. The reduction was smaller for fiber-reinforced concretes (Group C) than for plain concretes (Group D).

The toughness index reported in Table 8.2 was calculated as the area under the load-deflection curve up to 0.075 in. (1.9 mm), divided by the area under the load-deflection curve when the first crack appears. The toughness index of fiber concretes that exhibit high durability did not change considerably after freeze-thaw cycles. In most cases a slight decrease was observed, while in a few instances there was an increase (Table 8.2). However, fiber concretes that deteriorated under freeze-thaw cycles had much smaller toughness indices than virgin fiber concrete specimens. Hence, it can be concluded that if the fiber concrete is designed to withstand freeze-thaw conditions—

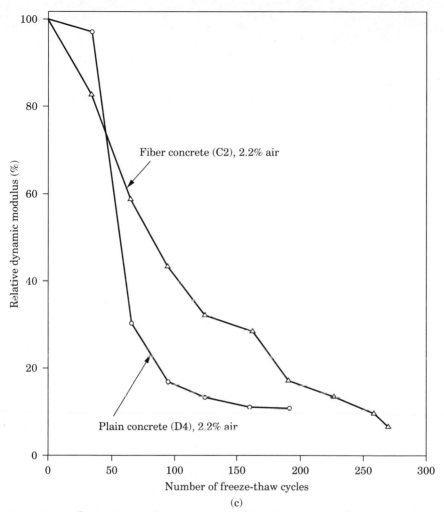

Relative dynamic modulus (%)

Number of freeze-thaw cycles

(c)

Figure 8-22c Comparison of freeze-thaw durability for plain and fiber-reinforced concretes: mixtures D4 and C2 [8.22].

by increasing its entrained air content, by raising its cement content, or by lowering its water-cement ratio—no appreciable change in the toughness index can be expected. Figure 8-23 shows load-deflection curves for two typical specimens, one at zero cycles and the other after 304 cycles of freeze-thaw loading. Both specimens behave in a similar manner, supporting the aforementioned observations.

In summary, we can say that

- Air content is the most significant parameter for freeze-thaw durability of fiber-reinforced concrete. The addition of entrained air improves freeze-thaw durability.

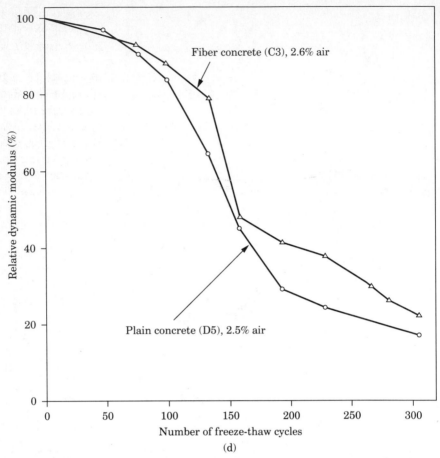

Figure 8-22d Comparison of freeze-thaw durability for plain and fiber-reinforced concretes: mixtures D5 and C3 [8.22]

- Under freeze-thaw cycling, the behavior of fiber-reinforced concrete is similar to that of plain concrete. For the same air content, the durability of fiber concrete and plain concrete are similar.

- For a water-cement ratio of more than 0.4 and a cement content of less than 700 lb/yd^3 (415 kg/m^3), a minimum of 6% air, preferably 8%, should be used to avoid deterioration under freeze-thaw cycling.

- The toughness index, which is an important property of fiber-reinforced concrete, does not change appreciably as a result of freezing and thawing if the mixture has been designed to prevent deterioration by entraining sufficient air into the mixture.

The results of Kobayashi [8.23] confirm those of Balaguru and Ramakrishnan [8.22]. In the former investigation freeze-thaw cycling was

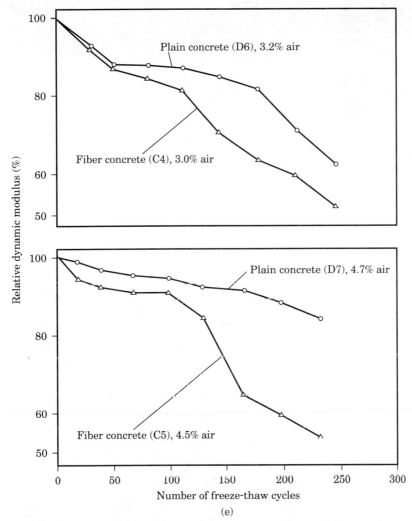

Figure 8-22e Comparison of freeze-thaw durability for plain and fiber-reinforced concretes: mixtures D6 and C4, and mixtures D7 and C5 [8.22].

done up to 1500 cycles. At large numbers of freeze-thaw cycles FRC was found to be more durable than plain concrete. The air content for the air-entrained plain and FRC was in the range of 5 to 6%. The amount of fibers used in this case [8.23] was much higher than that in Reference [8.22].

The results of freeze-thaw durability tests on polymeric fiber-reinforced concrete are not available in the published literature for fiber volume fractions less than 0.5%.

Limited studies on scaling resistance indicate that FRC with steel fibers fares better than plain concrete [8.24].

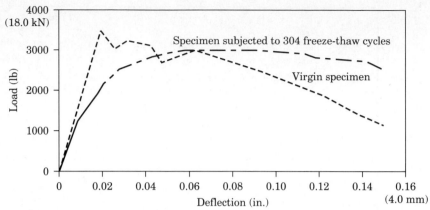

Figure 8-23 Comparison of load-deflection curves: a virgin specimen and a specimen subjected to 304 freeze-thaw cycles [8.22].

TABLE 8.2 Strength and Toughness Properties of Specimens Subjected to Freeze-Thaw Cycles [8.22]

Mixture desig-nation	Number of cycles completed	Dura-bility factor	Absorption coefficient (percent by weight) Zero cycle	After testing	Modulus of rupture (psi) Zero cycle	After testing	Toughness index Zero cycle	After testing
A1	307	45.1		6.5	782	386	4.7	6.1
A2	307	86.4		5.2	705	623	6.0	5.5
A3	300	98.3		5.9	903	922	4.9	7.6
A4	300	94.2		6.0	902	802	7.4	6.6
B1	300	98.8		5.3	1059	1313	5.3	4.7
B2	300	98.9		4.9	1022	1087	5.8	6.2
C1	270	19.7	5.4	7.6	614	109	7.3	1.0
C2	270	12.8	5.0	6.9	744	145	5.3	1.5
C3	304	27.9	5.2	6.7	647	439	6.0	1.8
C4	249	36.2	5.5	6.4	687	303	6.7	4.4
C5	235		5.1	6.5	686	205	6.6	4.4
D1	300	94.0		6.2	727	648	1.0	1.0
D2	300	98.2		5.6	791	939	1.0	1.0
D3	176	7.2	5.6	8.0	555	*	1.0	*
D4	191	6.1	5.1	7.8	649	*	1.0	*
D5	304	27.7	5.3	7.1	655	190	1.0	*
D6	249	49.3	5.6	6.7	647	173	1.0	1.0
D7	235		5.6	6.5	563	175	1.0	1.0

*Specimens disintegrated in the freeze-thaw cabinet.
1 psi = 0.0069 MPa

8.10   Moisture Absorption and Permeability

Limited results available indicate that FRC with steel fibers has lower water absorption and possibly lower permeability than plain concrete [8.22]. Permeability studies indicate that the permeability of polymeric fiber-reinforced concrete and plain concrete are about the same. A typical Darcy's coefficient for a concrete with a compressive strength of 3 ksi (20 MPa) was 10^{-10} cm/s. The fiber volume fraction was 0.1 %. When the samples were subjected to 1000 cycles of freezing and thawing, FRC specimens had a lower permeability than plain concrete.

8.11   References

8.1 Balaguru, P.; and Ramakrishnan, V. "Mechanical Properties of Superplasticized Fiber Reinforced Concrete Developed for Bridge Decks and Highway Pavements," *Concrete in Transporation* SP-93, American Concrete Institute, Detroit, Michigan, 1986, pp. 563–584.

8.2 Balaguru, P.; and Ramakrishnan, V. "Properties of Lightweight Fiber Reinforced Concrete", *Fiber Reinforced Concrete*, SP 105, American Concrete Institute, Detroit, Michigan, 1987, pp. 305–322.

8.3 Dipsia, M. "Mechanical Properties of Superplasticized Steel Fiber High Strength Semi-Lightweight Concrete," M.S. Thesis, Rutgers University, New Brunswick, New Jersey, 1987, 128 pp.

8.4 Balaguru, P. "Fiber-Reinforced Rapid-Setting Concrete," *Concrete International*, American Concrete Institute, Detroit, Michigan, Vol. 14, No. 2, 1992, pp. 64–67.

8.5 Balaguru, P. "Evaluation of New Synthetic Fiber for Use in Concrete," Civil Engineering Department Report No. 88-10, 1988, 93 pp.

8.6 Khajuria, A.; and Balaguru, P. "Behavior of New Synthetic Fiber Reinforced Concrete," Civil Engineering Department Report No. 89-13, October 1989, 89 pp.

8.7 Khajuria, A.; Chien, A.; and Balaguru, P. "Properties of Fiber Reinforced Lightweight Concrete," Civil Engineering Department Report No. 90-10, 1990, 70 pp.

8.8 Nanni, A. "Splitting-Tension Test for Fiber Reinforced Concrete," *ACI Materials Journal*, Vol. 82, No. 4, 1985, pp. 229–233.

8.9 Gopalaratnam, V. S.; Shah, S. P.; Batson, G.; Griswell, M.; Ramakrishnan, V.; and Wecharatana, M. "Fracture Toughness of Fiber Reinforced Concrete," *ACI Materials Journal*, Vol. 88, No. 4, 1991, pp. 339–353.

8.10 Patel, M. K. "Flexural Toughness Characteristics of Fiber Reinforced Concrete," M.S. Thesis, Rutgers University, New Brunswick, New Jersey, 1991, 145 pp.

8.11 ASTM, "Standard Method of Test for Flexural Toughness of Fiber Reinforced Concrete," *Annual Book of ASTM Standards,* Vol. 04.02, Concrete and Aggregates, 1991, pp. 507–513.

8.12 Khajuria, A.; Chien, A.; and Balaguru, P. "Toughness Characteristics of Fiber Reinforced Concrete," Civil Engineering Department Report No. 90-11, 1990, 46 pp.

8.13 ASTM, "Standard Test Method for Abrasion Resistance of Horizontal Concrete Surfaces," *Annual Book of ASTM Standards,* Vol. 04, No. 02, Concrete and Aggregates, 1991, pp. 365–369.

8.14 Nanni, A. "Abrasion Resistance of Concrete," *ACI Materials Journal,* Vol. 86, No. 1989, pp. 559–565.

8.15 Brandshaug, T.; Ramakrishnan, V.; Coyle, W. V.; and Schrader, E. K. "A Comparative Evaluation of Concrete Reinforced with Deformed End Fibers," Reports No. SDSM and T-CBS 7801, South Dakota School of Mines and Technology, Rapid City, South Dakota, May 1978, 52 pp.

8.16 Houghton, D. L.; Borge, D. E.; and Paxton, J. A. "Cavitation Resistance of Some Special Concretes," *ACI Journal,* Vol. 75, 1978, pp. 664–667.

8.17 Schrader, E. K.; and Much, A. V. "Fibrous Concrete Repair of Cavitation Damage," *ASCE, Journal of Construction Division,* Vol. 102, No. 2, 1976, pp. 385–399.

8.18 Schrader, E. K.; and Kaden, R. A. "Outlet Repairs at Dworshak Dam," *The Military Engineer,* Vol. 68, No. 443, 1976, pp. 254–259.

8.19 Mikkelmeni, M. R.; "A Comparative Study of Fiber Reinforced Concrete and Plain Concrete Construction," M.S. Thesis, Mississippi State University, Starkville, Mississippi, 1970.

8.20 Cook, D. J.; and Uher, C. "The Thermal Conductivity of Fiber Reinforced Concrete," *Cement and Concrete Research,* Vol. 4, 1974, pp. 497–509.

8.21 ASTM "Standard Test Method for Resistance of Concrete to Rapid Freezing and Thawing," *Annual Book of ASTM Standards,* Vol. 04, No. 02, Concrete and Aggregates, 1991, pp. 319–324.

8.22 Balaguru, P.; and Ramakrishnan, V. "Freeze-Thaw Durability of Fiber Reinforced Concrete," *ACI Journal,* Vol. 83, 1986, pp. 374–382.

8.23 Kobayashi, K. "Development of Fiber Reinforced Concrete in Japan," *International Journal of Cement Composites,* Vol. 5, 1983, pp. 27–40.

8.24 Ramakrishnan, V.; and Coyle, W. V. "Steel Fiber Reinforced Superplasticized Concretes for Rehabilitation of Bridge Decks and Highway Pavement," Report No. DOT/RSPA/DMA-50/84-2, Office of University Research, U.S. D.O.T., 1983, 408 pp.

9

FRC under Fatigue and Impact Loading

Fibers provide substantially improved resistance to fatigue and to impact loading. Improved fatigue performance is one of the primary reasons for the extensive use of FRC in pavements and bridge decks, where the composite is subjected to millions of cyclically varying loads during its lifetime. Likewise, the composite is subjected to some kind of impact loading during service. This chapter discusses the performance of FRC under these loading conditions. As in Chapters 6, 7, and 8, only concretes reinforced with a low volume fraction of steel and polymeric fibers are discussed. With respect to impact loading, both the drop-weight and the instrumented impact tests are discussed.

9.1 Fatigue Loading

Fatigue loading can be applied in all the three modes, namely, compression, tension (axial or flexural), or shear. Almost all investigations on FRC have been concerned with flexural fatigue loading because of its direct influence on the performance of pavement and industrial floor slabs. Typically, $4 \times 4 \times 14$ in. ($100 \times 100 \times 350$ mm) prisms were tested using loads placed at mid-third points over a simply supported span of 12 in. (300 mm). Variables that have been investigated include fiber type, fiber volume fraction, and matrix composition. In almost all cases the fiber addition resulted in improved fatigue (S-N curve) performance.

9.1.1   Basics of fatigue testing

Usually, fatigue performance of materials is presented in the form of stress range versus logarithm of the number of cycles to failure (S-N) curves, plotted on semilog paper. The stress range S_r is the difference between the maximum stress σ_{max} and the minimum stress σ_{min} at the maximally stressed (tension or compression) location of the extreme (tension or compression) fiber. In the case of concrete beams subjected to bending, failure always occurs in the (flexural) tension mode. Hence, if P_{max} and P_{min} are the maximum and minimum loads, the maximum σ_{max} and minimum σ_{min} stresses can be expressed as

$$\sigma_{max} = \frac{M_{max}}{Z} \text{ psi(Pa)} \tag{9.1}$$

$$\sigma_{min} = \frac{M_{min}}{Z} \text{ psi(Pa)} \tag{9.2}$$

where M_{max} = the moment caused by P_{max} (in. · lb, N · m)
  M_{min} = the moment caused by P_{min} (in. · lb, N · m)
  Z = the section modulus, $\frac{bd^2}{6}$ for rectangular sections (in.^{3}, mm^3)
  b = width of the cross section
  d = depth of the cross section

For $4 \times 4 \times 14$ in. beams tested under third-point loading,

$$\sigma_{max} = \frac{3P_{max}}{16} \text{ psi} \quad (1 \text{ psi} = 6.895 \text{ kPa}) \tag{9.3}$$

and

$$\sigma_{min} = \frac{3P_{min}}{16} \text{ psi} \tag{9.4}$$

where P_{max} and P_{min} are total maximim and minimum loads in pounds (N).

The stress range,

$$S_r = \sigma_{max} - \sigma_{min} \text{ psi(Pa)} \tag{9.5}$$

and the mean stress

$$\sigma_{mean} = \frac{\sigma_{max} + \sigma_{min}}{2} \text{ psi(Pa)} \tag{9.6}$$

A specimen can be subjected to a large variety of loading schemes. The simplest loading scheme consists of subjecting a beam to a chosen single maximum and (single) minimum load repeatedly. This type of loading is called constant amplitude loading. In this type of loading, once σ_{max} and σ_{min} (P_{max} and P_{min}) are chosen, S_r becomes constant. The number of cycles at which the beam fails is noted as fatigue life, N. Specimens made of the same material are tested at various levels of S_r and the failure cycles are noted. The plot of the variation of S_r with respect to log N becomes the S-N curve (Figure 9-1). Normally the experimental variation is very large. Hence, a number of samples have to be tested for each loading condition. The stress range could be plotted as ksi (MPa) or as a fraction of static strength. For example, in flexural loading the stress range can be expressed as a fraction (percentage) of the modulus of rupture f_r.

In flexural fatigue, σ_{min} can be negative, that is, the part of the beam experiencing maximum stress can be subjected to alternate compression and tension. This case is called reversed cyclic loading. With either plain or fiber-reinforced concrete prisms, very few researchers have attempted high-cycle reversed fatigue loading because such conditions rarely exist in the field; instead, in almost all cases the minimum stress was maintained positive. Usually the minimum stresses vary from 5% to 10% of the modulus of rupture. A common procedure to obtain S-N is as follows: (a) Test samples monotonically at a quasi-static rate of loading

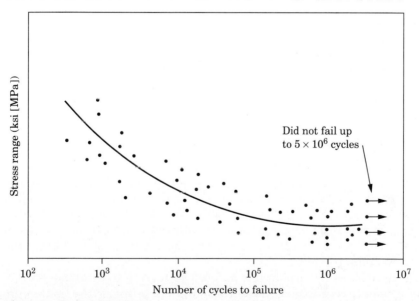

Figure 9-1 Typical S-N curve.

and obtain the average modulus of rupture f_r. (*b*) Choose a minimum and a maximum stress as a percentage of f_r. Typically $\sigma_{\min}$ is taken as 0.1 f_r and $\sigma_{\max}$ is taken as 0.9 f_r. (*c*) Test a few (at least three) samples at stresses between 0.1 and 0.9 f_r and obtain the number of cycles N at which failure occurs. These beams could fail in a few hundred to a few thousand cycles. (*d*) Reduce the maximum stress to, say, 0.8 f_r and repeat the tests. As the maximum stress and thus the stress range decreases, the number of cycles to failure will increase exponentially. Most researchers opt to stop the test after a certain number of cycles in cases where failure does not occur. The most popular number is 2 million cycles even though a few investigators have tested samples up to 5 or 10 million cycles.

Fatigue testing is a very time-consuming process. To obtain an S-N curve for a particular composite may take a few months. Hence, the variables to be tested should be chosen carefully to reduce the number of samples to be tested.

A number of factors affect fatigue performance [9.1], including:

- Minimum (mean) stress
- Compressive (flexural) strength
- Cement content
- Mix proportion
- Rate of loading
- Rest periods (hold times)

Fortunately, only the minimum or mean stress affects the fatigue strength considerably. Other factors, such as mixture design, were found to have little influence if fatigue strength is expressed as a fraction of the static strength. The rest period, if it is not prolonged, does not affect the fatigue strength considerably [9.1]. Rate of loading is important because it determines the time required for testing. Loading rates resulting in 70 to 900 cycles per minute (~1 to 15 Hz) have little effect on fatigue strength provided the maximum stress level is less than 75% [9.1]. Even though this observation was made for testing in compression, it can be considered valid for flexure because the behaviors under compression and flexure were found to be essentially the same [9.1]. FRC beams have been tested at rates varying from 2 to 20 cycles per second. The testing machine should have sufficient pump capacity to maintain the maximum and minimum loads at the chosen rate. For larger deformations it is often more difficult to maintain the loads at higher loading rates.

Since it takes a considerable amount of testing to establish an S-N curve, some researchers have chosen to determine the maximum stress the beams can sustain for 2 million cycles without failure. For example, if FRC beams subjected to a σ_{max} of $0.7f_r$ do not fail up to 2 million cycles, the fatigue strength of that particular material and fiber composition is taken as $0.7\ f_r$. Earlier investigators believed that concrete does not have an endurance limit up to 10 million cycles [9.1]. This concept is being reevaluated using recent data. If the endurance limit exists at 2 million cycles, then the maximum stress that the composite can sustain for 2 million cycles can be assumed to be the fatigue strength for design purposes.

9.1.2 Fatigue behavior of steel fiber–reinforced concrete

Flexural fatigue tests have been carried out on concretes containing straight and hooked-end steel fibers [9.2–9.5]. In all cases the fibers were found to increase the fatigue resistance compared with plain concrete.

Tests using straight steel fibers in a volume fraction range of 2.0% to 2.98% show that fiber-reinforced beams can sustain up to 83% of first crack stress for 2 million cycles [9.2]. The fibers used had aspect ratios varying from 75 to 90. The beam dimension was $4 \times 6 \times 102$ in. ($100 \times 150 \times 2590$ mm). The beams were tested at center-point loading over a simply supported span of 96 in. (2118 mm). The rate of loading was 9 cycles per second. The S-N curve for the fiber volume fraction of 2.98% is shown in Figure 9-2 [9.2]. The fatigue strength at 2 million cycles was 75% and 83% of first crack stress for reversed and nonreversed fatigue loading. The failures occurred owing to pullout of the fibers rather than fiber fracture.

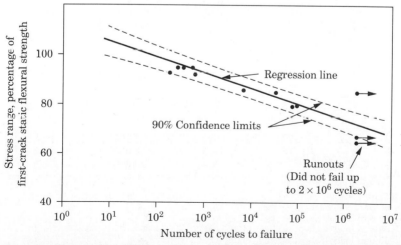

Figure 9-2   Fatigue life of FRC containing 2.98% straight steel fibers for reversed loading [9.2, 9.35].

Results in Reference [9.4] report the performance of hooked-end fibers. Two fiber lengths, of 50 and 60 mm, were evaluated. The diameters of the 50 and 60 mm fibers were 0.5 and 0.8 mm, respectively. Beams with two fiber volume fractions of 66 and 100 lb/yd^3 (40 and 60 kg/m^3) and control plain concrete beams were tested using third-point loading at the rate of 20 cycles per second. The beam dimension was $4 \times 4 \times 14$ in. ($100 \times 100 \times 350$ mm). Tests were run under load control. The results are presented in Table 9.1. Since most of the beams did not fail in 2 million cycles, a reasonable plot of S-N curve could not be obtained.

In Table 9.1, the last column indicates the maximum stress σ_{max} as a fraction of the modulus of rupture f_r. Since f_r values were obtained

TABLE 9.1   Fatigue Test Results on FRC Made with Steel Fibers [9.4]

Mix design[†]	Sample number	Flexural fatigue stress (psi)*		Number of cycles to failure	Percent of static flexural strength[‡]
		Minimum	Maximum		
OB1	1	80	640	2,000,000[§]	81.5
OB1	2	80	719	2,000,000	91.6
OB1	3	82	779	2,000,000	99.2
OB2	1	85	680	2,000,000	79.5
OB2	2	82	779	2,000,000	91.1
OB2	3	83	790	2,000,000	92.4
OB3	1	84	669	2,000,000	78.7
OB3	2	79	711	2,000,000	83.7
OB3	3	85	850	2,000,000	100.0
OB4	1	94	846	2,000,000	92.0
OB4	2	92	1009	2,000,000	109.7
OB4	3	89	1073	2,000,000	116.6
PL	1	71	356	2,000,000	49.4
PL	2	71	357	2,000,000	49.6
PL	3	71	393	2,000,000	54.6
PL	4	71	393	35,700	55.3
PL	5	74	406	13,400	56.4
PL	6	69	414	55,300	57.5
PL	7	73	679	800	93.6
PL	8	73	655	1,000	90.3
PL	9	73	577	126,400	79.5

*1 psi = 6.895 kPa
[†] PL = Plain concrete
OB1 = 60 mm long fibers at 40 kg/m^3 (66 lb/yd^3)
OB2 = 60 mm long fibers at 60 kg/m^3 (100 lb/yd^3)
OB3 = 50 mm long fibers at 40 kg/m^3 (66 lb/yd^3)
OB4 = 50 mm long fibers at 60 kg/m^3 (100 lb/yd^3)
[‡] $\sigma_{max}/f_r \times 100$
[§]The number 2,000,000 indicates that the specimens did not fail up to 2,000,000 cycles.

using different sets of specimens, fatigue strength for some specimens was more than 100%. The static strength of the specimens tested under fatigue loading could have been higher; however, the results indicate that specimens reinforced with hooked-end steel fibers can sustain a high level of flexural stress under fatigue loading. Note that plain concrete beams failed at about 55% of f_r. It was reported that FRC beams continued to sustain the loads even after cracking. Tests using a maximum stress, σ_{max}, of 0.8 times the modulus of rupture, f_r, were run up to 5 million cycles to see whether fatigue failure would occur. The specimens did not fail [9.4].

Fibers were also found to be effective in refractory concrete [9.5]. A premixed refractory aggregate and a calcium aluminate cement binder were used for the matrix. The fibers used were 1 in. (25 mm) long and 0.018 in. (0.46 mm) diameter and were made of melt-extract stainless steel. The fiber volume fractions investigated were 0, 0.5, 1.0, and 1.5%. Fatigue tests were run at a rate of 20 cycles per second using $4 \times 4 \times 14$ in. ($100 \times 100 \times 350$ mm) beams loaded at third-points over a simply supported span of 12 in. (300 mm). The load was varied from a minimum of about 10% of the static modulus of rupture to a number of maximum loads, resulting in maximum flexural stress in the range of 38% to 90% of modulus of rupture.

The results are shown in Figures 9-3 through 9-5. Figure 9-3 presents the variation of maximum stress with respect to the number of cycles to failure. Figures 9-4 and 9-5 provide comparisons among the various fiber volume fractions, plotted in terms of actual stress and of the ratio σ_{max}/f_r respectively. The bar graphs were plotted for a fatigue life of 2 million cycles. In Figure 9-5, f_r is taken

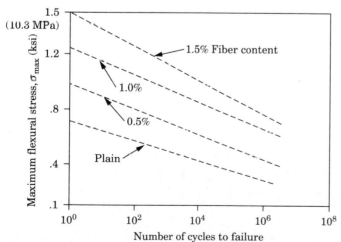

Figure 9-3 Comparison of the maximum fatigue flexural stress vs. log N, the number of cycles to failure, for plain and fiber concrete [9.5].

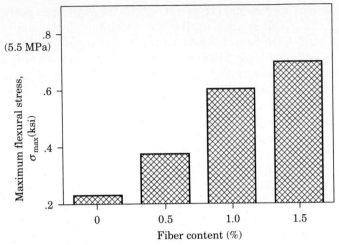

Figure 9-4 The effect of steel fiber content on fatigue strength: Maximum flexural stress that can be sustained for 2 million cycles [9.5].

as the modulus of rupture of plain matrix. Since the matrix strength increases with fiber addition, and if f_r of fiber-reinforced matrix is used for the ratio σ_{max}/f_r, then the ratios decrease considerably, as shown in Figure 9-6. Addition of fibers still results in some improvement.

9.1.3   Fatigue behavior of polymeric fiber–reinforced concrete

A limited number of fatigue tests were carried out using polymeric fibers made of polypropylene, polyethylene, and Nylon 6 fibers

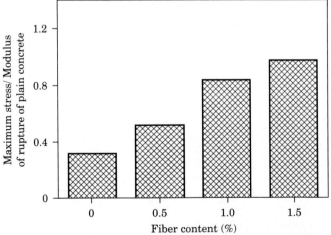

Figure 9-5 Comparison of plain concrete and FRC: Ratio of maximum stress/modulus of rupture of plain concrete for specimens that sustained at 2 million cycles [9.5].

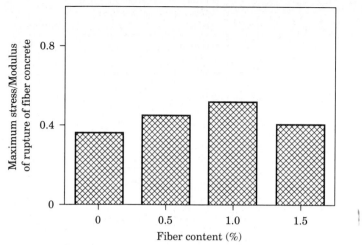

Figure 9-6 Comparison of plain concrete and FRC: Ratio of maximum stress/modulus of rupture of corresponding FRC specimens that sustained 2 million cycles [9.5].

[9.6–9.9]. Since the fiber volume fractions of these fibers were low, the fatigue strength increase is not so substantial as in steel fiber concrete. However, the fiber addition makes a definite contribution to the fatigue strength.

Fibrillated polypropylene fibers incorporated in concrete produce a noticeable effect on fatigue strength at a volume fraction of 0.2% [9.6]. The 2.25 in. (57 mm) long fibers were evaluated for volume fractions of 0, 0.1, 0.2, and 0.3%. The tests were conducted using $4 \times 4 \times 14$ in. ($100 \times 100 \times 350$ mm) beams loaded at third-points over a simply supported span of 12 in. (300 mm). The loading rate was 20 cycles per second. The minimum stress was kept at about 10% of the modulus of rupture. The maximum stress was varied from 40% to 94%.

The results are shown in Table 9.2. Since the same samples were run at different stress ranges, only qualitative information can be obtained from these results. At 2 million cycles, the plain concrete beams sustained a maximum stress of about 50% of modulus of rupture. This number consistently increased with fiber addition. At a volume fraction of 0.3%, the maximum sustainable stress increased to about 65%.

Results reported in Reference [9.7] provide data on fibrillated polypropylene fibers manufactured by a different company from that in Reference [9.6]. The fiber volume fractions used were 0, 0.1, 0.5, and 1.0%; hence, the influence of higher volume fraction was also studied. The details of the test setup were the same as those in previous investigations. Two matrices with two water-cement ratios were evaluated as well. The sustainable maximum stress at 2 million cycles was found

TABLE 9.2   Fatigue Test Results on FRC Made with Polypropylene Fiber [9.6]

Mix designation[†]	Sample beam number	Flexural fatigue stress (psi)[*] Minimum	Flexural fatigue stress (psi)[*] Maximum	Number of cycles to failure	Percent of static flexural stress	Comments[‡]
FR1	9	78	313	2,000,000	40.4	a
FR1	9	78	352	2,000,000	45.4	b
FR1	9	78	430	2,000,000	55.5	c
FR1	9	78	469	2,000,000	60.5	d
FR1	9	78	547	2,000,000	70.6	e
FR1	8	76	382	118,600	49.3	
FR1	7	83	454	610,400	58.6	
FR2	8	80	402	2,000,000	52.2	a
FR2	8	80	442	2,000,000	57.4	b
FR2	8	80	483	2,000,000	62.7	c
FR2	8	80	523	2,000,000	67.9	d
FR2	8	80	603	1,562,800	78.3	d
FR2	7	76	417	212,700	54.2	
FR2	9	78	469	192,600	60.9	
FR3	4	73	402	2,000,000	60.9	a
FR3	4	73	438	2,000,000	66.4	b
FR3	4	73	511	17,300	77.4	
FR3	9	71	423	2,000,000	64.1	a
FR3	9	71	458	2,000,000	69.4	b
FR3	9	71	529	27,500	80.2	
FR3	8	72	466	300	70.6	
F1	6	72	578	28,800	80.3	
F1	7	72	493	187,200	68.5	
F1	5	72	431	29,500	60.0	
F2	5	73	453	38,800	62.1	
F2	4	73	395	51,900	54.0	
F2	8	73	385	2,000,000	52.8	
F3	6	72	543	123,000	75.9	
F3	4	74	445	475,007	59.7	
F3	7	75	436	2,000,000	58.5	
PL	4	73	679	800	93.6	
PL	10	73	655	1,000	90.3	
PL	12	73	577	126,400	79.5	
PL	7	71	356	2,000,000	49.4	
PL	12	71	357	2,000,000	49.6	
PL	9	71	393	2,000,000	54.6	
PL	10	71	393	35,700	55.3	
PL	11	74	406	13,400	56.4	
PL	8	69	414	55,300	57.5	

[*]1 psi = 6.895 kPa
[†]PL = plain concrete, F1 and FR1 = 0.1% fibers,
    F2 and FR2 = 0.2% fibers, F3 and FR3 = 0.3% fibers
[‡]a—This specimen was tested at higher stress when it did not fail at 2×10^6 cycles.
  b—This specimen was tested for a second time with higher stress.
  c—This specimen was retested for a third time with higher stress.
  d—This specimen was retested for a fourth time with higher stress.
  e—This specimen was retested for a fifth time with higher stress.

to increase by about 25% and 30% over plain concrete for fiber volume fractions of 0.5% and 1.0%, respectively.

Fatigue tests using polyethylene and Nylon 6 fibers were conducted using the same test setups used for polypropylene fibers. These fibers were made of single filaments. The contribution of these fibers was found to be about the same as the polypropylene fibers.

9.1.4   Beams reinforced with continuous bars and discrete fibers

Fiber addition to normal concrete containing bar reinforcement also improves the fatigue resistance [9.10]. Constant amplitude fatigue tests were carried out using beams reinforced with three fiber types and three volume fractions. Straight steel fiber was used at a volume fraction of 1.27%, whereas the volume fractions for paddled and hooked-end fibers were 0.89% and 1.54%, respectively. Three bar reinforcement ratios of 0.17, 0.75, and 2.09 were investigated. The beam setup, dimensions of the beam, and the reinforcement details are shown in Figure 9-7 [9.10]. The load was applied at three cycles per second.

From Table 9.3 [9.10] one can see that the fatigue strength of beams with fibers was much higher than the strength of plain concrete beams. Longer fibers with a higher aspect ratio and higher fiber volume fractions provided better results. The influence of fibers was found to decrease with the increase in the amount of bar reinforcement. There was a discrepancy between calculated and measured stresses in the reinforcing bars (Table 9.3). The decrease in stress was assumed to be the result of the fiber contribution in the tension zone.

9.2   Impact Loading

Impact resistance of FRC can be measured by using a number of different test methods. These methods can be broadly grouped into the following categories [9.11]:

- Weighted pendulum Charpy type impact test
- Drop-weight test (single or repeated impact)
- Constant strain rate test
- Projectile impact test
- Split Hopkinson bar test
- Explosive test
- Instrumented pendulum impact test

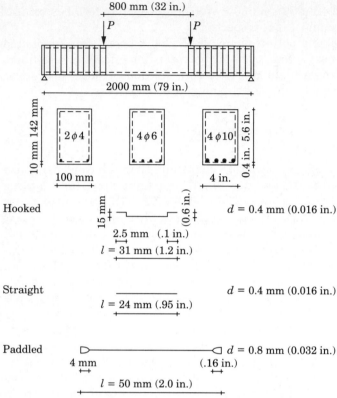

Figure 9-7 Beams and types of steel fibers for fatigue tests of reinforced-concrete beams with fibers (test results in Table 9.3) [9.10].

TABLE 9.3 Fatigue Strength of Beams Containing Bar and Fiber Reinforcement (Details of Beam in Figure 9-7) [9.10]

Fiber type	$\dfrac{pl}{d}$	P_{min}* (kN)	P_{max}* (kN)	Calculated $\sigma_{s,max}$ (MPa)	Cycles to failure (10^3)	Measured $\sigma_{s,max}$ (MPa)
None	0	3.42	19.10	402	265	380
Hooked end	69	3.99	19.52	411	453	340
Straight end	76	3.56	18.75	395	600	342
Paddled end	96	3.85	19.05	401	>1400	300

*Loads do not include dead weight.

1 kN = 224.809 lb
1 MPa = 145 psi
p = fiber volume (%)
l = length of fiber
d = diameter of fiber

The resistance of the material is measured using one of the following criteria:

- Energy needed to fracture the specimen
- The number of blows in a repeated impact test to achieve a specified distress level
- The size of the damage—measured using crater size, perforation, or scab—or the size and velocity of the spall after the specimen is subjected to a surface blast loading

The measured performance can be used to compare different material compositions or to design a structural system that should withstand certain kinds of impact loads. The second part is more critical because it assures field performance for the chosen material combination. The results from these tests should be interpreted very carefully because they depend on a number of factors including specimen geometry, loading configuration, loading rate, test system compliance, and the prescribed failure criteria [9.12].

The repeated impact (drop-weight) test and two forms of instrumented impact tests are briefly described in the following sections. More details regarding these test set-ups can be found in References [9.11 to 9.14]. Well-defined test procedures are not established for explosive and projectile impact tests because they are not used for routine material evaluation.

9.2.1 Test methods: drop-weight test

This is the simplest test for evaluating impact resistance. The test method cannot be used to determine basic properties of composites. Rather, the method is designed to obtain the relative performance of plain and fiber-reinforced concretes containing different types and volume fractions of fibers.

A 6 in. (150 mm) diameter, 2.5 in. (64 mm) thick concrete (or FRC) disc is subjected to repeated impact loads (blows) by dropping a 10 lb (4.54 kg) hammer from a height of 18 in. (460 mm). The load is transferred from the hammer to the specimen through a 2.5 in. (64 mm) steel ball placed at the center of the disc. The plan view and a cross section of the test equipment are shown in Figures 9-8 and 9-9 [9.14]. The test sample can be cast as a 6 in. (150 mm) diameter, 2.5 in. (64 mm) thick disc or cut from a standard 6 × 12 in. (150 × 300 mm) cylinder. For concrete containing fibers longer than 20 mm, specimens should be cut from the cylinder to avoid preferential fiber alignment.

The specimen is placed between four guide pieces (lugs) located 0.1875 in. (4.8 mm) away from the sample (Figures 9-8 and 9-9). A frame (positioning bracket) is then built in order to target the steel

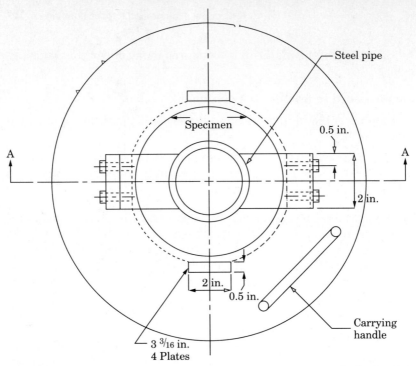

Figure 9-8 Plan view of test equipment for measuring impact strength. Section A-A is shown in 9-9 [9.14] (1 in. = 25.4 mm).

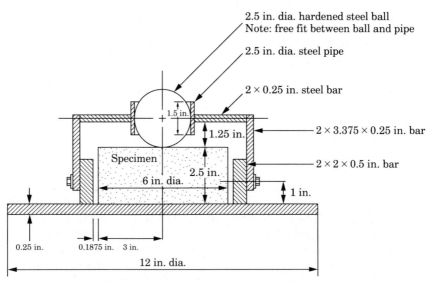

Figure 9-9 Section through test equipment for measuring impact strength shown in 9-8 [9.14] (1 in. = 25.4 mm).

ball at the center of the concrete disc. The disc should be coated at the bottom with a thin layer of petroleum jelly or a heavy grease to reduce friction between the specimen and the base plate.

The bottom part of the hammer unit is placed with its base upon the steel ball and the load is applied by dropping the 10 lb. (4.54 kg) weight. Standard equipment used for testing soils can be used for this test. Either manual or automated systems can be used for applying repeated blows. The number of blows that causes the first visible crack is recorded as the first crack strength. The loading is continued until the disc fails and opens up such that it touches three of the four positioning lugs. The number of blows that causes this condition is recorded as the failure strength.

Results of these tests exhibit high variability. A minimum of three and preferably more samples should be tested for each test condition.

9.2.2 Test methods: instrumented impact tests

By using instrumented impact tests, one can obtain additional information regarding material behavior, such as load-deflection response, strain and energy histories during the impact event, and the magnitude of ultimate strength. The disadvantage is the extensive instrumentation needed for the test.

Two types of systems are commonly employed. In the Charpy impact system, a swinging pendulum is used to impart the blow, whereas in the other system heavy weight attached to an instrumented tup is dropped to provide the impact. In both systems the instrumentation of the loading tup (or striker) and of the anvil that supports the beam (specimen) is needed to obtain the loading history.

A typical drop-weight system is shown in Figure 9-10 [9.11, 9.12, 9.14]. The heavy weight attached with the instrumented striker is dropped by gravity and guided by two columns. The weight and the height of the fall provide for the adjustment needed in terms of energy capacity and impact velocity.

In the Charpy system, shown in Figure 9-11 [9.14], the pendulum weight and the height of lift are used to obtain the required energy and impact velocity. Conventional machines can only be used to test small specimens because of the limitation on the maximum energy that can be attained in these machines.

Typical instrumentation needs are dynamic load cells, foil-type (resistance) strain gauges, and signal-recording systems such as amplifiers, oscilloscopes, and signal storage systems. The equipment should have a high-frequency response time because the load is impacted in about a millisecond.

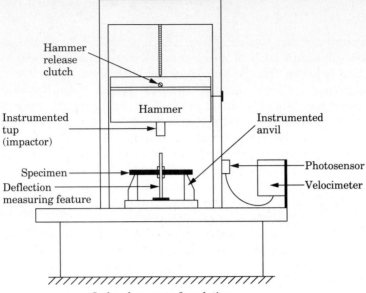

Figure 9-10 Block diagram of the general layout of the instrumented drop-weight system [9.14].

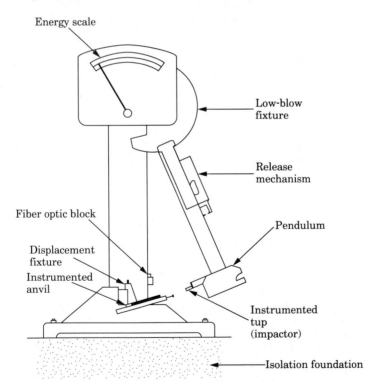

Figure 9-11 Block diagram of the general layout of the modified instrumented Charpy system [9.14, 9.35].

Anvil and striker loads should be recorded simultaneously in order to interpret inertial loads. The striking and support points should be rounded to avoid local damage at the contact points. Parasitic inertial loads in the responses recorded can cause problems. As a general guideline, the difference between the striker and anvil loads recorded during the test should be less than 5 percent. More detailed recommendations to reduce parasitic inertial loads can be found in References [9.11 and 9.15].

9.2.3 Impact resistance of steel fiber–reinforced concrete: drop-weight test

Addition of steel fibers to concrete leads to a substantial increase in impact strength [9.16 to 9.19]. Hooked-end fibers are more effective than straight fibers [9.16]. Tests using 50 mm (2 in.) long hooked-end fibers and 25 mm (1 in.) long straight fibers show that, at failure, hooked fibers can resist as much as seven times more blows than straight fibers (Figure 9-12). The straight fibers had an aspect ratio of 60, whereas hooked-end fibers had an aspect ratio of 100. The hooked-end fibers were used at low volume fractions of 80 and 105 lb/yd^3 (48 and 63 kg/m^3), while the dosage for straight fibers was 140 lb/yd^3 (84 kg/m^3). A matrix containing a pozzolan provided similar results [9.16].

The first-crack strength of concrete reinforced with straight steel fibers is only marginally higher than that of plain matrix, but the failure strength increases as much as 10-fold for FRC with hooked-end fibers over plain matrix concrete. The fibers are very effective in transferring stresses across the crack and in preventing shattering of the concrete.

Tests using various volume fractions of fibers show a consistent increase in impact resistance with increase in fiber content (Figure 9-13). Even a volume fraction of about 0.3% provides a fourfold increase in impact strength over plain concrete. The orientation of fibers along the diameter of the specimen seems to affect the impact strength considerably, as demonstrated in specimens fabricated using the shotcreting process. Shotcreted specimens having similar volume fractions of hooked-end fibers failed after receiving over a thousand blows, compared with fewer than 100 blows for molded specimens [9.18]. Since fibers tend to be deposited lengthwise in the shotcreting process, most of the fibers in the cores taken from these slabs had oriented along the diameter, providing better impact resistance.

Impact strength of FRC increases with maturity (age at testing) and follows the trend exhibited by compressive and flexural strengths

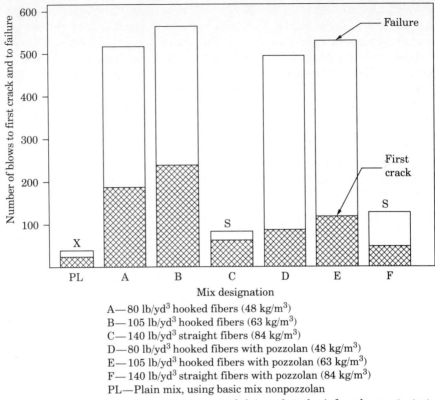

Figure 9-12 Comparison of impact resistance of plain and steel-reinforced concrete, tested at 28 days with the drop-weight test [9.16, 9.35].

[9.19]. Both first crack and failure strength increase with maturity. One should expect this change because with increasing maturity both the toughness of the matrix and the bond between the matrix and fibers improve, thus enhancing the failure strength. Note that the bond between fibers and matrix is the predominant factor that affects the transfer of load across the cracks.

9.2.4 Impact resistance of polymeric fiber–reinforced concrete: drop-weight test

Addition of polymeric fibers like Nylon 6, polypropylene, polyethylene, and polyester also improve impact strength [9.8, 9.9, 9.20]. The effects of fiber type, fiber volume fraction, and matrix composition have been investigated by Balaguru et al. In these studies polypropylene fibers were 19 mm (0.75 in.) long and fibrillated. Nylon 6 and polyester fibers were single filaments 19 mm (0.75 in.) long. The polyethy-

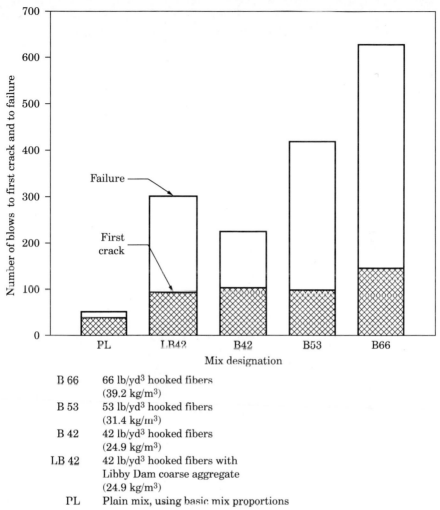

B 66   66 lb/yd³ hooked fibers
        (39.2 kg/m³)
B 53   53 lb/yd³ hooked fibers
        (31.4 kg/m³)
B 42   42 lb/yd³ hooked fibers
        (24.9 kg/m³)
LB 42   42 lb/yd³ hooked fibers with
        Libby Dam coarse aggregate
        (24.9 kg/m³)
PL   Plain mix, using basic mix proportions

Figure 9-13 Comparison of impact resistance of plain concrete reinforced with three different volume fractions of hooked-end steel fibers, tested at 28 days with the drop-weight test [9.17].

lene fibers were only a few millimeters long. The fiber dosage varied from 0.75 to 8 lb/yd³ (0.4 to 4.8 kg/m³) for longer fibers and 8 to 32 lb/yd³ (4.8 to 19.0 kg/m³) for shorter fibers. The composites with shorter fibers were cast using the shotcreting process. The matrix strength varied from 3 to 7 ksi (20 to 48 MPa).

The increases in first-crack and failure strength for matrices reinforced with Nylon 6 fibers are shown in Figures 9-14 and 9-15. Because of the high variability among replicates, averages were not computed. From Figures 9-14 and 9-15, one can see that the addition of fibers

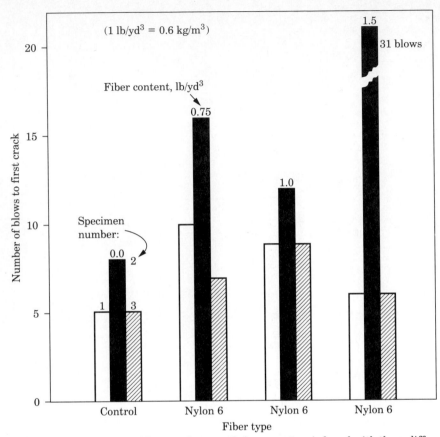

Figure 9-14 Comparison of first-crack strength for concrete reinforced with three different contents of Nylon 6 fibers [9.20].

provides a marginal improvement in first-crack strength and somewhat clearer improvement in failure strength. At a fiber dosage of 1.5 lb/yd³ (0.9 kg/m³), there is about 250% improvement. At a fiber volume fraction of 0.5% ($\approx$ 8 lb/yd³, 4.8 kg/m³) the improvement was reportedly about fourfold [9.20].

The variation of impact strengths among fiber types is shown in Figures 9-16 to 9-19. The difference in behavior between fiber types is not significant at a fiber content of 1.0 lb/yd³ (0.6 kg/m³). Polyester fiber, which has a higher modulus of elasticity, seems to provide slightly better performance at first crack. Polyethylene fibers provide good improvement in impact resistance at volume fractions of 1 and 2 percent. Their performance could not be compared with other fibers because they were not evaluated at these higher volume fractions.

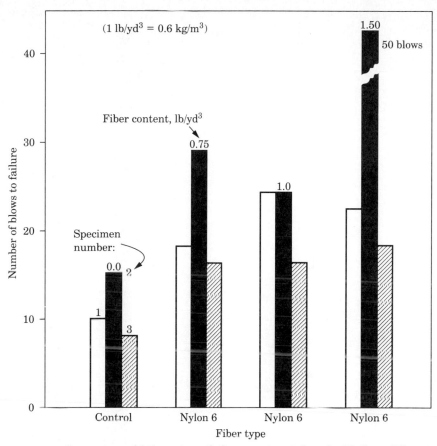

Figure 9-15 Comparison of failure strength for concrete reinforced with three different contents of Nylon 6 fibers [9.20].

9.2.5 Impact resistance of steel fiber–reinforced concrete: instrumented impact tests

Instrumented impact tests can provide information regarding the characteristics of brittle and quasi-brittle materials subjected to loads at high rates of strain. The results from these tests can be used for formulating constitutive stress-strain relations useful for structural design. If the basic behavior of the constitutent materials is known, then the structural elements can be designed to withstand an expected (or prescribed) external dynamic loading condition. Hence, the additional information provided by the instrumented tests makes the complexity of the procedure worthwhile.

The accuracy of instrumented impact tests depends largely on the accuracy of the recording systems and the proper analysis of the inertial effects. The load experienced by the beam is computed using

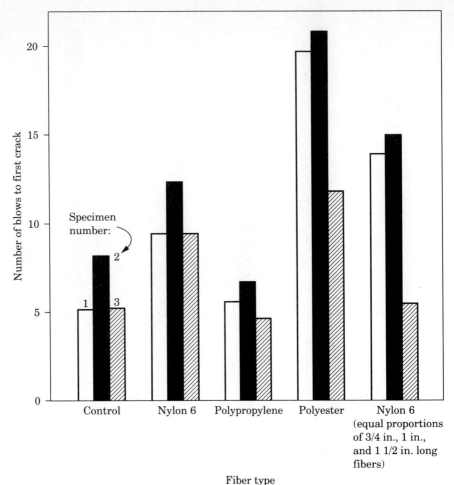

Figure 9-16 Comparison of first-crack strength for concretes reinforced by various polymeric fiber types at a fiber content of 1.0 lb/yd^3 (0.6 kg/m^3) [9.20].

the strain realized by the instrumented tup (striker) and/or the strain realized by the anvils (beam supports). Ideally, loads computed using strains recorded by the tup should exactly match the sum of the loads recorded by the anvils. However, because of inertial oscillations, these two loads may be significantly different. Reducing the velocity of impact, increasing the ratio of the tup (hammer) mass to beam mass, and increasing the ratio of the beam stiffness to the effective stiffness of the contact zone can minimize the difference [9.11].

Results obtained using the drop-weight and the Charpy test compare well in terms of both load-deformation behavior and energy absorption [9.12–9.15, 9.22]. Using the drop-weight test, Namaan and Gopalaratnam [9.22] compared three fiber volume fractions, three aspect ratios, and two matrix compositions. Mortar was used for both matrix com-

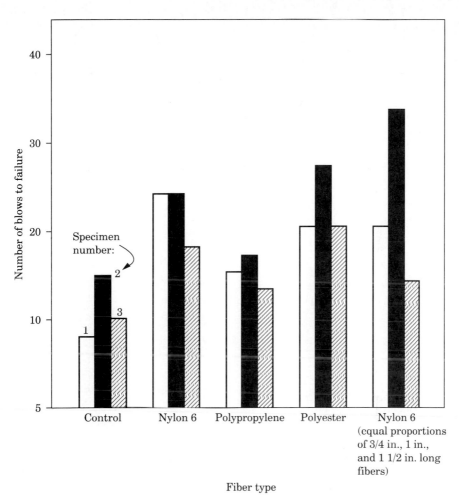

Figure 9-17 Comparison of failure strength for concretes reinforced by various polymeric fiber types at a fiber content of 1.0 lb/yd³ (0.6 kg/m³) [9.20].

positions. Fiber volume fractions of 1, 2, and 3% were evaluated for aspect ratios of 47, 62, and 100. The smooth fibers were made of either steel or brass.

Test specimens for flexural strength consisted of prisms that were 12 in. (300 mm) long, 3 in. (75 mm) wide and 0.5 in. (12 mm) thick. They were loaded at midspan using four loading rates. The contact velocities for the four loading rates were 0.1, 20, 1670, and 2360 in./min (0.0042, 0.85, 71, and 100 cm/s). The corresponding strain rates were 4.6×10^{-5}, 0.01, 0.8, and 1.2 in./in. (mm/mm) per second. Tests at the two lower velocities were conducted using an Instron machine and the others were conducted using the drop-weight system.

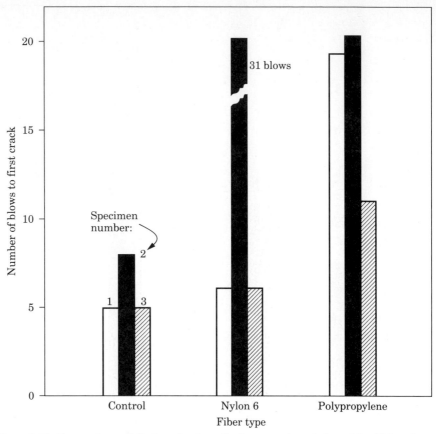

Figure 9-18 Comparison of first-crack strength for concrete reinforced by Nylon 6 or polypropylene at a fiber content of 1.5 lb/yd³ (0.9 kg/m³) [9.20].

Typical load-deflection curves at different loading rates are shown in Figure 9-20 [9.22]. From this figure one can see that both the peak load and energy absorption change significantly with increase in loading rate. The changes in modulus of rupture and energy absorption are shown in Figures 9-21 and 9-22 [9.22]. Energy absorption was calculated using the area under the curve up to a deflection of 0.5 in. (12 mm). From these figures one can note that changes are more noticeable in the strain rate region of 10^{-2} to 1.2 units/s than between 4.6×10^{-5} and 10^{-2} units/s. It can also be seen that fibers with higher aspect ratios are more rate sensitive. For a given aspect ratio, composites with higher volume fractions of fibers are more rate sensitive [9.22]. Composites with the weaker matrix were more rate sensitive than those with the stronger matrix.

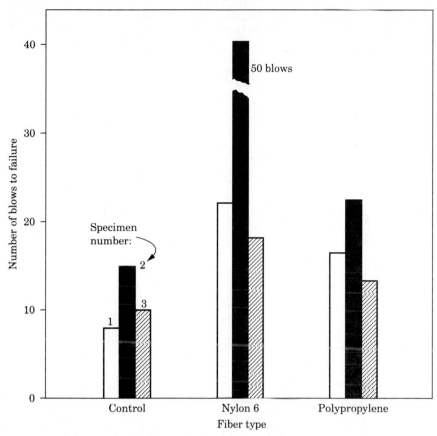

Figure 9-19 Comparison of failure strength for concrete reinforced by Nylon 6 or polypropylene at a fiber content of 1.5 lb/yd³ (0.9 kg/m³) [9.20].

For the unreinforced mortar matrix, surface energy to failure was about 0.22 lb · in./in.² (3.8 N · m/mm²) at a low rate of loading. This energy increased up to 32.6 lb · in./in.² (563 N · m/mm²) for composites reinforced with 3% fibers at the corresponding loading rate. A threefold increase in modulus of rupture and energy absorption was recorded when the strain rate increased from 4.6×10^{-6} to 1.2 in./in./s. The highest increase was recorded for the weak matrix containing fibers with the largest aspect ratio at a 3% fiber volume [9.22].

An extensive investigation carried out using the instrumented Charpy test indicates similar behavior of concrete reinforced with fiber volume fractions up to 1.5 percent [9.12]. The beams tested were 9 in. (225 mm) long, 1 in. (25 mm) wide, and 3 in. (75 mm) deep. Since the beams were relatively deep, differences in load-deformation characteristics were more marked than for the thin specimens reported in Reference [9.22].

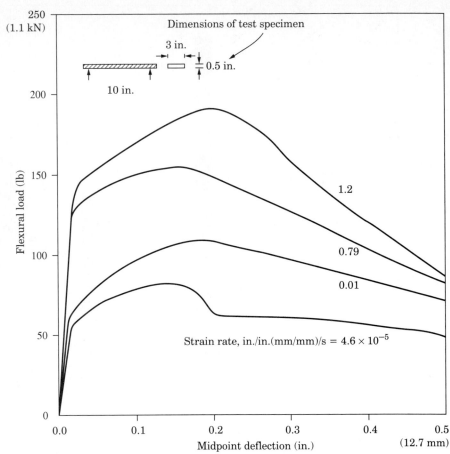

Figure 9-20 Load-deflection curves obtained using four strain rates [9.22] (1 in. = 25.4 mm).

Table 9.4 shows the different variables investigated in Reference [9.12]. Three fiber volume fractions and a control were evaluated using strain rates varying from 10^{-6} to 0.3 in./in./s. The beams were notched at the midspan. Typical load-deflection responses for plain and reinforced matrices are shown in Figures 9-23 and 9-24 respectively. In Table 9.4 and Figures 9-23 and 9-24, $\dot{\epsilon}$ represents the strain rate on the derivative of strain ϵ with respect to time. The average values for material properties at different strain rates are shown in Table 9.5. In both plain and fiber-reinforced beams, the loading rate increase results in increases in the peak load, the linear load deflection portion, and the deflection at peak load.

Results obtained by other investigators [9.23–9.25] are compared with results of Gopalaratnam and Shah [9.12] in Table 9.6. The following observations can be made based on these results.

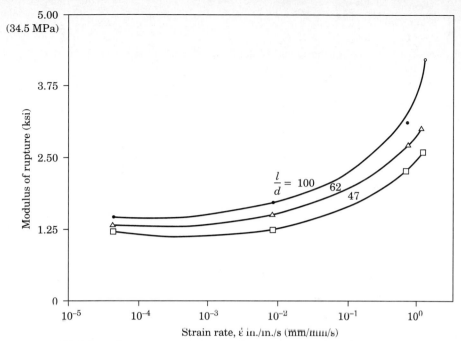

Figure 9-21 Modulus of rupture vs. strain rate for concrete reinforced with a fiber volume fraction of 2% steel fibers [9.22].

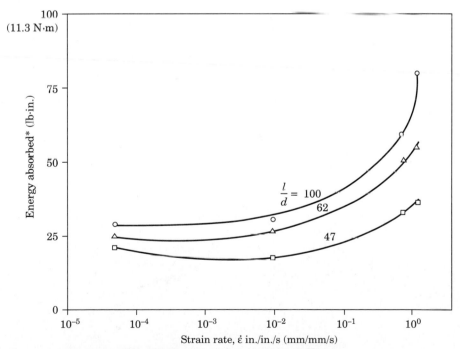

* Area under load-deflection curve up to 0.5 in. deflection.

Figure 9-22 Energy absorbed vs. strain rate for concrete reinforced with a fiber volume fraction of 2% steel fibers [9.22].

TABLE 9.4 Details of the Experimental Program Evaluating the Instrumented Charpy Test [9.12]

Mix*	Fiber content, V_f (percent)	Net cross section, depth × width (in. [mm])	Span (in. [mm])	$\dot{\epsilon}_1$	$\dot{\epsilon}_2$	$\dot{\epsilon}_3$	$\dot{\epsilon}_4$	$\dot{\epsilon}_5$	Compression tests‡ no. of specimens $\dot{\epsilon}_1$
1	0.0	3 × 1 (76 × 25)	8 (203)	4	4	4	4	4	4
		5/2 × 1 (64 × 25)	8 (203)	4	—	—	—	—	
2	0.5	5/2 × 1 (64 × 25)	8 (203)	4	4	4	4	4	4
3	1.0	5/2 × 1 (64 × 25)	8 (203)	4	4	4	4	4	4
4	1.5	5/2 × 1 (64 × 25)	8 (203)	4	4	4	4	4	4

The column headers "Number of specimens" span $\dot{\epsilon}_1$ through $\dot{\epsilon}_5$, and "Flexural tests†" spans from Net cross section through Number of specimens.

*Matrix 1 : 2 : 0 : 0.5 (cement : sand : coarse aggregate : water, by weight): fibers l = 1 in. (25 mm); d = 0.016 in. (0.4 mm).

†$\dot{\epsilon}_1 = 1 \times 10^{-6}$/s, $\dot{\epsilon}_2 = 1 \times 10^{-4}$/s, $\dot{\epsilon}_3 = 0.09$/s, $\dot{\epsilon}_4 = 0.17$/s, $\dot{\epsilon}_5 = 0.3$/s. Flexural tests at $\dot{\epsilon}_1$ and $\dot{\epsilon}_2$ were conducted using a closed-loop machine under displacement control. Flexural tests at $\dot{\epsilon}_3$, $\dot{\epsilon}_4$, and $\dot{\epsilon}_5$ were conducted using the modified instrumented Charpy setup.

‡Compression tests were conducted on 3 × 6 in. (76 × 152 mm) cylinders.

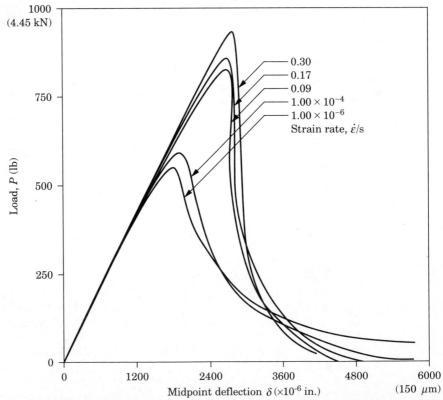

Figure 9-23 Load-deflection responses of unnotched plain mortar matrix beams at different rates of loading [9.12].

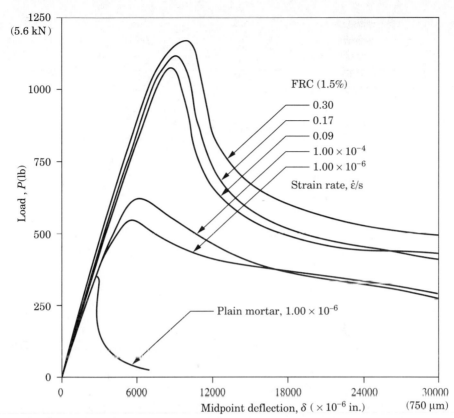

Figure 9-24 Load-deflection responses of notched beams (notch depth = 0.5 in. [12 mm]; width = 0.1 in. [2.5 mm]) at different rates of loading (fiber content = 1.5%, aspect ratio = 63) [9.12].

The ratio of the dynamic to the quasi-static modulus of rupture increased from 1.43 to 1.90 at the highest rate of loading included in Table 9.6. The corresponding numbers for FRC vary from 1.79 to 2.63, confirming that FRC is more rate sensitive than plain matrix. The corresponding numbers for energy absorption range from 1.35 to 2.35 for plain matrix and 1.52 to 1.86 for fiber-reinforced matrix. Even though the ratios are similar, the absolute values for energy absorption are higher for FRC specimens than for plain matrix specimens.

The lower the modulus of rupture, the higher the influence of strain rate on the properties of the composite. The tests conducted using different types of impact mechanisms and specimen sizes provide comparable results, indicating that the properties determined are directly related to the behavior of the basic material.

Energy absorption capacity as determined using conventional Charpy tests is higher than that determined by instrumented drop-weight setups. This increase may be due to the energy absorbed by

TABLE 9.5 Average Values for Material Properties at Different Strain Rates Determined with the Instrumented Charpy Test [9.12]

Material	Fiber content, V_f (percent)	Property†	Strain rate‡, ϵ/s				
			$\dot\epsilon_1$	$\dot\epsilon_2$	$\dot\epsilon_3$	$\dot\epsilon_4$	$\dot\epsilon_5$
Mortar	0.0	MOR, psi (MPa)*	747 (5.15)	800 (5.52)	1113 (7.68)	1153 (7.95)	1240 (8.55)
		G_f, lb/in. (N/m)	0.43 (75)	0.42 (73)	0.52 (92)	0.56 (97)	0.58 (101)
	0.5	MOR, psi (MPa)	806 (5.56)	864 (5.96)	1334 (9.20)	1392 (9.60)	1450 (10.00)
		G_f, lb/in. (kN/in.)	8.01§ (1.40)	8.09§ (1.42)	12.09 (2.12)	12.89 (2.26)	13.54 (2.37)
FRC	1.0	MOR, psi (MPa)	864 (5.96)	989 (6.82)	1603 (11.06)	1670 (11.52)	1718 (11.85)
		G_f, lb/in. (kN/m)	9.77§ (1.71)	10.00§ (1.76)	15.24 (2.67)	15.93 (2.78)	17.20 (3.01)
	1.5	MOR, psi (MPa)	1056 (7.28)	1181 (8.15)	2054 (14.16)	2150 (14.83)	2237 (15.43)
		G_f, lb/in. (kN/m)	14.32§ (2.50)	15.18§ (2.66)	21.20 (3.71)	21.83 (3.82)	24.50 (4.29)

*Average of four specimens for each reported value of MOR and G_f, defined below

†Modulus of rupture (MOR) computed using elastic theory and beam depth. Fracture energy G_f computed as area under the load deflection curve (for fracture of plain mortar and up to a central deflection of 0.1 in. for FRC) for unit net cross-sectional area

‡$\dot\epsilon_1 = 1.0 \times 10^{-6}/s$; $\dot\epsilon_2 = 1.0 \times 10^{-4}/s$; $\dot\epsilon_3 = 0.09/s$; $\dot\epsilon_4 = 0.17/s$; $\dot\epsilon_5 = 0.3/s$

§ Stopped test at $\delta = 0.075$ in., G_f^* up to $\delta = 0.1$ in. estimated by extrapolating P-δ curve between $\delta = 0.075$ in. and $\delta = 0.1$ in.

TABLE 9.6 Results from Instrumented Impact Tests on Mortar Concrete and FRC [9.12]

Reference	Setup	Specimen size: Net cross section depth × width (in.)	Span (in.)	Material	C:S:A:W (by weight)	Static properties§ MOR (psi)	G_f (lb/in.)	Impact properties¶ Strain rate	$\dfrac{MOR_d}{MOR_s}$	$\dfrac{G_{fd}}{G_{fs}}$
Zech and Wittmann	Instrumented drop-weight	0.79 × 0.79	7.9	Mortar	1:4.7:0:0.57	1813	—	1.00	1.50	—
				Mortar	1:0.2:0:0.90	1030	—	1.00	1.90	—
				Mortar	1:2:0:0.5	1060	0.43	0.27	1.67	2.35
				Concrete	1:2:3:0.5	1430	0.59	0.27	1.43	—
Suaris and Shah	Instrumented drop-weight	3 × 1.5	15	SFRC* V_f = 1 percent l/d = 100	1:2:0:0.5	1370	15.83	0.27	2.02	1.86
				Mortar	—	1030	0.59	0.20	1.10 (1.74†)	1.56
Kayanagi et al.	Instrumented drop-weight	3 × 3	24	SFRC* not reported Mortar	—	1262	7.97	0.20	2.63	1.52
				Mortar	1:2:0:0.5	747	0.43	0.30	1.65	1.35
		3 × 1	8	Concrete	1:2:2:0.5	1400	0.32	0.30	1.50	1.47
				SFRC* V_f = 0.5 percent l/d = 63	1:2:0:0.5	806	8.01	0.30	1.79	1.69
Gopalaratnam and Shah [9.12]	Instrumented Charpy	2.5 × 1	8	SFRC* V_f = 1.0 percent l/d = 63	1:2:0:0.5	864	9.77	0.30	1.99	1.76
				SFRC* V_f = 1.5 percent l/d = 63	1:2:0:0.5	1056	14.32	0.30	2.11	1.71

*G_f for all FRC specimens is reported up to a 0.1 in. midpoint deflection   †When sufficient potential energy is available in the impact
‡C—cement; S—sand; A—coarse aggregate; W—water   §MOR—modulus of rupture; G_f—fracture energy
¶MOR$_d$—the MOR for dynamic loading; MOR$_s$—the MOR for static loading; G_{fd}—the G_f for dynamic loading; G_{fs}—the G_f for static loading
1 in. = 2.54 cm; 1 psi = 0.0069 MPa; 1 lb/in. = 175.2 N/m

the support system, which was deducted from the observed total energy absorption.

Comparison of results obtained from (*a*) instrumented impact tests, (*b*) repeated impact tests, (*c*) conventional Charpy tests, and (*d*) toughness under static bending tests indicates that the static bending test provides the best estimate of energy absorption under impact loading [9.12]. This observation is based only on results obtained using straight smooth steel fibers. Fiber geometry might influence the validity of the above statement.

Researchers have recorded different magnitudes of change in impact resistance, depending on the method used for impact. The reported changes in modulus of rupture range from 35% to 1000% [9.26–9.28]. The very high increase could be attributed to inertial effects [9.11].

9.2.6 Impact resistance of polypropylene fiber–reinforced concrete: instrumented impact tests

A few investigators have studied the contribution of polypropylene fibers to impact resistance using the instrumented drop-weight test [9.29–9.32]. The primary variables investigated were fiber volume fraction, fiber lengths, and matrix composition. Fibers made by two manufacturers were tested. Lengths varied from 0.5 to 2.0 in. (12 to 50 mm). Fiber volume fraction ranged from 0 to 1.0%. Cement paste, mortar, concrete containing just cement, and concrete containing admixtures of fly ash or silica fume were the matrix composition variables.

The tests were conducted using a drop-weight hammer weighing 345 kg (760 lb), dropped from a height of 200 mm (8 in.). An accelerometer was mounted at the midspan of the beam. The overall beam dimensions were $100 \times 100 \times 350$ mm ($4 \times 4 \times 14$ in.). The hammer was instrumented so as to obtain the loads. The load-deflection curve was used to compute the fracture energy, which was taken as the area under the load-deflection curve up to a deflection at which the load dropped to one-third of peak load.

At a fiber volume fraction of 0.1% the improvement in both peak load and fracture energy was not significant. The peak load actually decreased in some instances.

For cement paste, the peak load decreased from 6.7 to 6.0 kN (1.5 to 1.3 kips) and the fracture energy increased from 8.5 to 9.7 N · m (6.3 to 7.2 ft · lb). For mortar the peak load increased from 9.1 to 11.7 kN (2.0 to 2.6 kips) and fracture energy increased from 14.2 to 17.9 N · m (10.5 to 13.2 ft · lb). In concrete the fiber addition resulted in a reduction of peak load from 8.2 to 7.8 kN (1.8 to 1.7 kips) and an increase of fracture energy from 9.5 to 10.4 N · m (7.0 to 7.7 ft · lb).

Concrete containing coarse aggregate registered the lowest increase in fracture energy. An increase in fiber length from 0.5 to 2 in. (12 to 50 mm) did not improve the fracture properties considerably. Overall, we can assume that the effect of fibrillated polypropylene fibers at a volume fraction of 0.1 percent is negligible. This assumption, of course, only pertains to plain cement or concrete beams tested at high strain rates.

Table 9.7 presents the effect of volume fraction on peak load and fracture energy. The effect of the incorporation of fibers on the peak load is negligible up to a volume fraction of 0.7%. The fracture energy increases up to the volume fraction 0.5%. The decrease in fracture energy at 0.7% volume fraction compared to 0.5% may be due to incomplete compaction or entrapment of excess air because of high fiber content. These results are similar to the other published results [9.29 to 9.31], even though the magnitudes of improvement reported vary considerably. For example, at a fiber volume fraction of 0.5%, the reported improvement in fracture energy varies from 32% [9.29] to 80% [9.30, 9.31]. The fibers were found to be more effective in normal-strength concrete than in high-strength concrete [9.29].

In summary, polypropylene fibers are not effective under dynamic loading at the normally recommended volume fraction levels of 0.1%. Improvement is not substantial even at 0.5% volume. The reasons for this limited improvement are not yet established. Even though these fibers were found not to be effective in plain matrix, they were effective in reinforced concrete, as discussed in the next section.

Tests conducted using polyethylene fibers and an Instron machine with crosshead velocity in the range of 1 to 200 mm/min (0.04 to 8 in./min) showed that the incorporation of 4% fiber provides a 25% increase in the modulus of rupture [9.33]. This report indirectly confirms the results discussed so far, even though the impact velocity was low.

9.2.7 Impact resistance of beams reinforced with continuous bars and discrete polypropylene fibers: instrumented impact tests

Tests have also been carried out using reinforced concrete beams containing polypropylene fiber reinforcement [9.30–9.32]. The primary variables investigated were fiber volume fraction and strength of concrete.

Typical results are shown in Table 9.8 [9.32]. The maximum load and fracture energy were obtained using 1200 mm (47 in.) long, 100

TABLE 9.7 The Effect of Polypropylene Fiber Content on the Impact Strength of Concretes* [9.32]

Fiber content (% volume)	Peak load (kN)	Fracture energy (Nm)
0	8.2	9.5
0.1	7.8 (−5%)	10.4 (+9%)
0.3	8.1 (−1%)	11.3 (+19%)
0.5	8.4 (+2%)	13.9 (+46%)
0.7	8.3 (+1%)	12.0 (+26%)

*The numbers in parenthesis indicate the relative improvement (%) over the plain matrix. Fiber length: 2.5 in. (63 mm) 1 kip = 4.4 kN 1 in. · lb = 0.11 N · m

mm (4 in.) wide, and 125 mm (5 in.) deep beams reinforced with two No. 10 (Canadian Standard) bars and various amounts of fiber. The reinforcement area was 100 mm^2 (0.16 in^2) and, hence, the reinforcement ratio was 1.0%. The beams were tested by dropping a 345 kg (760 lb) hammer from a height of 500 mm (20 in.).

From Table 9.8 we can see that the fibers are very effective in improving the performance of normal-strength concrete under impact load. The energy absorbed increases by almost threefold up to a fiber volume content of 0.5%. The increase for high-strength concrete is not significant, however, the absolute value of energy absorbed is still very high. The only drawback is that considerably higher amounts of fibers than the currently recommended levels must be used to obtain substantial increases in energy absorption.

Results reported in Reference [9.34] indicate that fibers are as effective in high-strength concrete as they are in normal-strength concrete. Fracture energy was found to increase threefold even for high-strength concrete. The primary difference from the other reports [9.30–9.32] was the age at testing, which was one year for the high-strength concrete, compared with a few weeks for normal-strength concrete.

TABLE 9.8 The Effect of Polypropylene Fiber Reinforcement on the Impact Resistance of Conventionally Reinforced Normal- and High-Strength Concretes [9.32]

Type of concrete	Fiber content (% volume)	Maximum load (kN)	Energy (Nm)
Normal-strength 30.4 MPa	0	40.8	286
	0.1	39.4	350
	0.3	44.3	444
	0.5	45.2	840
	1.0	34.0	644
High-strength silica fume, 73.7 MPa	0	47.2	866
	0.3	51.3	941
	1.0	55.5	1020

1 ksi = 6.9 MPa
1 kip = 4.4 kN
1 in. · lb = 0.11 N · m

Since fibers are not so effective in plain concrete as in reinforced concrete, one may hypothesize that fibers provide a confining effect, resulting in better bonding between reinforcement and concrete. Since the presence of fibers reduces the shattering effect, the bars can maintain higher pull-out loads even at large deformations. This hypothesis also explains the larger load-carrying capacity of beams at higher deflections.

9.3 References

9.1 ACI Committee 215. "Considerations for Design of Concrete Structures Subjected to Fatigue Loading," ACI Manual of Concrete Practice, 1990, Part I, pp. 215-1 to 215-25.

9.2 Batson, G.; Ball, C.; Bailey, L.; Landers, E.; and Hooks, J. "Flexural Fatigue Strength of Steel Fiber Reinforced Concrete Beams," *ACI Journal,* Vol. 69, No. 11, 1972, pp. 673–677.

9.3 Ramakrishnan, V.; and Josifek, C. "Performance Characteristics and Flexural Fatigue Strength on Steel Fiber Composites," Proceedings of the International Symposium on Fiber Reinforced Concrete, Oxford and IBH Publishing Company, New Delhi, December 1987, Madras, India, pp. 2.73–2.84.

9.4 Ramakrishnan, V.; Oberling, G.; and Tatnall, P. "Flexural Fatigue Strength of Steel Fiber Reinforced Concrete," *Fiber Reinforced Concrete Properties and Applications,* SP-105, American Concrete Institute, Detroit, Michigan, 1987, pp. 225–245.

9.5 Wu, G. Y.; Shivaraj, S. K.; and Ramakrishnan, V. "Flexural Fatigue Strength, Endurance Limit, and Impact Strength of Fiber Reinforced Refractory Concretes," *Fiber Reinforced Cements and Concretes: Recent Developments,* Elsevier, New York, 1989, pp. 261–273.

9.6 Ramakrishnan, V.; Gollapudi, S.; and Zellers, R. "Performance Characteristics and Fatigue Strength of Polypropylene Fiber Reinforced Concrete," *Fiber Reinforced Concrete Properties and Applications,* SP-105, American Concrete Institute, Detroit, Michigan, 1987, pp. 159–177.

9.7 Vondran, G. L.; Nagabhusharam, M.; and Ramakrishnan, V. "Fatigue Strength of Polypropylene Fiber Reinforced Concrete," *Fiber Reinforced Cement and Concretes: Recent Developments,* Elsevier, New York, 1989, pp. 533–543.

9.8 Balaguru, P. "Properties of Concrete Reinforced with Polyethylene Fibers," Civil Engineering Department Report 90-20, Rutgers University, New Brunswick, New Jersey, 1990, 320 pp.

9.9 Balaguru, P. "Mechanical Behavior of Nylon 6 Fiber Reinforced Concrete," Civil Engineering Department Report 90-15, Rutgers University, New Brunswick, New Jersey, 1990, 106pp.

9.10 Kormeling, H. A.; Reinhardt, H. W.; and Shah, S. P. "Static and Fatigue Properties of Concrete Beams Reinforced with Continuous Bars and with Fibers," *ACI Journal,* Vol. 77, No. 1, 1980, pp. 36–43.

9.11 Gopalaratnam, V. S.; Shah, S. P.; and John, R. "A Modified Instrumented Charpy Test for Cement-Based Composites," *Experimental Mechanics,* Vol. 24, No. 2, June 1984, pp. 102–111.

9.12 Gopalaratnam, V. S.; and Shah, S. P. "Properties of Fiber Reinforced Concrete Subjected to Impact Loading," *ACI Journal,* Vol. 83, No. 1, 1986, pp. 117–126.

9.13 Suaris, W.; and Shah, S. P. "Test Methods for Impact Resistance of Fiber Reinforced Concrete," *Fiber Reinforced Concrete—International Symposium,* SP-81, American Concrete Institute, Detroit, Michigan, 1984, pp. 247–265.

9.14 ACI Committee 544; "Measurement of Properties of Fiber Reinforced Concrete," American Concrete Institute, Detroit, Michigan, 1989, 11 pp.

9.15 Suaris, W.; and Shah, S. P. "Inertial Effects in the Instrumented Impact Testing of Cementitious Composites," *Cement, Concrete and Aggregates,* Vol. 3, No. 2, Winter 1981, pp. 77–83.

9.16 Ramakrishnan, V.; Brandshaug, T.; Coyle, W. V.; and Shrader, E. K. "A Comparative Evaluation of Concrete Reinforced with Straight Steel Fibers and Fibers with Deformed Ends Glued Together into Bundles," *ACI Journal,* Vol. 77, No. 3, 1980, pp. 135–143.

9.17 Ramakrishnan, V.; Coyle, W. V.; Kulandaisamy, V.; and Schrader, E. K. "Performance Characteristics of Fiber Reinforced Concrete with Low Fiber Contents," *ACI Journal,* Vol. 78, No. 5, 1981, pp. 388–394.

9.18 Ramakrishnan, V.; Coyle, W. V.; Dahl, L. F.; and Schrader, E. K. "A Comparative Evaluation of Fiber Shotcrete," *Concrete International: Design and Construction,* Vol. 3, No. 1, 1981, pp. 56–59.

9.19 Balaguru, P.; and Ramakrishnan, V. "Mechanical Properties of Superplasticized Fiber Reinforced Concrete Developed for Bridge Decks and Highway Pavements," *Concrete in Transportation,* SP-93, American Concrete Institute, Detroit, Michigan, 1986, pp. 563–584.

9.20 Khajuria, A.; and Balaguru, P. "Behavior of New Synthetic Fibers for Use in Concrete," Civil Engineering Department Report No. 89-13, Rutgers University, New Brunswick, New Jersey, 1989, 89 pp.

9.21 Balaguru, P. "Evaluation of New Synthetic Fiber for Use in Concrete," Civil Engineering Department Report No. 88-10, Rutgers University, New Brunswick, New Jersey, 1988, 93 pp.

9.22 Namaan, A. E.; and Gopalaratnam, V. S. "Impact Properties of Steel Fiber Reinforced Concrete in Bending," *International Journal of Cement Composites and Lightweight Concrete,* Vol. 5, No. 4, 1983, pp. 225–233.

9.23 Zech, B.; and Wittmann, F. A. "Variability and Mean Value of Strength of Concrete as a Function of Load," *ACI Journal,* Vol. 77, No. 5, 1980, pp. 358–62.

9.24 Suaris, W.; and Shah, S. P. "Properties of Concrete Subjected to Impact," *Journal of Structural Engineering,* ASCE, Vol. 109, No. 7, 1983, pp. 1727–1741.

9.25 Koyanagi, W.; Rokugo, K.; Uchida, Y.; and Iwase, H. "Energy Approach to Deformation and Fracture of Concrete under Impact Load," *Transactions,* Japan Concrete Institute, Tokyo, Japan, Vol. 5, 1983, pp. 161–168.

9.26 Hibbert, A. P. "Impact Resistance of Fiber Concrete," Ph.D. Thesis, University of Surrey, England, 1977.

9.27 Radomski, W. "Application of the Rotating Impact Machine for Testing Fiber Reinforced Concrete," *International Journal of Cement Composites and Lightweight Concrete,* Vol. 3, No. 1, 1981, pp. 3–12.

9.28 Butler, J. E.; and Keating, J. "Preliminary Data Derived Using a Flexural Cyclic Loading Machine to Test Plain and Fibrous Concrete," RILEM *Materials and Structures,* Vol. 14, No. 79, 1981, pp. 25–33.

9.29 Banthia, N. P. "Impact Resistance of Concrete," Ph.D. Thesis, University of British Columbia, Canada, 1987.

9.30 Banthia, N. P.; Mindess, S.; and Bentur, A. "Impact Behavior of Concrete Beams, RILEM, *Materials and Structures,* Vol. 20, 1987, pp. 293–302.

9.31 Mindess, S.; and Vondran, G. "Properties of Concrete Reinforced with Fibrillated Polypropylene Fibers Under Impact," *Cement and Concrete Research,* Vol. 8, 1988, pp. 109–115.

9.32 Bentur, A.; Mindess, S.; and Skalny, J. "Reinforcement of Normal and High Strength Concretes with Fibrillated Polypropylene Fibers," *Fiber Reinforced Cements and Concretes: Recent Developments,* Elsevier, New York, 1989, pp. 229–239.

9.33 Kobayashi, K.; and Cho, R. "Flexural Behavior of Polyethylene Fiber Reinforced Concrete," *International Journal of Cement Composites and Lightweight Concrete,* Vol. 3, No. 1, February 1981, pp. 19–25.

9.34 Mindess, S.; Banthia, N.; and Bentur, A. "The Response of Reinforced Concrete Beams with a Fiber Matrix to Impact Loading," *International Journal of Cement Composites and Lightweight Concrete,* Vol. 8, 1986, pp. 165–170.

9.35 Shah, S. P.; and Skarendahl, A. (eds.). *Steel Fiber Concrete,* US–Sweden Joint Seminar (NSF-STU), Swedish Cement and Concrete Research Institute, Stockholm, Sweden, 1985, 520 pp.

10

Creep, Shrinkage, and Long-Term Performance

Creep and shrinkage are important parameters associated with the long-term performance of cement composites. For plain concrete the ultimate shrinkage strains vary from 200 to 1000 μin./in. (200 to 1000 μm/m). The creep strains can be more than twice the initial elastic strain at working load levels. Fiber-reinforced concrete is used in a number of applications in which shrinkage and creep strains are important in the design and performance of structures. This chapter deals with fiber-reinforced concrete containing steel and polymeric fibers at comparatively low volume fractions. The behavior of thin-sheet products, which typically contain larger fiber volume fractions, is discussed in Chapters 13 and 14.

Fiber durability is another factor to be considered in long-term performance. For steel fibers, the primary concern is corrosion. The results of a few investigations show that fiber corrosion is not a problem. In the area of polymeric fibers, the primary concern is the durability and effectiveness of the fibers themselves in the alkaline environment present in concrete. Durability tests carried out using an accelerated aging process show that some fibers fare better than others.

10.1 Creep and Shrinkage of Steel Fiber–Reinforced Concrete

Creep and shrinkage characteristics of steel fiber concrete were studied by a limited number of investigators [10.1–10.9]. The results available are even more limited in the area of creep [10.1–10.4]. In general, the presence of fibers reduces shrinkage to a limited extent, whereas the effect that their presence has on creep is still not known conclusively. It should be noted that the fiber volume fractions used in most of these studies are less than 2 percent by volume.

10.1.1   Creep behavior of steel fiber–reinforced concrete

The primary variables investigated are the fiber type, the fiber volume fraction, the stress-to-strength ratio, the relative humidity of exposure, and the age at loading [10.1, 10.2]. The results obtained using hooked-end fibers are reported in Reference [10.1.]. Two mix proportions containing 0 and 75 lb/yd³ (45 kg/m³) 2 in. long × 0.02 in. diameter (50 mm long, 0.5 mm diameter) steel fibers were subjected to about 20% of their compressive strengths. The mix proportions used are shown in Table 10.1. The plain concrete mixtures are designated as P1 and P2, and the corresponding fiber mixtures are designated as F1 and F2. Mixtures P1 and F1 had a lower cement content and higher water-cement ratio. The compressive strengths for these mixtures averaged about 6000 psi (40 MPa). Mixtures P2 and F2, which had a higher cement content and lower water-cement ratio, had an average compressive strength of about 8000 psi (55 MPa). Both shrinkage and creep measurements were made for all four mixes. The creep study also included creep-recovery measurements.

Creep tests were conducted using ASTM C512 procedures. Three 6×12 in. (150 × 300 mm) test cylinders that were moisture-cured for 28 days were subjected to a sustained load using a spring system. The tests were conducted in a room where a temperature of 23°C (73.4° F) and a relative humidity of 50% were maintained. All the samples were subjected to a sustained load of 40,000 lb (178 kN) resulting in approximate stress-strength ratios of 0.22 and 0.19 for the two mixtures tested. The strains were measured over a gauge length of 10 in. (250 mm), and the measurements were continued for a minimum of 337 days.

TABLE 10.1   Mixture Proportions: Creep and Shrinkage Study [10.1]

Mixture designation	F1*	F2*	P1	P2
Cement content (lb/yd³)†	611	799	611	799
Water-cement ratio by weight	0.4	0.3	0.38	0.28
High-range water-reducing admixture, percentage by weight of cement	1.0	1.2	1.0	1.2
Air-entraining agent, percentage by weight of cement	0.15	0.20	0.1	0.13

*Fiber-reinforced concrete mixtures, F1 and F2, had a fiber content of 75 lb/yd³ (45 kg/m³).
†1 lb/yd³ = 0.6 kg/m³

In the creep-recovery study, strain measurements were taken for an unloaded period of 12 days, and the specimens were reloaded again. Creep measurements were also taken for another unloaded period after another 12 days of sustained loading.

The creep strains and the creep-recovery behavior are shown in Figures 10-1 and 10-2 respectively [10.1]. The increase in stress-strength ratio results in increased creep strains. This is consistent with the results obtained on plain concrete. However, the strains of 0.46 and 0.59×10^{-6} in./in./psi (67 and 86×10^{-3} mm/mm/Pa) measured after 375 days are smaller than the values generally reported for structural concrete. In both instances involving the two concrete mixtures, the fiber-reinforced concrete had higher creep strains (Figure 10-2).

The results reported in Reference [10.2] deal with the effects of fiber volume fraction, age at loading, and the relative humidity of the surrounding environment on the creep behavior. The concrete had an average compressive strength of about 33 MPa (4800 psi) at 28 days. Straight carbon steel fibers with a shallow notch on one side were used at volume fractions of 1% and 2%. The specimens were loaded on 4, 7, and 28 days after coating. Test environmental conditions included rooms with 50% and 100% relative humidity and a temperature of 35° C (95° F). The details of the various combinations of testing are shown in Table 10.2 [10.2]. The tests were run using 150 × 300 mm (6 × 12 in.) cylinders loaded using a spring system. Shrinkage strains were also monitored under similar environmental conditions.

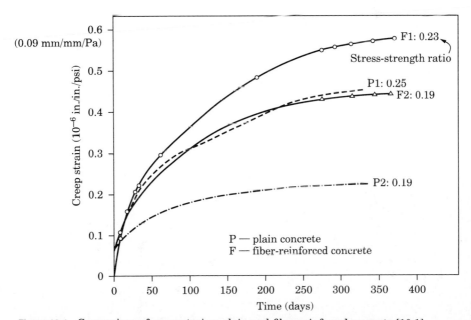

Figure 10-1 Comparison of creep strains: plain and fiber-reinforced concrete [10.1].

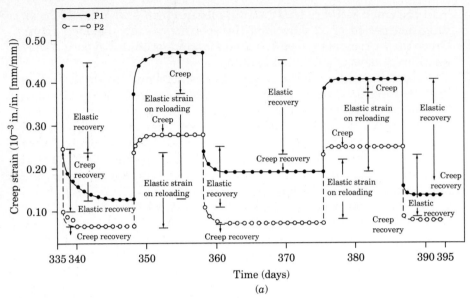

Figure 10-2a Elastic and creep recovery: plain concrete [10.1]

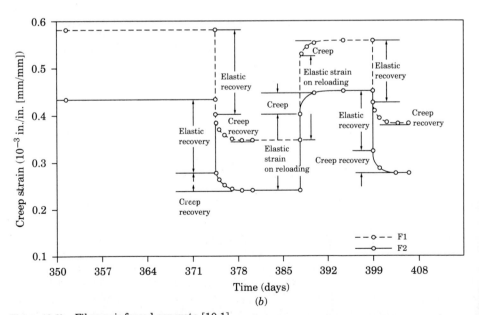

Figure 10-2b Fiber-reinforced concrete [10.1].

TABLE 10.2   Experimental Program for Creep Study [10.2]

Designation of specimen	t' (days)	v (%)	$f'c$ (kg/cm²)	E ($\times 10^5$) (kg/cm²)	Applied stress (kg/cm²)	Test environment*
W07	7	0	248.4	1.93	48.1	W
D07	7	0	248.4	1.93	48.1	D
W028	28	0	288.8	2.35	67.9	W
D028	28	0	288.8	2.35	67.9	D
H028	28	0	288.8	2.35	67.9	H
W14	4	1	215.2	1.94	45.3	W
D14	4	1	215.2	1.94	45.3	D
W17	7	1	268.8	2.22	62.3	W
D17	7	1	268.8	2.22	62.3	D
W128	28	1	331.0	2.58	79.2	W
D128	28	1	331.0	2.58	79.2	D
H128	28	1	331.0	2.58	79.2	H
W27	7	2	274.5	2.23	62.3	W
D27	7	2	274.5	2.23	62.3	D
W228	28	2	345.2	2.61	79.2	W
D228	28	2	345.2	2.61	79.2	D
H228	28	2	345.2	2.61	79.2	H

*Note: D: Dry room (23°C, 50% relative humidity), W: Fog room (23°C, 100% relative humidity), H: (35°C, 95% relative humidity). There was an average of two specimens for each test.

1 kg/cm² = 14.2 psi; °C = (°F − 32)/1.8

The effect of fibers on creep strains loaded at 7 and 28 days is shown in Figure 10-3 [10.2]. The figure also presents the variations of creep strains with respect to the relative humidity of the surrounding environment. As expected, dryer conditions increase the creep strains considerably. Higher fiber contents reduce the creep strains consistently for both 50% and 100% relative humidity, as shown in Figure 10-4. Specimens tested at 35°C (95°F) and 95% relative humidity recorded higher creep strains than specimens tested at 23°C (73°F) and 100% relative humidity, but much lower strains compared to specimens tested at 50% relative humidity (Figure 10-5). The fiber-reinforced specimens recorded lower creep strains under all testing conditions.

The magnitude of creep strain decreased faster at later ages of loading and for specimens tested at 35°C (95°F). The temperature effect is shown in Figure 10-6. Higher maturity could have possibly improved the bond strength and reduced the creep strains by the process of stress redistribution between the concrete and the fibers.

The results reported in Reference [10.3] provide information for stress-strength ratios varying from 0.3 to 0.9. The variables investigated include fiber types, fiber volume fraction, and variation in matrix composition. These results provide more comprehensive understanding of the creep behavior of FRC. However, only a few mixtures were similar to the ones used in the field. The matrix consisted of

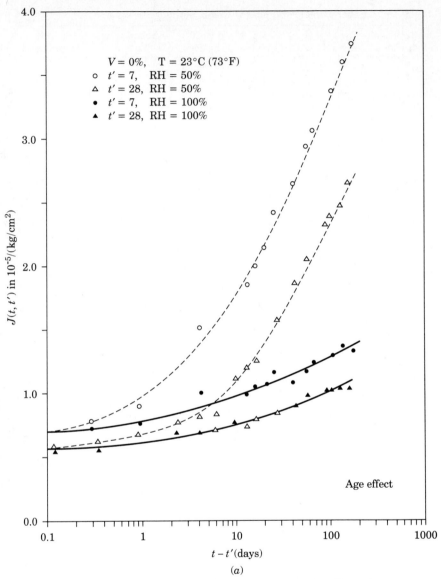

Figure 10-3a Effect of age and relative humidity on creep strains: fiber volume fraction = 0%.

cement paste, cement mortar, and concrete. The melt-extract fibers were used at volume fractions of 1.5% and 3%. The hooked-end fibers were used at volume fractions of 1% to 3%. The tests were conducted using $100 \times 100 \times 500$ mm ($4 \times 4 \times 20$ in.) prisms. In most cases the prisms were removed from their molds after 24 hours and were cured in a room maintained at 20°C (68°F) and 55% relative humidity.

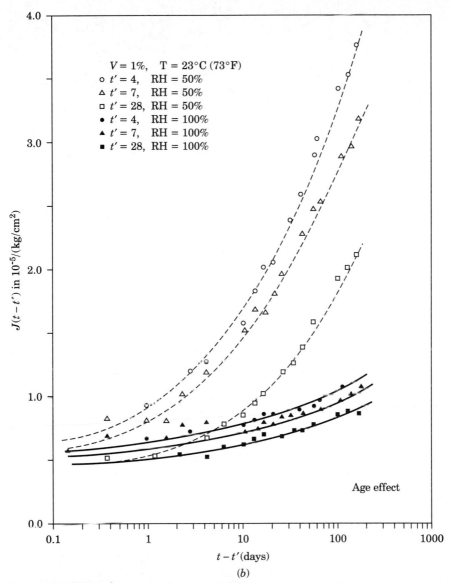

Figure 10-3b Effect of age and relative humidity on creep strains: fiber volume fraction = 1%.

Two types of creep tests were conducted depending on the magnitude of the stress-strength ratio. For stress-strength ratios of 0.30, 0.41, and 0.55, test specimens were placed in creep rigs. Creep strains were measured for a period of 90 days followed by creep-recovery measurements for 60 days. For stress-strength ratios varying from 0.64 to 0.90, strains were recorded up to 432 minutes.

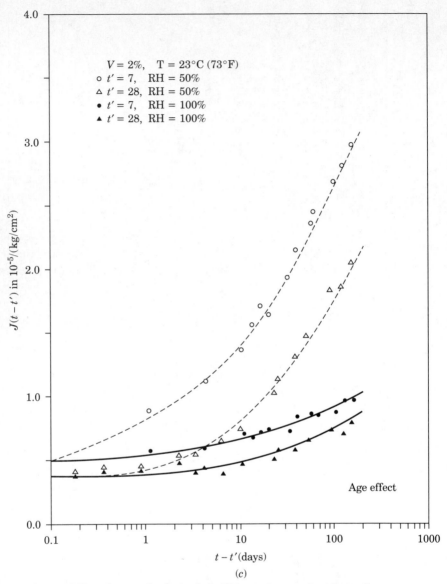

Figure 10-3c Effect of age and relative humidity on creep strains: fiber volume fraction = 2%.

The results show that cement paste has the highest creep strains, followed by mortar and concrete. The addition of fibers results in a consistent decrease in creep strains both at lower and higher stress-strength ratios. The difference between the fiber types, namely melt-extract and hooked-end fibers, is not significant. The elastic (instantaneous) recovery was found to be independent of fiber addition and

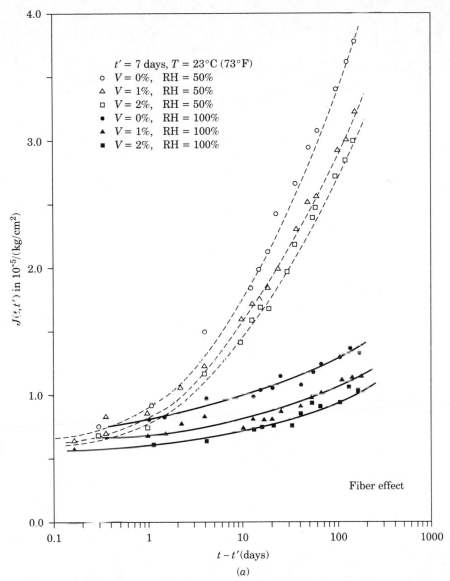

Figure 10-4a Effect of fiber volume fraction on creep strains: age at loading = 7 days.

averaged about 90% of the initial elastic deformation. Elastic recovery is typically less than the initial elastic deformation because of the increase in stiffness that comes with maturity. Fibers have no influence on the creep-recovery process. Specimens that had undergone higher creep strains had larger recovery strains, irrespective of the fiber presence.

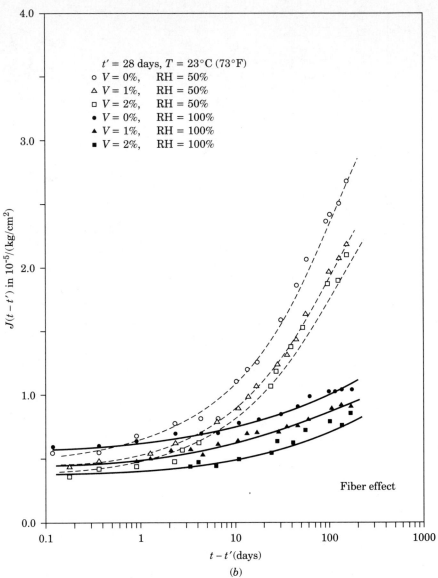

Figure 10-4b Effect of fiber volume fraction on creep strains: age at loading = 28 days.

The effect of fibers on concrete containing silica fume are reported in Reference [10.4]. The test results also include a mixture with no silica fume. The fibers used were steel fibers with hooked ends at a volume fraction of about 1%. The silica fume content was 0%, 5%, and 10% by weight of cement. Silica fume was considered as a replacement for cement. The compressive strengths averaged from 37 to 43.6 MPa (5000 to 6000 psi).

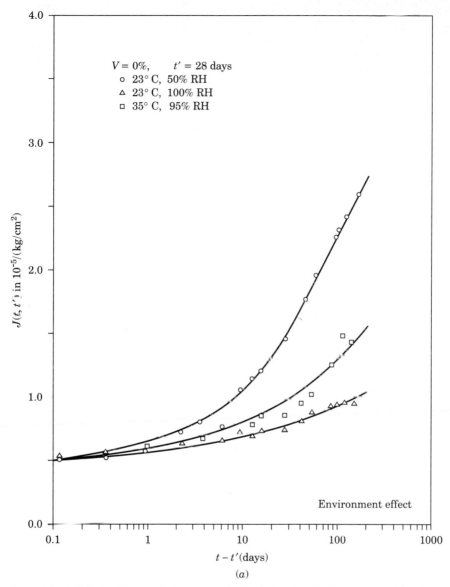

Figure 10-5a Effect of temperature exposure and relative humidity on creep strains: fiber volume fraction = 0%.

Creep tests were conducted using 150×300 mm (6×12 in.) cylinders loaded to a sustained stress of 0.4 times the compressive strength. The specimens were moisture-cured for 7 days. The creep test started after 28 days. The temperature and relative humidity in the test room were 21°C (70°F) and 50%.

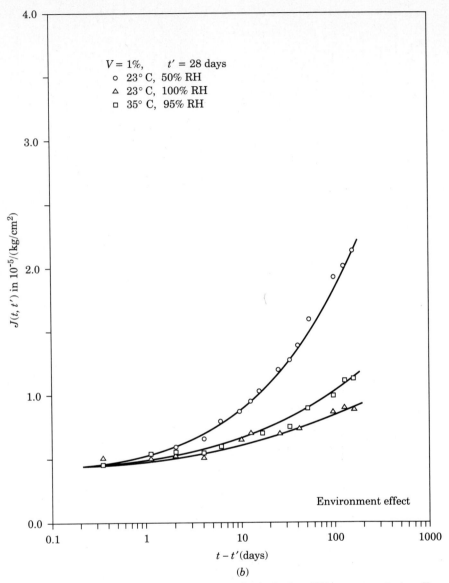

Figure 10-5b Effect of temperature exposure and relative humidity on creep strains: fiber volume fraction – 1%.

The test results show that fiber concrete had higher creep strains in all three cases, with 0%, 5%, and 10% silica fume. This tends to agree with the results reported in Reference [10.1] but contradicts the results of References [10.2] and [10.3]. It should be noted that the fiber volume used was relatively low for results reported in References [10.1] and [10.4]. They also contained much higher aggregate content

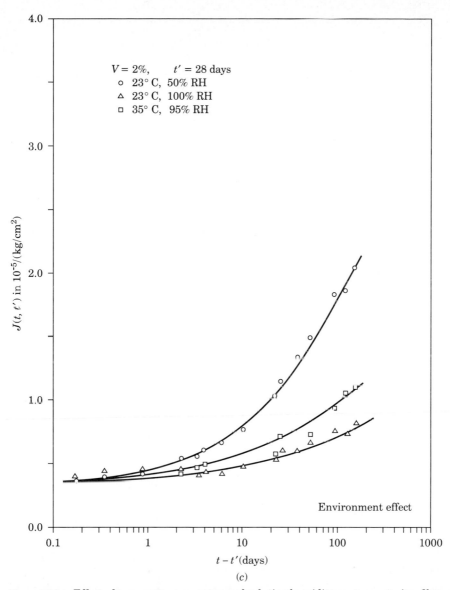

Figure 10-5c Effect of temperature exposure and relative humidity on creep strains: fiber volume fraction = 2%.

compared to mixes used in References [10.2] and [10.3]. The results available from creep studies can be summarized as follows.

Steel fibers can be expected to reduce creep strains at volume fractions higher than 1%. The fibers are also more effective in matrices that undergo larger creep strains. For concrete containing less than 1%

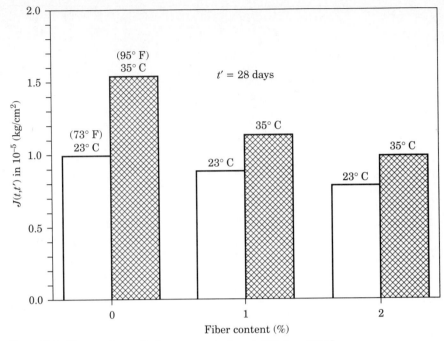

Figure 10-6 Temperature effect on creep strains at 170 days [10.2].

steel fibers, the creep strains are slightly higher. The other characteristics such as creep recovery can be assumed to be unaltered.

10.1.2 Shrinkage behavior of steel fiber–reinforced concrete

In general, the addition of fibers reduces the shrinkage strains. Typical parameters that affect shrinkage are mixture proportions; curing conditions, including the type and length of curing; age at which measurements are made; and the volume-to-surface ratio. For steel fiber-reinforced concrete, the additional primary variables are type, volume fraction, and aspect (length/diameter) ratio of the fibers. All these variables have been investigated to a limited extent. The following sections provide a summary of the results of the pertinent investigations.

Typically, cement paste has the highest shrinkage compared to mortar and concrete. Even though cement paste is rarely used by itself, investigations using it provide basic information. Limited test results on the behavior of fiber-reinforced paste indicate that the addition of fibers decreases the shrinkage strains. The tests were conducted using $100 \times 100 \times 500$ mm ($4 \times 4 \times 20$ in.) prisms placed at $20°C$ ($68°F$) and 55% relative humidity. The water-cement ratio was 0.35, and the cube strength was 59 MPa (8560 psi). The addition of 3% volume of melt-extract fibers that were 22.5 mm (0.9 in.) long and 0.4 mm (0.016 in.) in diameter had reduced the shrinkage strains by about 50% at 500 days.

Shrinkage strains obtained using similar test conditions for plain mortar matrix are only about 50% of the strains obtained for paste after a duration of about 500 days. The addition of fibers reduces the shrinkage strain by about 25% in the mortar mixture. The fiber volume fraction was 2% for both the mixtures containing melt-extract and hooked fibers. Hooked fibers seem to be a little more effective than melt-extract fibers, but the difference is not significant.

Fiber contribution to the shrinkage reduction of concrete is shown in Figures 10-7 and 10-8 [10.1]. The results were obtained using 75 × 75 × 292 mm (3 × 3 × 11.5 in.) prisms cured in water for 28 days at 23°C (73.4°F) temperature. During shrinkage strain measurements, the temperature and relative humidity were maintained at 23°C (73.4°F) and 50%. The mixture proportions for the two concretes are shown in Table 10.1. The shrinkage strains of the plain matrix in these cases are much smaller compared to results reported in Reference [10.5]. Consequently, the reduction provided by the fibers is also smaller than that reported in Reference [10.5].

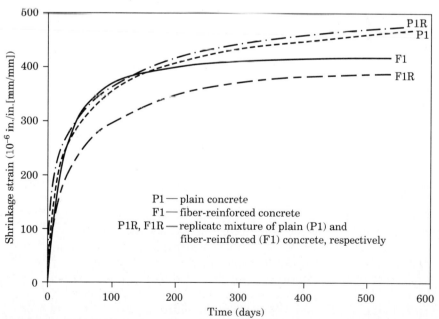

Figure 10-7 Shrinkage of plain and hooked-end steel fiber concrete: low cement content [10.1].

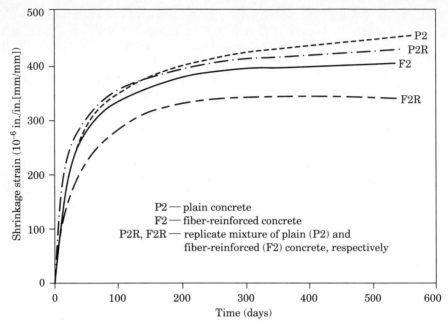

Figure 10-8 Shrinkage of plain and hooked-end steel fiber concrete: relatively high cement content [10.1].

The age and size effects are reported in References [10.2] and [10.6]. The age effect was studied by starting the shrinkage measurements at 4, 7, 28, and 90 days after casting. The size effect was studied by using two different size prisms and cylinders. The two prism sizes were $75 \times 75 \times 300$ mm ($3 \times 3 \times 12$ in.) and $50 \times 50 \times 300$ mm ($2 \times 2 \times 12$ in.), and the cylinder size was 150×300 mm (6×12 in.). The concrete consisted of 1 part cement, 2 parts fine aggregate, and 2 parts coarse aggregate. The 28 day compressive strength was about 40 MPa (5800 psi). The fibers used were straight carbon steel fibers with a shallow notch on one side. They were 19 mm (0.75 in.) long and had an equivalent diameter of 0.43 mm (0.017 in.). The fiber volume fractions were 1%, 2%, and 4%. The tests were conducted in a room maintained at 23 ± 0.5°C temperature and 50 ± 2 percent relative humidity.

The results show that shrinkage strains reduce as the age at initial measurement increases. Fiber incorporation reduces the shrinkage strains, but the magnitude is substantial only for higher fiber volume fractions. The improvement provided by the fiber decreases with increase in initial (measurement) age. The contribution of fibers are more evident after about 100 days. In addition, the shrinkage process stabilizes earlier for steel fiber–reinforced concrete compared to plain concrete.

Irrespective of the initial age, an increase in surface area for a given volume results in larger shrinkage strains in the initial stages. But the curves seem to converge, resulting in the same ultimate shrinkage strains. This observation is true for both plain and fiber-reinforced matrices. This is logical because stabilization and the movement of moisture take a longer time for specimens with smaller surface areas, and shrinkage therefore occurs at a slower rate.

The shrinkage behavior of fiber-reinforced lightweight concrete is shown in Figure 10-9 [10.9]. The concrete was made using 287, 123, 560, and 696 kg/m^3 (486, 208, 950, and 1180 lb/yd^3) of cement, fly ash, sand, and Pytag lightweight aggregate respectively. The concrete had a 28 day compressive strength of about 44 MPa (6400 psi). Fiber inclusion again reduces the shrinkage by about 20%. There is no significant difference between the fiber types: crimped, hooked, and paddle fibers.

In summary, the following observations can be made regarding the shrinkage behavior of steel fiber–reinforced composites.

- The fiber addition decreases the shrinkage strain. An increase in fiber content can be expected to produce a consistent decrease in shrinkage.

- The shrinkage reduction provided by fibers is the maximum for cement paste, followed by mortar, rich mixtures containing more cement, and lean mixtures containing less cement.

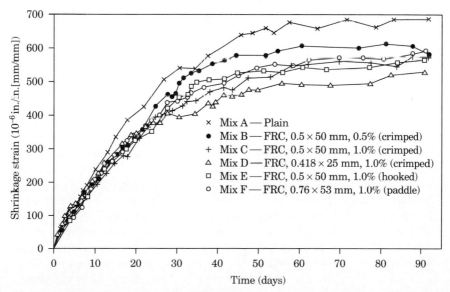

Figure 10-9 Shrinkage of lightweight FRC [10.9] (1 in. = 25.4 mm; 1 lb/yd^3 = 0.6 kg/m^3).

- The differences among fiber types tested are not significant. The types tested include crimped, surface-deformed, hooked-end, paddle, and melt-extract fibers.

- The contribution of fibers becomes more pronounced at later ages of drying shrinkage. The shrinkage process also stabilizes faster for fiber composites compared to plain matrices.

- Age (time of initial measurement), size (surface-volume ratio), and curing conditions have the same effect on both plain and fiber-reinforced composites.

10.2   Creep and Shrinkage of Polymeric Fiber–Reinforced Concrete

The results available for concrete containing polymeric fibers is even more limited than those for steel fiber concrete. The addition of 3 kg/m^3 (5 lb/yd^3) of polypropylene fibers to concrete resulted in an increase in creep strains [10.4]. The increase, however, was found to be less for polypropylene fiber concrete compared to steel fiber-reinforced concrete tested under similar conditions. The addition of silica fume was found to decrease the creep strains. However, for mixes containing 5% and 10% silica fume, plain concrete specimens always had lower creep strains than fiber-reinforced specimens.

The results reported in Reference [10.11] are in contradiction with the results of Reference [10.4]. Shrinkage tests conducted using 0.1% polypropylene fibers showed that fibers provide a reduction of 7% and 5% for concretes with compressive strengths of 3000 and 4000 psi (21 and 31 MPa). Tests conducted using accelerated drying conditions indicate that fibers provide a substantial reduction in drying shrinkage strains [10.12].

10.3   Long-Term Performance

The primary concerns associated with long-term performance are durability, including resistance to freezing and thawing cycles and frost; dimensional stability involving creep and shrinkage; and the capability of the composite to retain its original properties. The freeze thaw durability of fiber-reinforced concrete is presented in Chapter 8. Creep and shrinkage are discussed in the previous two sections. This section deals with the long-term durability and effectiveness of the fibers themselves. As mentioned earlier, corrosion is the primary concern for steel fibers, whereas the durability of the polymeric fibers in alkaline environment is the concern for polymeric fiber-reinforced concrete.

10.3.1 Corrosion of steel fibers

In crack-free concrete, the corrosion of fibers is limited to the surface skin of the concrete. In situations where carbonation occurs, the corrosion was found to occur up to the depth of carbonation [10.13]. The performance of uncracked FRC beams was studied by placing $100 \times 100 \times 2134$ mm ($4 \times 4 \times 84$ in.) beams in a tidal zone. The bottoms of the specimens were immersed in sea water, and the tops exposed to air at all times. The middle part was subjected to wetting and drying by tidal waves. The concrete had a cement content of 580 kg/m^3 (983 lb/yd^3) and a water-cement ratio of 0.51. The fiber content was 160 kg/m^3 (270 lb/yd^3). After the exposure period of 2.5, 5.0, and 10.0 years, the beams were cut into six sections and evaluated for carbonation, chloride penetration, fiber corrosion, and flexural strength variation.

The average flexural strengths were 11.4, 12.6, and 11.4 MPa (1700, 1800, and 1700 psi) after 2.5, 5.0, and 10.0 years of exposure, indicating no loss of flexural strength.

The carbonation depth was about 1 mm (0.04 in.) after 5 years of exposure. After 10 years of exposure, the carbonation depth was 1 mm (0.04 in.) at the immersed sections, 1.5 to 2.0 mm (0.06 to 0.08 in.) at the tidal zone sections, and up to 3 mm (0.12 in.) at the sections exposed to air. In all instances, the corrosion of fibers was limited to the carbonation depth.

Chloride ion penetration does not affect the fiber corrosion. Even in sections with high chloride ion concentration fibers were not corroded. This is explainable because the chloride ion concentration gradient, rather than the amount of ions, was found to cause corrosion in metals. Within the fiber length there may not be sufficient ion gradient to cause corrosion.

Surface staining was found to occur in the first month. Staining was found to be independent of the cover (provided by the concrete) or amount of fibers near the surface.

The fibers were found to be corrosion-resistant even in mixes with a relatively high water-cement ratio. Tests of FRC beams made using 428 kg/m^3 (725 lb/yd^3) cement and a water-cement ratio of 0.58 after 2000 cycles of marine water spray indicate that fibers are fully effective [10.14]. The load-deflection curves of flexural specimens indicate that both strength and ductility (toughness) characteristics are fully retained. The fibers used in this case were either stainless steel melt extract fibers or corrosion-resistant fibers at a volume fraction of 3% and 2.2% respectively.

The corrosion of fibers could be more intense if the concrete is cracked. Studies have been conducted by exposing cracked FRC beams to a chloride environment. An earlier study indicated that fibers bridging the crack do not corrode if crack widths are less than 0.1 mm (0.004 in.) [10.15]. A more recent investigation shows that the permissible crack widths could be as high as 0.15 mm and 0.2 mm (0.006 in. and 0.008 in.) for low carbon steel fibers and melt extract fibers respectively [10.16]. This conclusion was based on the strength and toughness test results of FRC beams exposed to wetting-and-drying marine cycles. The following are the pertinent details of the test program [10.16].

Two concrete mixtures, two fiber types, and two volume fractions were evaluated using $100 \times 100 \times 500$ mm ($4 \times 4 \times 20$ in.) prisms under flexural loading. The details of the test variables are shown in Table 10.3 [10.16]. The specimens were cast using a table vibrator and were left in their molds for 24 hours, followed by exposure to a laboratory environment for 14 days. At the end of this exposure period, cracks were induced by bending specimens under third-point loading. The crack widths ranged from 0.3 to 1.73 mm (.012–0.053 in.). One-third of the cracked specimens were sealed by applying 3 coats of silicone rubber over the cracks and a thick layer of bituminous paint. The sealed specimens and one-third of the cracked unsealed specimens were subjected to marine spray cycles. The prisms were tested under third-point loading after 650 and 1450 cycles of marine spray. The remaining one-third of the prisms were stored in the laboratory to compare their performance with the other test samples.

The influence of initial crack width on modulus of rupture after 1450 marine cycles is shown in Figure 10-10. The figure also shows the average flexural strengths of uncracked specimens at equivalent age. From Figure 10-10, it can be seen that the flexural strength of cracked and weathered specimens do not decrease if crack widths are less than 0.15 mm (0.006 in.). Unsealed specimens performed better than sealed specimens. The healing process that occurred in unsealed specimens might have contributed to their improved performance. Specimens containing fly ash performed better than the specimens made using portland cement alone (Figure 10-11). In this case, specimens with crack widths as high as 0.25 mm (0.01 in.) had flexural strengths comparable to uncracked prisms. These results are in agreement with two other investigations, reported in References [10.15] and [10.17].

Typical load deflection curves of uncracked and cracked specimens exposed to laboratory and marine cycle environments are shown in Figure 10-12. From this figure, it can be seen that the ductility is not affected by the marine cycle exposure. This indicates that both fiber integrity and its bonding to the matrix were not affected by marine exposure. The crack widths for these specimens ranged from 0.05 mm to 0.6 mm (0.002 to 0.024 in.). The area under the load-deflection curve decreases considerably when the crack widths are greater than 0.9 mm (0.036 in.).

TABLE 10.3 Details of Mixes, Fibers, and Curing Conditions: Durability Study under Marine Environment [10.16]

Mix	Mix proportions (by weight)					Cement content (kg/m^3)	Age at test		Fiber details				Fiber type	Curing conditions
	PFA	OPC	F agg.	C agg.	w/c		Days	Marine cycles	l (mm)	d (mm)	l/d	V_f (%)		
A_{ME}	0.00	1.00	1.50	0.86	0.40	590	900	1450	25.0	0.51	49	3.0	Melt extract (ME)	Marine spray (unsealed) Marine spray (unsealed)
B_{ME}	0.26	0.74	1.51	0.84	0.40*	435	450	650	26.5	0.44	60	1.8	Melt extract (ME)	Marine spray (unsealed) Marine spray (sealed) Lab air-cured (unsealed)
B_{MS}	0.26	0.74	1.51	0.84	0.40*	435	450	650	28.2	0.48	59	1.7	Low carbon steel hooked (MS)	Marine spray (unsealed) Marine spray (sealed) Lab air-cured (unsealed)

*Water/(OPC + PFA) = 0.40   OPC—ordinary portland cement   PFA—pulverized fly ash   F agg.—fine aggregate
C agg.—coarse aggregate,   1 in. = 25.4 mm, $1 lb/yd^3 = 0.6 kg/m^3$

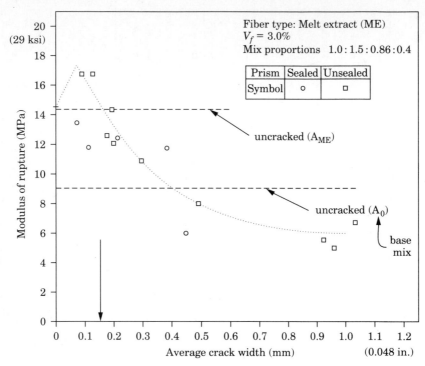

Figure 10-10 Influence of initial crack width and marine spray on modulus of rupture [10.16].

Corrosion characteristics of fibers that are exposed to less-aggressive environments such as pavements and industrial floors have not been systematically investigated. The general inspection of existing installations of FRC indicate that the corrosion of fibers is not a problem.

10.3.2   Durability of polymeric fibers

As mentioned earlier, the primary concern with polymeric fibers is their durability in the alkaline environment present in portland cement–based matrices. Polypropylene fiber is the only polymeric fiber tested for a field exposure duration of 10 years [10.18]. Accelerated aging tests have been used to evaluate other fiber types.

The investigation on polypropylene fibers was conducted using flexural test specimens reinforced with 54 layers of continuous networks of fibrillated polypropylene film [10.18]. The primary independent variables were exposure conditions, the condition of the composite, and the degree of ultraviolet stabilization. The exposure conditions consisted of natural outdoor weathering and indoor storage. Cracked and

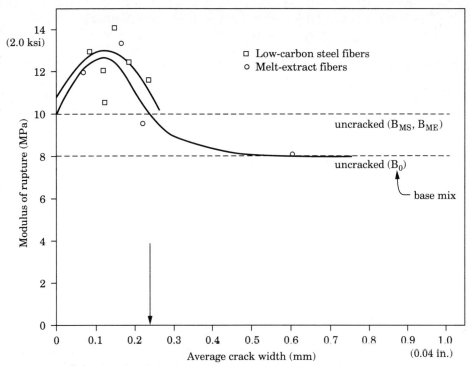

Figure 10-11 Behavior of sealed prism specimens after exposure to 650 marine cycles [10.16].

uncracked samples constituted two specimen conditions. Minute cracks were induced by subjecting the specimens to a tensile strain of 0.003 mm/mm (in./in.) by bending the prisms at one-third points. In the area of ultraviolet stabilization, natural and high-UV stabilization conditions were evaluated.

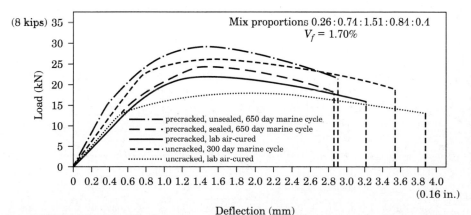

Figure 10-12 Comparison of flexural behavior of uncracked and precracked beams after exposure to marine cycles [10.17].

The durability and effectiveness of the fibers were evaluated by testing the exposed samples in third-point bending over a simple span of 135 mm (5.4 in.). Information on both strength and ductility was obtained by recording the load deflection curves. Test ages were 1, 6, and 12 months and 2, 3, 5, and 10 years.

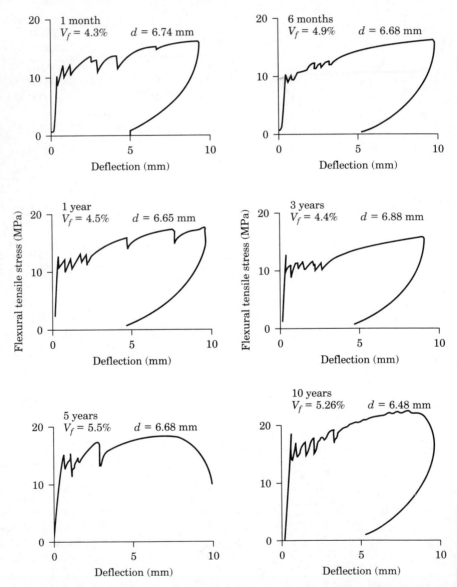

Figure 10-13 Load (tensile stress) deflection behavior of polypropylene fiber-reinforced beams exposed to natural weathering conditions, $V_f = 4$ to 5% [10.18] (1 MPa = 145 psi).

The load deflection responses of uncracked and precracked samples are shown in Figures 10-13 and 10-14. The figures show that for both uncracked and precracked conditions, weathering has little effect on both initial bend-over point and ultimate flexural strength.

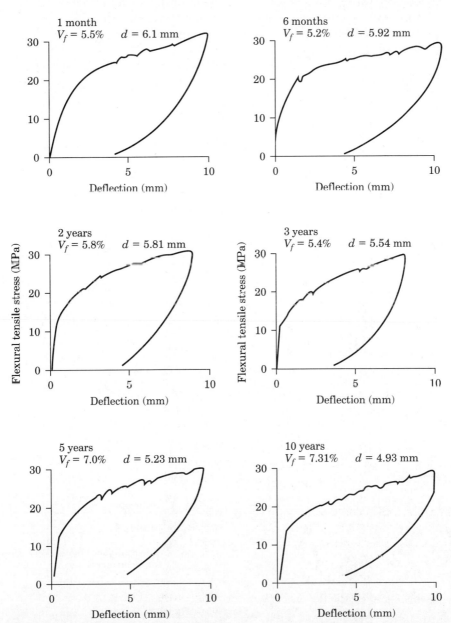

Figure 10-14 Load-(tensile-stress) deflection behavior of polypropylene fiber reinforced beams exposed to natural weathering conditions, $V_f = 5\text{--}7\%$ [10.18].

The samples stored inside the laboratory had lower BOP strength compared to specimens stored outside (Figure 10-15). The ultimate flexural strength was similar for both exposure conditions. The change in load capacity around working load levels was also negligible, as shown in Table 10.4, which shows the bending stress at a deflection of 2 mm (0.08 in.) at various exposure times. Overall, fibers were durable and they provided effective reinforcement, indicating a good bond between matrix and fibers.

Results available on the durability of Nylon 6, polyester, and polypropylene fibers are based on an accelerated aging test [10.19]. Since polypropylene fiber durability is established using long-term data, its durability under accelerated aging could be compared with the durability of other fibers. Results reported in Reference [10.19] indicate that Nylon 6 fibers are durable in an alkaline environment. The accelerated aging was done by immersing the test specimens in lime-saturated water maintained at 50°C (112°F). This procedure is similar to the one used for glass fibers. More details about the development and validity of the test are presented in Chapter 13.

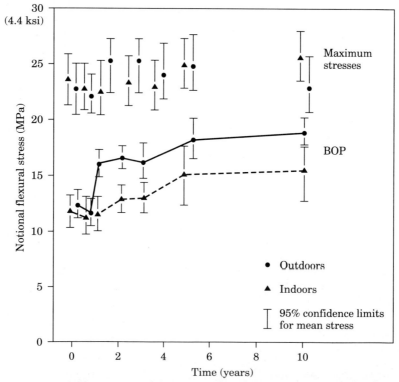

Figure 10-15 Effect of exposure time versus flexural properties of polypropylene fiber-reinforced beams [10.18].

TABLE 10.4   Average Stress at 2 mm Deflection at the Different Ages: Durability Study for Polypropylene Fibers [10.18]

Age	Exposure	Mean sample thickness (mm)	Mean fibre vol (%)	Notional stress (MPa)	Standard deviation (MPa)
28 days	Curing room	6.51	5.17	15.3	2.2
6 mos	Inside	6.69	5.12	14.3	1.88
1 year	″	6.73	5.20	13.8	1.79
2 years	″	6.65	5.32	15.5	2.04
3 years	″	6.63	5.19	15.2	1.65
5 years	″	6.53	5.85	16.9	2.66
10 years	″	6.76	5.50	18.0	2.69
28 days	Curing room	6.55	5.39	14.8	2.05
6 mos	Outside	6.56	5.25	14.7	2.06
1 year	″	6.65	5.44	17.5	1.74
2 years	″	6.6	5.37	17.8	1.97
3 years	″	6.53	5.21	17.0	2.38
5 years	″	6.65	5.90	18.8	2.25
10 years	″	6.88	5.42	17.8	1.95

1 in. = 25.4 mm   1 ksi = 6.895 MPa

The test specimens consisted of $4 \times 4 \times 14$ in. ($100 \times 100 \times 350$ mm) prisms reinforced with 0.5% volume of fibers. All three types of fibers (Nylon 6, polypropylene, and polyester) were 0.75 in. (19 mm) long. Polypropylene fibers were fibrillated, whereas the other two fibers were made of single filaments. The specimens were cast using a Plexiglas mold and were moisture-cured for 28 days before subjecting them to accelerated aging. The matrix (concrete) had a cement content of 517 lb/yd^3 (307 kg/m^3) and a water-cement ratio of 0.57. The average 28 day compressive strength measured using 6×12 in. (150×300 mm) cylinders was 3000 psi (20 MPa).

The durability and effectiveness of the fibers were measured using flexural strength and flexural toughness. The ASTM test method C1018 was used for flexural testing. The beams were tested under third-point loads using a simply supported span of 12 in. (300 mm). Tests were conducted after 0, 4, 8, 16, 32, and 52 weeks of accelerated aging.

The flexural strengths obtained after various aging periods are shown in Table 10.5. The toughness index values are shown in Table 10.6. The load deflection responses after 0, 16, and 32 weeks are shown in Figures 10-16, 10-17, and 10-18 for Nylon 6, polypropylene and polyester fibers respectively. The variations of toughness indices and the ratios of toughness indices I_{30}/I_{10} with respect to aging are shown in Figures 10-19 and 10-20 respectively. The following observations can be made using the results shown in the tables and figures.

TABLE 10.5   Modulus of Rupture of Aged Specimens: Accelerated Aging Study [10.19]

Acc. Aging (weeks)	Modulus of rupture (psi)		
	Nylon 6	Polypropylene	Polyester
0	313	262	450
4	309	274	282
8	438	381	356
16	350	431	378
32	406	400	456
52	543	444	481

1 psi = 6.895 Pa

Polyester fibers provide a higher modulus of rupture at the beginning of aging (Table 10.5), but the values slightly decrease or remain about the same with accelerated aging. Specimens with Nylon 6 and polypropylene fibers showed an increase in strength. The magnitude of increase is not significant, but the fact that postcrack strengths were higher than precrack strengths is significant. This behavior indicates

TABLE 10.6   Toughness Indices at Various Stages of Accelerated Aging [10.19]

Toughness index		Age (weeks)					
		0	4	8	16	32	52
I_5	N6*	3.9	3.6	5.1	5.0	4.4	4.1
	PP†	4.2	4.6	4.8	4.8	4.6	4.9
	PY‡	3.9	4.3	3.7	4.0	4.3	4.0
I_{10}	N6	6.0	5.8	11.3	9.6	9.0	7.5
	PP	7.3	7.9	10.5	10.2	9.3	10.2
	PY	5.4	6.8	5.1	5.5	6.7	5.7
I_{30}	N6	16.4	15.8	38.9	27.0	27.9	20.8
	PP	20.7	24.6	38.2	36.1	32.6	31.1
	PY	10.1	10.9	7.5	7.8	9.7	7.4
I_{10}/I_5	N6	1.54	1.61	2.22	1.92	2.05	1.85
	PP	1.74	1.72	2.19	2.12	2.02	2.10
	PY	1.38	1.58	1.38	1.38	1.56	1.44
I_{30}/I_{10}	N6	2.73	2.72	3.44	2.81	3.10	2.77
	PP	2.84	3.11	3.64	3.54	3.51	3.04
	PY	1.87	1.60	1.47	1.42	1.45	1.29

*N6—Nylon 6
†PP—polypropylene
‡PY—polyester

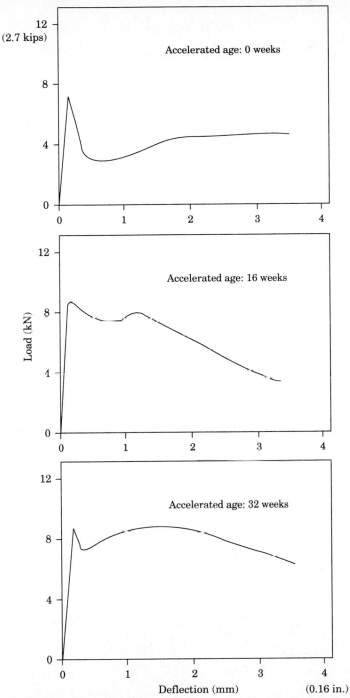

Figure 10-16 Load deflection behavior of beams subjected to accelerated aging: Nylon 6 fibers [10.19].

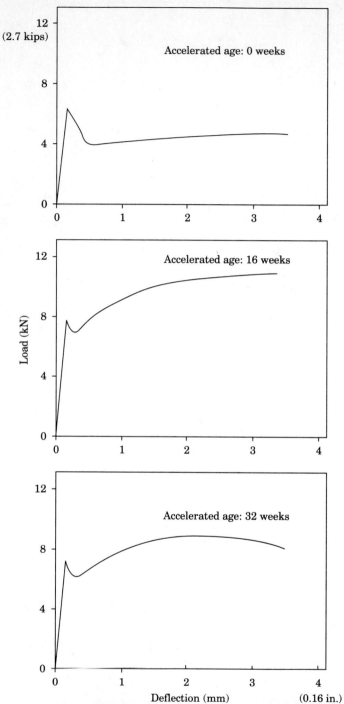

Figure 10-17 Load deflection behavior of beams subjected to accelerated aging: polypropylene fibers [10.19].

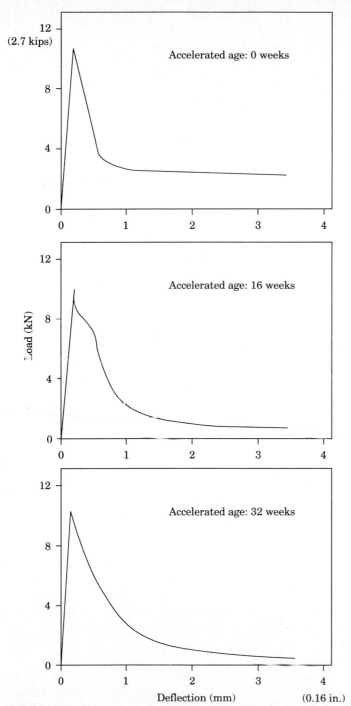

Figure 10-18 Load deflection behavior of beams subjected to accelerated aging: polyester fibers [10.19].

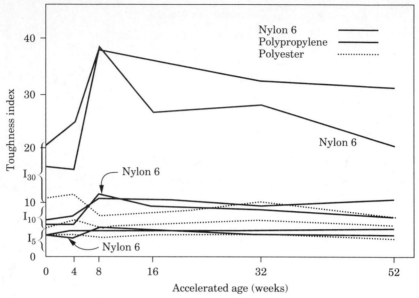

Figure 10-19 Flexural toughness versus accelerated aging [10.19].

that the fibers not only did not deteriorate but also bonded well within the matrix to provide the postcrack resistance. Since the fibers do not gain strength with age, the increase can be assumed to be the contribution of the better bonding that is provided by the more complete hydration of cement, compared to hydration at 28 days.

The toughness index I_5 values are about the same for all three types of fibers at the beginning of aging. The differences among fiber types are not significant up to 52 weeks of accelerated aging (Table 10.6 and Figure 10-19). The I_{10} values are higher for Nylon 6 and polypropylene fibers compared to polyester fibers. Accelerated aging resulted in an increase of I_{10} for Nylon 6 and polypropylene fibers, whereas the numbers remained about the same for polyester fibers (Table 10.6 and Figure 10-19). The I_{30} values were considerably higher for Nylon 6 and polypropylene fibers. The values also increased from about 16 to 38 for Nylon 6 and from 20 to 38 for polypropylene fibers. As mentioned earlier, postcrack strength was higher than precrack strength, resulting in an I_{30} value higher than 30. Note that, for ideal elastoplastic behavior, the I_{30} value should be 30 (Chapter 4).

The ratios I_{10}/I_5 and I_{30}/I_{10} provide an indication of ductility at higher deformations. The ratio I_{30}/I_{10} is significant because deflection up to 15.5 δ (δ is the first-crack deflection) is considered for I_{30}. Variation of this index with respect to accelerated aging, shown in Figure 10-20, indicates that the performance improves for Nylon 6

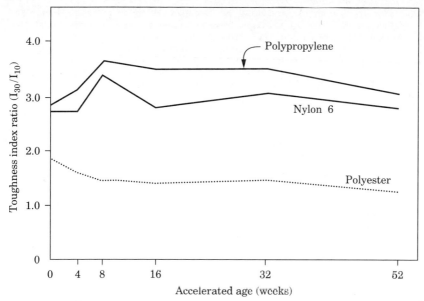

Figure 10-20 Flexural toughness ratios versus accelerated aging [10.19].

and polypropylene fibers. Polyester fiber had a lower ratio at the big-inning of aging. This value decreased within the first eight weeks of aging and continued to be the same afterward.

Overall, it can be said that Nylon 6 and polypropylene fibers provide the same and sometimes even more effective reinforcement with accelerated aging. Both these fiber types could be assumed to have long-term durability.

Accelerated aging and normal exposure tests have shown that polyvinylalcohol (PVA) fibers are durable in an alkaline environment. The durability tests were conducted using thin (6 mm thick) specimens reinforced with a 3% volume of fibers. The maximum duration of exposure was seven years. The fibers were found to retain both strength and deformation characteristics.

10.4   References

10.1 Balaguru, P.; and Ramakrishnam, V. "Properties of Fiber Reinforced Concrete: Workability, Behavior under Long-Term Loading, and Air-Void Characteristics," *ACI Materials Journal*, Vol. 85, No. 3, 1988, pp. 189–196.

10.2 Chern, J. C.; and Young, C. H. "Compressive Creep and Shrinkage of Steel Fiber Reinforced Concrete," *The International Journal of Cement Composites and Lightweight Concrete*, Vol. 11, No. 4, 1989, pp. 205–214.

10.3 Mangat, P. S; and Azari, M. M. "Compression Creep Behavior of Steel Fiber Reinforced Cement Composites," *Materials and Structures*, RILEM, Vol. 19, 1986, pp. 361–370.

10.4 Houde, J.; Prezeau, A.; and Roux, R. "Creep of Concrete Containing Fibers and Silica Fume," *Fiber Reinforced Concrete Properties and Applications,* SP-105, American Concrete Institute, Detroit, Michigan, 1987, pp. 101–118.

10.5 Mangat, P. S.; and Azari, M. "Shrinkage of Steel Fiber Reinforced Cement Composites," *Materials and Structures,* RILEM, Vol. 21, 1988, pp. 163–171.

10.6 Chern, J. C.; and Young, C. H. "Study of Factors Influencing Drying Shrinkage of Steel Fiber Reinforced Concrete," *ACI Materials Journal,* Vol. 87, 1990, pp. 123–129.

10.7 Grzybowski, M.; and Shah, S. P. "Shrinkage Cracking of Fiber Reinforced Concrete," *ACI Materials Journal,* Vol. 87, 1990, pp. 138–148.

10.8 Grzybowski, M.; and Shah, S. P. "A Model to Predict Cracking in Fiber Reinforced Concrete Due to Restrained Shrinkage," *Magazine of Concrete Research,* Vol. 41, No. 148, 1989, pp. 125–135.

10.9 Swamy, R. N.; "Steel Fiber Concrete for Bridge Deck and Building Floor Applications,"*Steel Fiber Concrete,* Elsevier, New York, 1986, pp. 443–478.

10.10 Swamy, R. N.; and Theodorakopoulos, D. D.; "Flexural Creep Behavior of Fiber Reinforced Cement Composites," *The International Journal of Cement Composites,* Vol. 1, 1979, No. 1, pp. 37–48.

10.11 Litvin, A. "Report to Wire Reinforcement Institute on Properties of Concrete Containing Polypropylene Fibers," CTL, Portland Cement Association, Skokie, Illinois, 1985.

10.12 Zollo, R. F.; Ilter, J. A.; and Bouchacourt, G. B. "Plastic and Drying Shrinkage in Concrete Containing Collated Fibrillated Polypropylene Fiber," Third International Symposium on Developments in Fiber Reinforced Cement and Concrete, RILEM, 1986.

10.13 Schupack, M. "Durability of SFRC Exposed to Severe Environments," *Steel Fiber Concrete,* Elsevier, 1986, pp. 479–496.

10.14 Mangat, P. S.; Molloy, B. T.; Gurusamy, K. "Marine Durability of Steel Fiber Reinforced Concrete of High Water/Cement Ratio," *Fiber Reinforced Cements and Concretes–Recent Developments,* Elsevier, 1989, pp. 553–562.

10.15 Morse, D. C.; and Williamson, G. R. "Corrosion Behavior of Steel Fibrous Concrete," NTIS, Springfield, Virginia, May 1977, 36 pp.

10.16 Mangat, P. S.; and Grusamy, K. "Permissible Crack Widths in Steel Fiber Reinforced Marine Concrete," *Materials and Structures,* RILEM, Vol. 20, 1987, pp. 338–347.

10.17 Hannant, D. J. "Additional Data on Fiber Corrosion in Cracked Beams and Theoretical Treatment of the Effect of Fiber Corrosion on Beam Load Capacity," RILEM Symposium on Fiber Reinforced Cement and Concrete, 1975, Vol. II, pp. 533–538.

10.18 Hannant, D. J. "Ten Year Flexural Durability Tests on Cement Sheets Reinforced with Fibrillated Polypropylene Networks," *Fiber Reinforced Cements and Concretes–Recent Developments,* Elsevier, 1989, pp. 572–563.

10.19 Khajuria, A.; Bohra, K.; and Balaguru, P. "Long-Term Durability of Synthetic Fibers in Concrete," *Durability of Concrete,* SP-126, American Concrete Institute, Detroit, Michigan, 1991, pp. 851–868.

Plastic and Early Drying Shrinkage

One of the primary contributions of fibers is their ability to improve the strain capacity of the composite. The improved strain capacity in tension helps reduce cracking under restrained conditions. For example, a large plain-concrete slab that is cast on a dry, windy day could develop extensive cracking. These cracks, called plastic shrinkage cracks, occur because the top of the slab shrinks faster than the bottom. Since the bottom of the slab is restrained, tension is created at the top of the slab, resulting in cracks. The addition of a small quantity of polymeric fibers, as small as 1 lb/yd^3 (0.6 kg/m^3), was found to reduce this plastic cracking.

The fibers' contribution was also found to be critical in the first few days after casting if the composite is subjected to tension. In most instances the tensile stress develops because of restraints provided either at the edges or at the bottom of the slab. The mechanism that causes the stress is primarily the change in length caused by drying shrinkage or temperature variation.

This chapter deals with the effect of fibers on plastic and early drying shrinkage. Almost all the plastic shrinkage occurs in the first four to six hours after casting. The effect of drying shrinkage is considered for about 40 days. The long-term shrinkage is discussed in Chapter 10. In addition, the focus of this chapter is crack reduction provided by the fibers under restrained conditions. Test methods and the results available on various types of fibers are discussed. Emphasis is placed on concrete containing low volumes of fibers (0.1 to 2%) cast in relatively thick sections, as opposed to thinsheet products.

11.1 Plastic Shrinkage

As mentioned earlier, freshly cast concrete shrinks primarily because of the evaporation of water. The effect of this shrinkage is widely evident in slabs that have large exposed areas, such as pavement slabs. The differential shrinkage between the top and bottom of the slab and the restraint provided at the bottom lead to cracks on the surface. Some of these cracks eventually extend to full depth. The development of these cracks leads to the fast deterioration of the slabs when they are exposed to dryness and moisture or freezing and thawing conditions. The addition of small quantities of polymeric fibers was found to decrease cracking considerably. This reduction in plastic shrinkage cracking is the primary motive for using polymeric fibers in concrete.

The testing procedure to determine the effectiveness of the fibers is still under development. There are two primary types of setups, one involves the use of plate elements subjected to rapid drying on the top [11.1], and the other procedure involves the use of ring-type samples, again subjected to rapid drying on the top surface [11.2]. In both cases, restraints are provided at the outer edges. Details of these test procedures are presented in the following sections.

11.1.1 Test procedures to evaluate the plastic shrinkage-reduction potential of fibers

Plate test. In this test rectangular slabs that are 2×3 ft (600×900 mm) and 0.75 in. (19 mm) or 2 in. (50 mm) thick are cast using either wooden or Plexiglas molds [11.1]. Normally the base is made of 0.5 in. (12 mm) thick plywood. A tile board could be glued to the top of the plywood to obtain a smooth surface. The sides could also be made using wood or Plexiglas. In most cases the specimens are 0.75 in. (19 mm) thick, but in some instances 2 in. (50 mm) thick concrete specimens containing coarse aggregate have also been evaluated [11.3]. End restraint is provided by a strip of 0.5×1 in. (12×25 mm) hardware cloth placed along the perimeter (Figure 11-1). The hardware cloth is held in position by nails. Other types of restraints such as projecting nails have also been used. The mold is usually lined with a polyethylene sheet that is placed on the top of the tile board to provide for free movement of the slab.

The test slabs cast using these molds are subjected to rapid evaporation using either fans or wind tunnels. In some cases the temperature of the surrounding environment is raised to hasten the evaporation process. A reduction of relative humidity also increases the evaporation rate. For 0.75 in (19 mm) thick slabs, the mix usually consists of cement mortar. The mortar should be relatively rich, having fewer than two parts sand for each part of cement in order to develop plastic

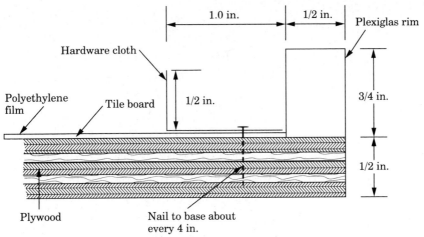

Figure 11-1   Cross-section of mold [11.1, 11.3] (1 in. = 25.4mm).

shrinkage cracks. The cracks start developing about three hours after the casting. The cracking process is normally complete in about eight hours.

The effectiveness of the fibers is measured using the crack area of fiber-reinforced and plain mortar specimens. The crack area in a plain mortar matrix is taken as 100 percent, and the crack area of the fiber-reinforced specimens is expressed as a percentage of the plain matrix crack area. The amount of crack area reduction is taken as an indicator of the effectiveness of fiber.

Ring Tests.   Ring specimens have been used to a limited extent for evaluating the plastic shrinkage–reduction potential of fibers (Figure 11-2) [11.2]. A hollow cylindrical ring of matrix is cast using two annular rings. The inner and outer rings have a diameter of 280 mm (11 in.) and 580 mm (23 in.) respectively, resulting in a concrete ring that is 150 mm (6 in.) wide. The height of the specimen is typically 80 mm (3.1 in.). Twelve steel ribs welded to the outer ring provide the restraint (hold the matrix to the outer ring). The specimens are dried on the top using an air stream with a speed of 4 m/s (13.1 ft/s). The temperature and relative humidity have to be controlled. The crack areas of plain and reinforced matrices are compared to evaluate the effectiveness of the fiber, as in the case of plate test.

11.1.2   Contribution of polymeric fibers to plastic shrinkage crack reduction

Plate specimens have been used to evaluate the effectiveness of polymeric fibers in crack reduction by a few investigators [11.1, 11.3–11.6].

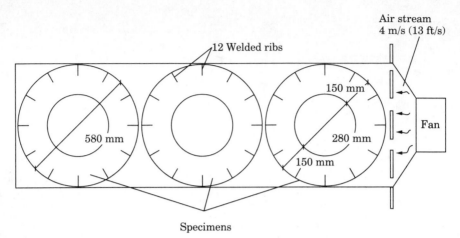

Specimens

Figure 11-2 A ring-type experimental setup for plastic shrinkage cracking (1 in. = 25.4 mm) [11.2].

The variables investigated include fiber type, fiber length, fiber volume fraction, and matrix composition. The types of fibers tested include Nylon 6, polyester, polyethylene, and polypropylene. The fiber lengths varied from a fraction of a millimeter (pulp) to 1.5 in. (38 mm). Both single filament and fibrillated fibers have been tested. Fiber volume fractions were varied from 0.025% to 0.5%. In most cases the matrix consisted of cement mortar with cement-to-sand ratios varying from 1:1 to 1:3. A few tests have also been conducted using concrete containing coarse aggregates.

The panel sizes were 2×3 ft and 0.75 in. thick ($600 \times 900 \times 19$ mm) in most cases. Some slabs measuring 2×3 ft and 2 in. thick ($600 \times 900 \times 50$ mm) and 3×3 ft and 0.75 in. thick ($900 \times 900 \times 19$ mm) were also used for testing.

Figures 11-3, 11-4, and 11-5 show some typical results obtained for Nylon 6, polypropylene, and polyester fibers. The fibers were all 0.75 in. (19 mm) long, except in one case, in which three fiber lengths, 0.75, 1.0, and 1.5 in. (19, 25, 37 mm), were mixed together. The polypropylene fibers were fibrillated, whereas the other two fiber types were made of single filaments. The matrix consisted of cement and sand in the ratio of 1:1.5, and the water-cement ratio was 0.5.

Figure 11-3 shows the variation of crack area with respect to fiber content. It can be seen that an increase in fiber content leads to a consistent reduction in crack area. This trend was found to continue up to a volume of 0.5% fiber [11.6]. In some cases the visible cracking was completely eliminated.

Figures 11-4 and 11-5 compare the effectiveness of fiber types at 1.0 and 1.5 lb/yd³ (0.6 and 0.9 kg/m³) fiber loadings. For a given fiber

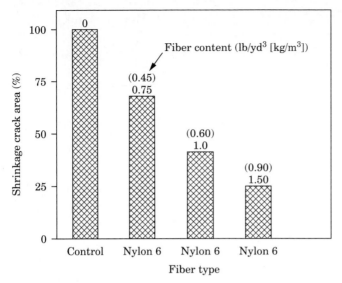

Figure 11-3 Comparison of shrinkage crack–reduction potential. Nylon 6 fibers [11.6].

volume, finer fibers and those with low modulus were found to provide better results. The presence of more fibers could provide more restraint leading to less cracking. The better results that are provided by lower-modulus fibers are not readily explainable; however, the results are not extensive, and solid conclusions could not therefore be drawn.

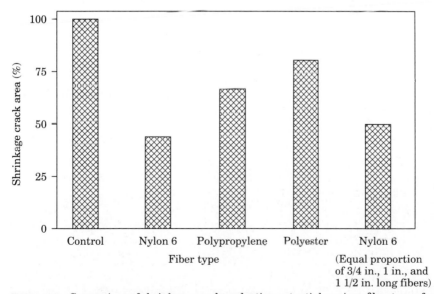

Figure 11-4 Comparison of shrinkage crack–reduction potential: various fiber types for a fiber content of 1.0 lb/yd³ (0.6 kg/m³) [11.6].

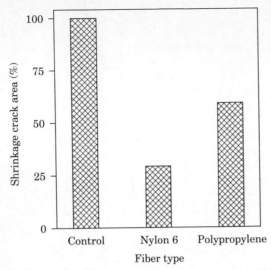

Figure 11-5 Comparison of shrinkage crack–reduction potential: various fiber types for a fiber content of 1.5 lb/yd^3 (0.9 kg/m^3) [11.6].

Figure 11-6 shows the results obtained using very fine polyethylene fibers and rich matrix. Again an increase in fiber content leads to a decrease in crack area. These fibers were found to be more effective for richer mixes than for leaner mixes, as shown in Figure 11-7. Fiber count and stiffness were found to have the same influence for shorter and longer fibers. Coarser and stiffer fibers had smaller crack reductions.

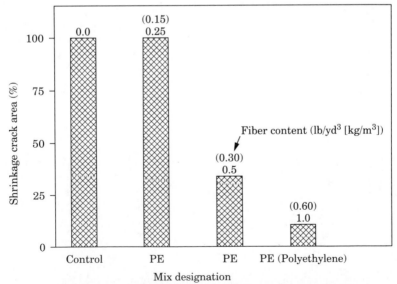

Figure 11-6 Fiber content versus crack reduction: polyethylene fiber [11.3].

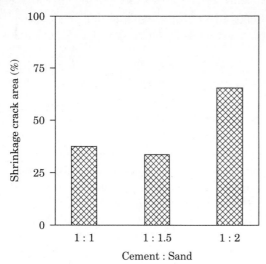

Figure 11-7 Influence of sand content on crack reduction: polyethylene fiber [11.3].

Longer fibers were found to be more effective in leaner mixes. Limited tests on concrete indicate that longer fibers (lengths of about 10 mm) are more effective than shorter (lengths of about 1 mm) fibers. This was found to be true even though shorter fibers had much higher length-diameter ratios. It should be noted that these observations are based on very limited tests and hence should be considered as a trend rather than as conclusive results.

Tests conducted using ring specimens also show that plastic shrinkage cracking could be reduced by using polyproplyene fibers. The field tests done using 8 × 30 ft. and 8 in. thick (2438 × 9144 × 200 mm) concrete slabs also show that the fibers are effective in reducing plastic shrinkage cracking. The test that was conducted in hot weather at low humidity was again limited in scope.

11.1.3   Contribution of steel fibers to plastic shrinkage crack reduction

Typically, steel fibers are used to improve the ductility of the hardened matrix and other mechanical properties such as fracture toughness. The reduction in plastic shrinkage is an added advantage. Limited tests conducted using steel fibers with hooked ends show that cracking could be considerably reduced by using 100 lb/yd^3 (60 kg/m^3) of fibers or more. Tests were conducted using 30 mm (0.5 mm diameter), 50 mm (0.5 mm diameter), and 60 mm (0.8 mm diameter) long fibers. The 50 mm fibers were found to be more effective in reducing the crack area.

Thinner and shorter steel fibers are being developed for specific use in crack reduction. The results on the effectiveness of these fibers are not yet available.

Steel fibers were found to be more effective for restrained drying shrinkage crack reduction as discussed in the next section.

11.1.4 Theoretical models for the prediction of plastic shrinkage crack widths

At present, a theoretical model is not available for the prediction of crack widths and fiber contribution to a reduction in plastic shrinkage cracking. The fibers could possibly aid the release of surface tensions and provide enhanced tensile strength during the early part of hydration. The basic phenomena have not been studied in enough depth to propose even an empirical approach for the prediction of crack reduction provided by the fibers.

11.2 Drying Shrinkage

The drying shrinkage of concrete has been studied extensively. It is an established fact that concrete shrinks when subjected to a dry environment. The various factors that affect the extent of drying shrinkage (such as the cement and the temperature and humidity of the environment) have been investigated. The discussion presented here is focused on cracking caused by restrained shrinkage and the use of fibers for minimizing crack widths.

Normally, buildings and other concrete structures (such as bridges and tunnel linings) are built in stages. Concrete cast at different time periods may shrink at different rates. If provisions are not made, high tensile stresses can be induced because of differential movement. This can result in the cracking of concrete. Similarly, concrete walls cast directly on rock could crack at the bottom (Figure 11-8), because free movement (owing to shrinkage) of the wall is restricted at the bottom. Cracking can also occur because of differential shrinkage occurring at repaired surfaces. If the repaired patch shrinks and the movement is restricted, then the patch can crack at the interface between the old and new material. If the interface strength is high, then the patch itself will crack at the top.

One way to reduce the shrinkage cracking is to provide reinforcement for resisting tensile forces. In recent years, researchers have investigated the use of discrete fibers for reducing restrained shrinkage cracks [11.7–11.11]. The fibers could also be used in combination with regular continuous reinforcement.

As in the case of plastic shrinkage, test methods for evaluating the effectiveness of fibers are still in the developmental stage. However,

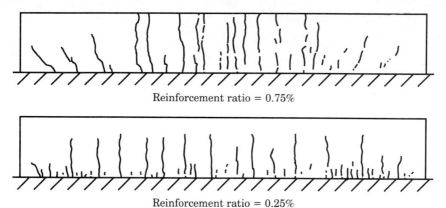

Reinforcement ratio = 0.75%

Reinforcement ratio = 0.25%

Figure 11-8   Observed crack pattern in a wall subjected to restrained shrinkage [11.15].

the methods that researchers have developed thus far (presented in the following section) could provide good quantitative measurements on the effectiveness of various types and volume fractions of fibers.

11.2.1   Test procedures to evaluate the contribution of fibers to drying shrinkage crack reduction

Three different shapes (linear, plate, and ring) have been tried for measuring the contribution of fibers to shrinkage crack reduction. Ring specimens seem to have better potential because they can be provided with good restraint. The arrangement is also conducive to the development of mathematical models, as explained in the later sections.

Linear Specimens.   The test specimens consist of long prisms with flared ends [11.10]. The lateral dimensions 0.2 × 4.8 in. (5 × 120 mm) are small compared to the longitudinal dimension 60 in. (1500 mm) (Figure 11-9). Restraints are applied at the flared ends, inducing cracks in the middle (uniform) section. The method was successfully used for studying the contribution of steel fibers.

Another way to provide restraint is to use reinforcing bars in the center of the matrix (Figure 11-10) [11.12]. The center part of the bar was debonded using a rubber tube. Hence the midsection of the specimen is subjected to tension created by the shrinkage. The specimen size was 2 × 2 × 12 in. (50 × 50 × 300 mm).

In another study, prismatic concrete specimens were glued to a stiff steel frame (Figure 11-11) [11.13]. The specimen dimensions were 500 × 80 × 20 mm (20.0 × 3.2 × 0.8 in.). The ends were held in position by the rigid frame. When the composite started to shrink, the reduction in length created the tensile stresses.

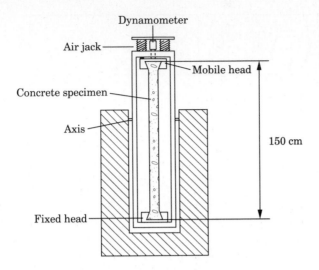

Cross-section 0.5 cm × 12 cm

Figure 11-9 Restrained shrinkage test: Restraints applied using external grips [11.10] (1 in. = 2.54 cm).

In these three types of specimens, stress distribution at the restrained ends is rather complex. Therefore, it is difficult to formulate an analytical model for the prediction of tensile stresses and crack widths that develop because of shrinkage strains.

Plate Specimens.   Plate specimens, similar to the ones used for plastic shrinkage, have also been tried for measuring cracks caused by drying shrinkage [11.14]. The restraints were provided by means of stirrups attached to rigid steel frames. The primary difficulty with this kind of setup is to estimate the actual extent of restraint provided by the stirrups. Hence, the tests could be used only to make qualitative judgments among the various types of fibers.

Ring specimens.   Ring specimens were used by a number of investigators for evaluating fiber-reinforced cement composites under restrained drying shrinkage [11.7–11.12, 11.15, 11.16]. Essentially, a ring of concrete is cast around a stiff steel ring (Figure 11-12). As the composite shrinks, it induces stresses on the steel ring. Since the steel ring is stiff and undergoes very little deformation, the outer cement composite ring is subjected to tension. If the concrete ring is thin in relation to the internal diameter, then the stresses across the thickness can be considered uniform. The compressive stress developed at the interface between the steel ring and the concrete ring is also negligible. The researchers used various external diameters for steel rings.

1	Center reinforcing bar, 3/8"ϕ or 1/4"ϕ
2	Channel
3	Rigid bar of the frame, 3/4"ϕ
4	End plate
5	Bonding bar, 3/16"ϕ
6	Fasten screw
7	Fasten screw
8	Rubber tube, 3/4" (with strain gauges inside)
9	Specimen $2 \times 2 \times 12$ – in. prism

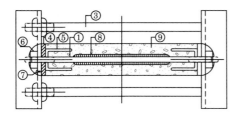

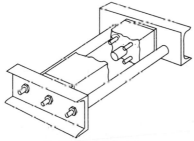

Figure 11-10 Restrained shrinkage test: Restraints applied using bar embedded in the specimen [11.12] (1 in. = 25.4 mm).

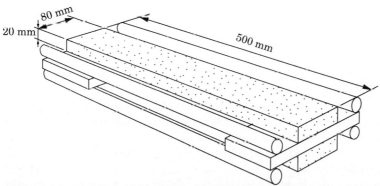

Figure 11-11 Test set up for linearly restrained shrinkage [11.13] (1 in. = 25.4 mm).

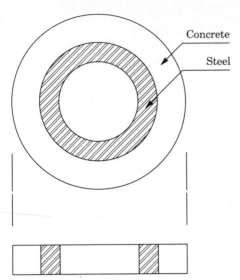

Concrete

Steel

Figure 11-12 Schematic view of a restrained ring shrinkage setup [11.16].

The thickness of the cement composite was also varied depending on the composition of the matrix. Typically, thicker sections were used with concrete containing coarse aggregates.

As mentioned earlier, this setup shows the most promise because of the uniform restraint provided by the steel ring. The restraining force is imposed by the steel ring across the perimeter of the concrete, instead of two or four locations as with linear and plate specimens. The method is described in detail in the following paragraphs. More details can be found in Reference [11.16].

The variation of stresses across the thickness of the concrete ring depends on the internal diameter of the ring. For the dimensions shown in Figure 11-13, the difference between the values of tensile hoop stress on the outer and inner surface is only 10% [11.16]. In addition to hoop stress, the concrete ring is also subjected to radial compressive stress when the steel ring exerts radial pressure. Since the diameter of the ring is relatively large, this radial compressive stress is only 20% of maximum hoop stress. Since cement composites are an order of magnitude stronger in compression, the maximum compressive stress in the ring is only about 2% of the compressive strength. Hence, the effect of compressive stresses can be neglected.

The concrete is sealed at the top using a silicone rubber sealer, allowing it to dry evenly only at the outer edge. A relatively large ratio of the width (exposed surface) to the thickness (4 or higher) can provide uniform drying across the thickness.

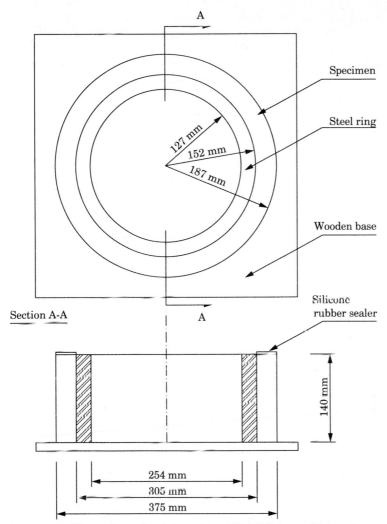

Figure 11-13 Dimensions of a ring specimen [11.16] (1 in. = 25.4 mm).

The cement composite can be cast between a steel ring and an annular outer mold. The outer mold can be made of cardboard or plastic. Provisions should be made to remove the outer mold without causing disturbance to the young cement composite. Care should also be taken to place the outer ring concentrically with the inner ring to avoid nonuniform thickness of the cement composite ring.

The outer mold can be removed as soon as the concrete hardens. The minimum period required is about 2.5 hours. Once the outer mold is removed the specimen can be subjected to the desired drying scheme. It can be exposed to drying right after the removal of the outer mold or it can be cured for 7 or 28 days before exposing to the drying scheme. The drying should be done in a controlled environment at a chosen temperature and relative humidity.

The shrinkage of the unrestrained specimen, known as free shrinkage, should also be measured for estimating the amount of stresses and crack widths. Measurements taken using an unrestrained ring (in which the inner steel tube was removed) and prismatic specimens show very little difference (Figure 11-14) [11.6]. Bar specimens recorded slightly larger strains, which should be expected, because of larger exposed areas. The dimensions of the bar specimen were 225 × 75 × 35 mm (9.0 × 3.0 × 1.4 in.).

Figure 11-15 shows the microscope setup for measuring the cracks that develop because of restrained shrinkage [11.16]. The microscope is attached to the center of the steel ring and can rotate 360°. It can also travel up and down, facilitating the crack-width measurements across the 140 mm (5.5 in.) width of the exposed concrete surface.

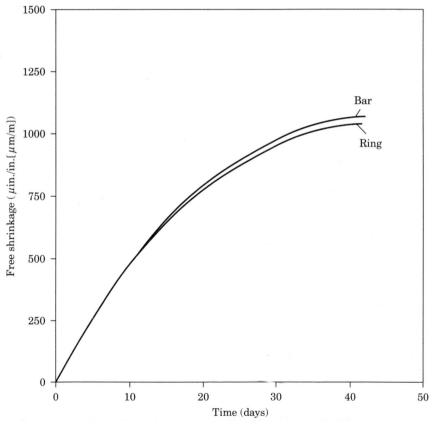

Figure 11-14 Comparison of the results between ring and prismatic bar specimens for plain concrete [11.16].

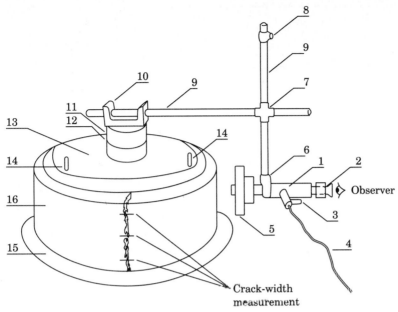

Figure 11-15   Test setup to measure crack width using a microscope [11.16].

1   Microscope
2   Filar eyepiece
3   Focusing knob
4   Internal illumination
5   Fluorescent lamp
6, 7, 8, 9, 10, 11   Microscope, mounting, hardware
12   Preloaded bearing pivot
13   Mounting plate
14   Dowel pins
15   Wooden base
16   Specimen

The strains in concrete can be measured by using strain gauges. These gauges should be placed along the circumference. If cracks develop near a strain gauge, then that gauge reading may not be useful. The amount of stresses imposed on the concrete ring can be measured by monitoring the strains developed in the steel ring. Sensitive gauges are needed to measure the strains in the steel ring because it undergoes very little deformation.

11.2.2   Contribution of polymeric fibers to drying shrinkage crack reduction

When plain concrete is subjected to restrained shrinkage using ring specimens, after the development of one crack the ring shrinks freely resulting in no further cracks. This phenomenon is shown in Figure 11-16 [11.16]. The test was conducted using a ring

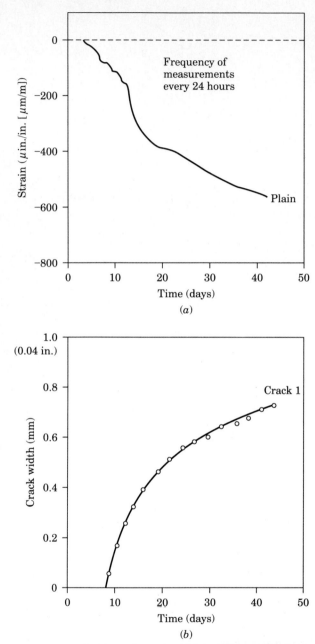

Figure 11-16 Strain and crack-width measurements for a plain concrete specimen [11.16].

specimen described in the previous section. The concrete had 1 part cement, 2 parts sand, and 2 parts coarse aggregate (9 mm [0.375 in.] maximum size). The water-cement ratio was 0.5. As the concrete shrinks, the tensile stresses develop in the concrete ring and a crack develops when the stresses produced (by shrinkage) exceed the tensile strength of the concrete. From Figure 11.16b it can be seen that a crack developed around the seventh day. The crack continued to widen as the concrete ring split by the crack was free to shrink. The strains measured in the concrete ring show a gradual increase of shrinkage strains up to 40 days (Figure 11-16a). The width of the crack also continues to increase with time.

If fibers are added to the matrix, the free shrinkage strains decrease (Chapter 10). However, this decrease is not substantial, especially at the early stages, as shown in both Chapter 10 and Figure 11-17. Results shown in Figure 11-17 were obtained using prism specimens

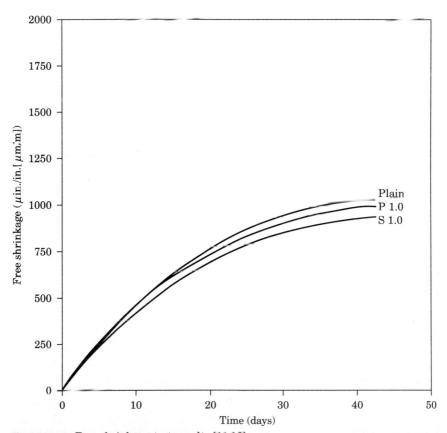

Figure 11-17 Free-shrinkage test results [11.16].

made of the same matrix as the ring specimens. The major contribution of fibers comes from the bridging of cracks and force transfer across the cracks.

When cracks form in fiber-reinforced concrete, the fibers that bridge the crack do not allow for free opening of the crack. As the ring shrinks further, the fibers transmit the forces across the crack and thus create tensile stress along the ring. If the force transmitted by the fibers is very small, say, in composites containing very low volumes of fibers, then a second crack will not form because the force transmitted across the crack is smaller than the tensile strength of the ring. If the fibers can transmit sufficient force across the crack, then a second crack will form. The process will continue, resulting in multiple cracking.

In Figure 11-18, which shows the crack widths and strains, it can be seen that strains do not increase monotonically (Figure 11-18a), as in the case of plain concrete specimens. There is a decrease in compressive strain in around 11 days. This decrease in strain can be attributed to the tension created by the forces developed by fibers bridging the crack. When the volume fraction of fibers is increased, more cracks and less maximum crack width can be expected. The experimental results do confirm this trend. The results obtained using fibrillated polypropylene fibers at volume fractions of 0.1%, 0.25%, 0.5% and 1.0% are shown in Table 11.1 and Figures 11-19 and 11-20.

From Table 11.1 it can be seen that both the total crack width and maximum crack width decrease with the addition of fibers. Even though specimens containing 0.25% and 0.50% fiber developed only one crack, the crack width reduced considerably. Since the free shrinkage strain does not show corresponding reduction, the reduction in crack width has to come from the resistance provided by the fibers across the crack. Even though the fibers bridging the crack could not develop sufficient force to induce a second crack, they could transfer some of the shrinkage strains to the uncracked part of the ring, creating tensile stress and strain. At a fiber volume fraction of 1%, the fibers were able to generate sufficient tensile force to induce a second crack.

Overall it can be said that polymeric fibers reduce maximum and total crack widths under restrained conditions. A fiber volume of about 1% is needed to develop multiple cracking.

11.2.3   Contribution of steel fibers to drying shrinkage crack reduction

Steel fiber contributes to both free shrinkage reduction and crack reduction under restrained conditions. The shrinkage reduction provided by steel fibers at early ages is shown in Figure 11-17 [11.16]. Even though steel fibers seem to be more effective than polypropylene

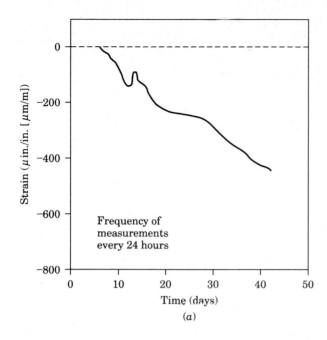

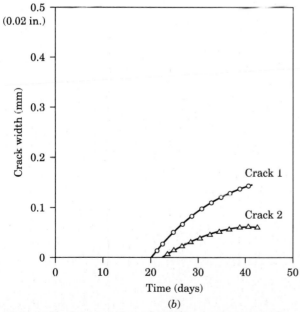

Figure 11-18 Strain and crack-width measurements for a specimen reinforced with a 1.0% volume of polypropylene fibers [11.16].

TABLE 11.1 Experimental Results and Comparison with Computed Results [11.16]

Specimen†	Number of cracks	Crack 1		Crack 2		Crack 3		Total crack width (mm)	Theoretical total crack width (mm)
		Crack width (mm)	Circumferential position* (mm)	Crack width (mm)	Circumferential position (mm)	Crack width (mm)	Circumferential position (mm)		
Plain	1	0.900	875	–	–	–	–	0.900	1.050
S0.25	2	0.300	890	0.045	190	–	–	0.345	0.410
S0.5	3	0.100	800	0.065	0	0.050	420	0.215	0.252
S1.0	1	0.075	87	–	–	–	–	0.075	0.145
S1.5	3	0.011	210	0.010	395	0.009	690	0.03	0.080
P0.1	1	0.875	0	–	–	–	–	0.875	1.030
P0.25	1	0.480	900	–	–	–	–	0.480	0.610
P0.5	1	0.230	1110	–	–	–	–	0.230	0.400
P1.0	2	0.150	720	0.065	110	–	–	0.215	0.250

*Total circumferential length = 1170 mm. (1 mm = 0.04 in.).
†S—steel fiber P—polypropylene fiber

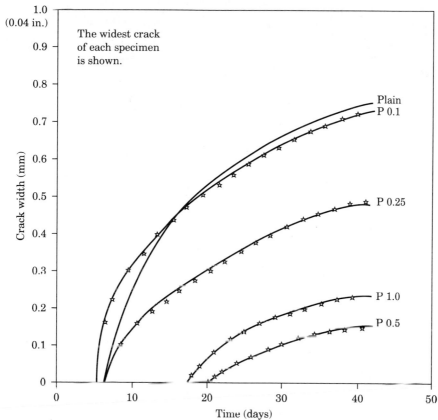

Figure 11-19 Crack width versus time for various volume percentages of polypropylene fibers [11.16].

fibers, the reduction in shrinkage is not significant. Steel fibers are more effective in reducing restrained shrinkage cracking because of their higher stiffness, resulting in better load transfer across the crack. Experimental results obtained using ring tests are shown in Table 11.1 and Figure 11-21 [11.16]. The matrix composition was the same as the one used for polypropylene fibers described in the previous section. The steel fibers used were 25 mm (1 in.) long and 0.4 mm (.016 in.) in diameter. The addition of steel fibers decreased the crack width considerably even at a volume fraction of 0.25%. An increase in fiber content produces a consistent decrease in both maximum and total crack widths. The formation of a first crack is also delayed in time with an increase in fiber content. In addition, crack width stabilizes at earlier ages for specimens having higher fiber content. The magnitude of crack reduction is large in ranges of 0–0.25% and 0.25–0.5%. Overall, steel fibers provide effective crack reduction under restrained conditions.

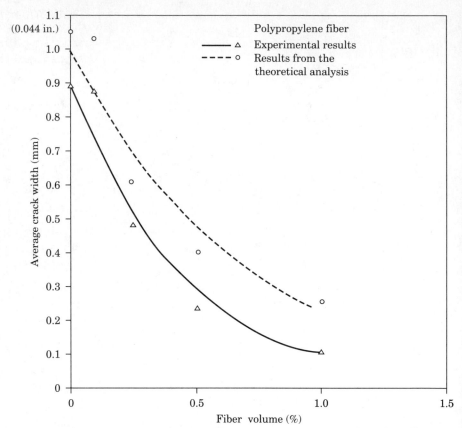

Figure 11-20 Average crack width versus fiber volume: polypropylene fibers [11.16].

11.2.4 Theoretical model for the prediction of crack widths under restrained drying shrinkage conditions

The physical concept of the formation of cracks under restrained conditions can be described as follows (see Ref. [11.17] for further details). At a given time, the composite undergoes a certain shrinkage strain $\epsilon_{Sh}(t)$. If the movement needed for this shrinkage is prevented by restraint, then a tensile stress $\sigma(t)$ is developed. This tensile stress $\sigma(t)$, which is a function of both strain $\epsilon_{Sh}(t)$ and the elastic modulus of the composite $E(t)$ at time t, can be expressed as

$$\sigma(t) = E(t)\epsilon_{Sh}(t) - \sigma^R(t) \tag{11.1}$$

where $\sigma^R(t)$ is the stress reduction provided by relaxation effects. Note that stress, strain, and elastic modulus are time-dependent variables.

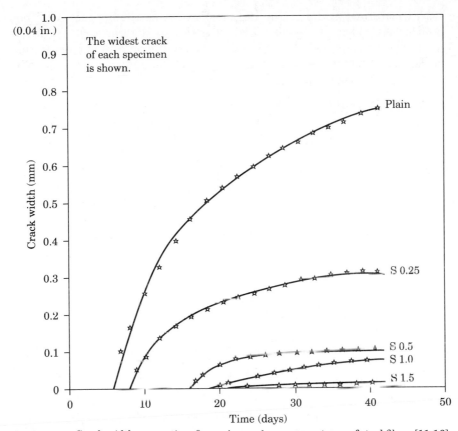

Figure 11-21   Crack width versus time for various volume percentages of steel fibers [11.16].

When the resulting stress $\sigma(t)$ reaches the tensile strength of the composite at t, then a crack will form. After the crack forms, the uncracked portion will continue to shrink, resulting in a widening of the crack, as shown in Figure 11-16.

For a second crack to form, tensile force has to be transmitted across the crack. In plain cement (concrete) matrix, the magnitude of force that can be transferred across the crack is very much limited, and hence a second crack does not form. In the case of fiber-reinforced matrix, there is a force transfer across the crack. The magnitude of this force transfer depends on the fiber type, the fiber volume fraction, and the matrix composition. As the uncracked part shrinks more, part of the strain is relieved by an increase in crack width, and another part produces tensile stresses in the ring. If the tensile stress exceeds the tensile strength, another crack forms and the process continues.

A theoretical model is available for predicting the number of cracks and crack widths; it is described in detail in Reference [11.17]. The model can be applied to a three-dimensional structure subjected to a given drying environment with a specified boundary restraint. The outline of the model is as follows.

Step 1. For a chosen time step, estimate the free shrinkage of the matrix. Free shrinkage can be estimated using the matrix composition and established equations available for cement matrices. Experimentally measured shrinkage values for the particular matrix composition provide the best estimate. If the element is thick, a layered approach can be used to estimate the shrinkage strains at various locations (depths) of the member. The problem can be solved using time increments. For each time increment, shrinkage strains and other parameters have to be computed.

Step 2. Compute the restraining forces. In the case of a ring specimen, the steel ring is assumed to be infinitely rigid, providing no change in circumference.

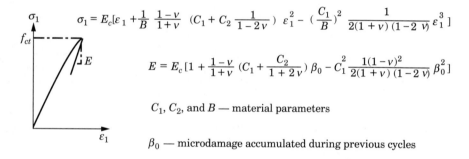

$$\sigma_1 = E_c[\varepsilon_1 + \frac{1}{B}\frac{1-v}{1+v} \ (C_1 + C_2 \frac{1}{1-2v}) \ \varepsilon_1^2 - (\frac{C_1}{B})^2 \ \frac{1}{2(1+v)(1-2v)} \varepsilon_1^3]$$

$$E = E_c [1 + \frac{1-v}{1+v} \ (C_1 + \frac{C_2}{1+2v}) \ \beta_0 - C_1^2 \ \frac{1(1-v)^2}{2(1+v)(1-2v)} \ \beta_0^2]$$

C_1, C_2, and B — material parameters

β_0 — microdamage accumulated during previous cycles

(*a*) Stress–strain relationship before the peak

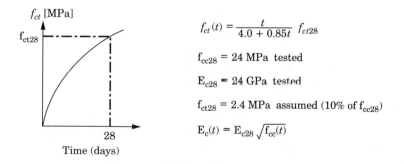

$$f_{ct}(t) = \frac{t}{4.0 + 0.85t} \ f_{ct28}$$

$f_{cc28} = 24$ MPa tested

$E_{c28} = 24$ GPa tested

$f_{ct28} = 2.4$ MPa assumed (10% of f_{cc28})

$E_c(t) = E_{c28} \sqrt{f_{cc}(t)}$

(*b*) Development of the tensile strength and E-modulus in time

Figure 11-22 Constitutive model for uncracked concrete [11.16] (1 ksi = 6.895 MPa).

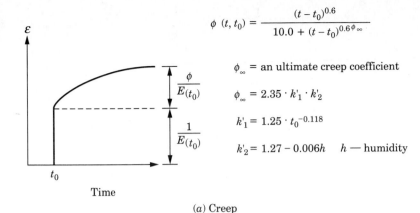

$$\phi\,(t, t_0) = \frac{(t - t_0)^{0.6}}{10.0 + (t - t_0)^{0.6}\phi_\infty}$$

ϕ_∞ = an ultimate creep coefficient

$\phi_\infty = 2.35 \cdot k'_1 \cdot k'_2$

$k'_1 = 1.25 \cdot t_0^{-0.118}$

$k'_2 = 1.27 - 0.006h \quad h$ — humidity

(a) Creep

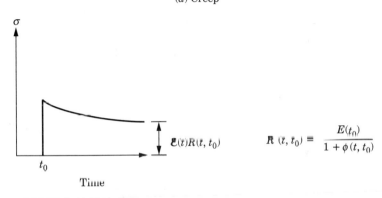

$$R\,(t, t_0) \equiv \frac{E(t_0)}{1 + \phi\,(t, t_0)}$$

(b) Relaxation

Figure 11-23 Creep and relaxation functions [11.16].

Step 3. For the estimated shrinkage strain that is restrained, compute the tensile stress. For accurate results, a nonlinear stress-strain response such as the one shown in Figure 11-22a is recommended. It is also important to include the change in strength and stiffness with time (Figure 11-22b).

Step 4. The stress developed due to shrinkage is partially reduced by the relaxation [Equation (11.1)]. The reduction in stress can be computed using the relaxation curves available in the literature. A typical relation is shown in Figure 11-23 for a linear viscoelastic behavior.

Step 5. As time progresses, the tensile stress keeps increasing. When this stress exceeds tensile strength of the composite, a crack forms.

Step 6. The shrinkage process continues after the formation of the first crack. The stress development in the element will depend on the force transfer across the crack. Even plain concrete was found

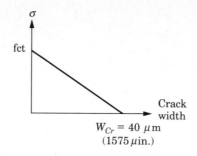

(a) Stress transfer in a cohesive crack

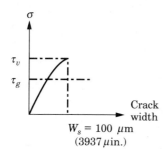

$$\sigma = \frac{4}{\pi} \frac{l}{d} V_f \tau_v [\frac{W}{w_s} - (\frac{W}{w_s})^2 + \frac{1}{2} \frac{\tau_g}{t_v} (\frac{W}{w_s})^2]$$

	Steel fiber	PP fiber
τ_v [MPa]	4.0	3.5
τ_g [MPa]	3.0	0.7
E [GPa]	210	4.83

l = length of the fibers
d = diameter of the fibers
V_f = volume of the fibers
W = crack width
W_s = critical crack width
τ_v = interfacial shear strength
τ_g = frictional shear strength

(b) Interfacial behavior in the bond

Figure 11-24 Constitutive model for the cracked concrete [11.16] (1 ksi = 6.895 MPa).

to resist tension until the crack is too wide [11.18]. The relation between traction carried across the crack and crack width can be assumed to be linear, as shown in Figure 11-24a.

For fiber-reinforced composites, the traction across the crack is much higher. The amount of traction depends on the fiber type, geometry, and volume fraction. The resistance is contributed by fiber debonding, pull-out, and stretching. The stress–crack width relationships for straight steel fibers and fibrillated polypropylene fibers are shown in Figure 11-24b.

The tensile stress in the cracked element has to be computed using the appropriate traction developed by the crack. Part of the shrinkage strain will contribute to an increase in the existing crack width, and the remaining shrinkage strain will induce tensile forces.

Step 7. If the stresses developed in the cracked element exceed the tensile strength, a second crack will form. The development of further cracks have to be monitored using resultant tensile forces at various time periods.

Another factor to be considered for the development of second and subsequent cracks is the effect of cyclic loading. Before the first crack forms, the entire element is subjected to tension, and the crack develops in the statistically weak point. When the uncracked part of the ring is subjected to tension again, it may not sustain the static tensile strength. Owing to cyclic loading effect, the tensile strength decreases. Hence the uncracked part of the element could develop a crack at a stress less than the virgin tensile strength. The effect of this cyclic loading can be accounted for using the following equation [11.17–11.19]

$$\epsilon_d^{pk} = \left(-\frac{1}{k}L_n\left(\frac{f_t^d}{f_t}\right)\right)^{1/\lambda} \tag{11.2}$$

where f_t is the tensile strength of concrete, f_t^d is the reduced tensile strength of concrete, and ϵ_d^{pk} is the reduced peak value of strain; k and λ are material constants. Based on the experimental results (Figure 11-25) [11.18], λ and k can be assumed as 1.01 and 0.071 respectively.

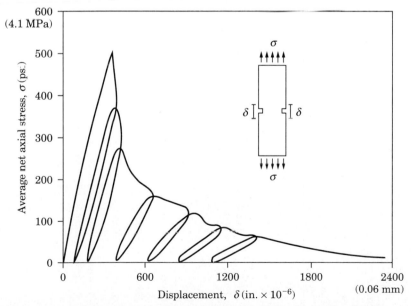

Figure 11-25   Response of concrete subjected to uniaxial cyclic loading [11.18].

TABLE 11.2 Results of the Analysis for a Ring Specimen Containing a 1% Volume of Steel Fibers [11.17]

Time t (days)	$\Delta\epsilon_i^{sh}$	$\Delta\epsilon_i^d$	$\Delta\epsilon_i^{cr}$	σ (MPa)	f_t (MPa)	f_t^{d*} (MPa)	Description	Number of cracks	Total crack width (mm)
5.50	0.00003	0.00003	0.0	1.14	1.77	1.77		0	0.0
6.00	0.00003	0.00003	0.0	1.76	1.85	1.85		0	0.0
6.50	0.00003	0.000011	0.000019	1.42	1.91	1.86	First crack forms	1	0.013
—									
7.00	0.00003	0.000025	0.000005	1.75	1.97	1.93		1	0.017
7.50	0.00003	0.000012	0.000018	1.49	2.02	1.93	Second crack forms	2	0.028
—									
9.50	0.000023	0.000019	0.00004	2.00	2.20	2.13		2	0.045
10.00	0.000023	0.000006	0.000017	1.67	2.24	2.10	Third crack forms	3	0.057
—									
12.00	0.000023	0.000014	0.000009	1.87	2.37	2.13		3	0.075
—									
26.00	0.000027	0.000027	0.000010	2.26	2.79	2.65		3	0.134
—									
46.00	0.000018	0.000011	0.000007	2.53	2.99	2.85	End of six weeks' drying	3	0.185

*f_t^d is the reduced tensile strength of the matrix. (1 ksi = 6.895 MPa; 1 in. = 25.4 mm)

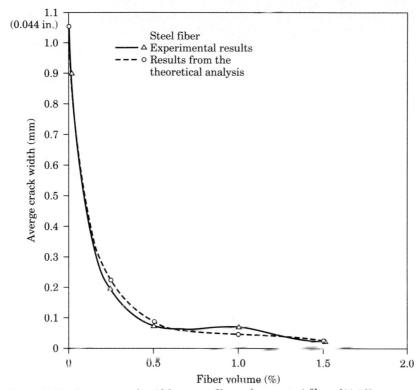

Figure 11-26   Average crack width versus fiber volume: steel fibers [11.16].

The comparisons of experimental and the results predicted using the above procedure are shown in Figures 11-20 and 11-26 for polypropylene and steel fibers respectively. From these figures it can be seen that the model accurately predicts both the number of cracks and the crack widths.

The estimated strains, stresses, and crack widths at various time intervals for another test sample are shown in Table 11.2. The term ϵ_i^d is the strain in the uncracked zone, whereas ϵ_i^{cr} is the strain in the cracked zone. For this sample, 30 mm long and 0.44 mm diameter fibers with hooked ends were used. The traction force across the crack was increased to account for the resistance provided by the hooks at the end of the fibers.

11.3   References

11.1   Kraai, P. P. "A Proposed Test to Determine the Cracking Potential Due to Drying Shrinkage of Concrete," *Concrete Construction,* September 1955, pp. 775–778.

11.2   Dahl, P. A. "Plastic Shrinkage and Cracking Tendency of Mortar and Concrete Containing Fiber Mesh," Report No. STF65A85039, SINTEF Div. FCB, Trondheim, Norway, September 1985.

11.3   Balaguru, P. "Plastic Shrinkage Characteristics of Fiber Reinforced Cement Composite," Rutgers, New Brunswick, New Jersey, August 1989.

11.4   Shaeles, C. A.; and Hover, K. C. "Influence of Mix Proportions and Construction Operations on Plastic Shrinkage Cracking in Thin Slabs," *ACI Materials Journal,* Vol. 85, 1988, pp. 495–504.

11.5   Balaguru, P. "Evaluation of New Synthetic Fibers for Use in Concrete," Civil Engineering Report No. 88-10, Rutgers, New Brunswick, New Jersey, November 1988, 93 pp.

11.6   Khajuria, A.; and Balaguru, P. "Evaluation of New Synthetic Fibers for Use in Concrete," Civil Engineering Report, Report No. 89-13, October 1989, 89 pp.

11.7   Swamy, R. N.; and Stavrides, H. "Influence of Fiber Reinforcement on Restrained Shrinkage and Cracking," *ACI Journal,* Vol. 76, No. 3, 1979, pp. 443–460.

11.8   Malmberg, B.; and Skarendahl, A. "Method of Studying the Cracking of Fiber Concrete under Restrained Shrinkage," *RILEM Symposium on Testing and Test Methods of Fiber Cement Composites,* 1978, pp. 173–179.

11.9   Krenchel, H.; and Shah, S. P. "Restrained Shrinkage Tests with PP-Fiber Reinforced Concrete," *Fiber Reinforced Concrete Properties and Applications,* SP-105, American Concrete Institute, Detroit, Michigan, 1987, pp. 141–158.

11.10   Paillere, A. M.; Buil, M.; and Serrano, J. J. "Effect of Fiber Addition on the Autogeneous Shrinkage of Silica Fume Concrete," *ACI Materials Journal,* Vol. 86, No. 2, March-April 1989, pp. 139–144.

11.11   Hahne, H.; Karl, S.; and Worner, J. "Properties of Polyacrylonitrile Fiber Reinforced Concrete," *Fiber Reinforced Concrete Properties and Applications,* SP-105, American Concrete Institute, Detroit, Michigan, 1987, pp. 211–223.

11.12   Pan, K. W., et al. "A Study on Restrained Shrinkage Cracking of Fly Ash, Cementitious Materials," Private communication, 1987.

11.13   Pihlajavaara, S. E.; and Pihlman, E. "Results of Long-Term Deformation Tests of Glass Fiber Reinforced Concrete," NORDFORSK-FRC Project 1974–76, Delrapport, Technical Research Center of Finland, Otaniemi, 1978.

11.14   Opsahl, O. A.; and Kvam, S. E. "Betong med EE-stal-fiber (Concrete with EE-Steel Fibers)," Report No. STF 65A82036, SINTEF Div. FCB, Trondheim, Norway, June 1982.

11.15   Grzybowski, M. "Determination of Crack Arresting Properties of Fiber Reinforced Cementitious Composites," Publication TRITA-BRO-8908, ISSN 1100-648X, Royal Institute of Technology, Stockholm, Sweden, June 1989, 190 pp.

11.16   Grzybowski, M.; and Shah, S. P. "Shrinkage Cracking of Fiber Reinforced Concrete," *ACI Materials Journal,* Vol. 87, 1990, pp. 138–148.

11.17   Grzybowski, M.; and Shah, S. P. "A Model to Predict Cracking in Fiber Reinforced Concrete Due to Restrained Shrinkage," *Magazine of Concrete Research* (London), Vol. 41, No. 148, 1989, pp. 125–135.

11.18   Gopalaratnam, V. S.; and Shah, S. P. "Softening Response of Plain Concrete in Direct Tension," *ACI Journal,* Vol. 82, No. 3, 1985, pp. 310–323.

11.19   Reinhardt, H. W.; and Cornelissen, H. A. W. "Post-Peak Cyclic Behavior of Concrete in Uniaxial Tensile and Alternating Tensile and Compressive Loading," *Cement and Concrete Research,* Vol. 14, 1984, pp. 263–270.

12

Fiber-Reinforced Shotcrete

Shotcrete is a portland cement-based mortar or concrete placed in position by pneumatically projecting the mix at high velocity onto a surface. There are basically two methods used for this process. In the first, called dry shotcreting, a premoistened mix of concrete is mixed with water at the nozzle of the pneumatic gun. In the other, all the constituents including the water are premixed and shot through the pneumatic gun. In both cases air is used to carry the mix. More details regarding equipment and the shotcreting process can be found in References [12.1, 12.2].

Fiber-reinforced shotcrete (FRS) contains discrete fibers in addition to the other constituent materials. The fibers used most often are steel fibers. Polymeric fibers are also being used, but to a limited extent. Spray-up processes used for making glass fiber–reinforced concrete (GFRC) panels can also be considered as an aspect of shotcrete. GFRC is presented in a separate chapter (Chapter 13) and is not discussed here. This chapter deals with fiber-reinforced mortar and concrete that is applied using standard equipment that is used for plain (concrete) shotcrete. The primary applications include tunnel linings, stabilization of rock slopes, steel pipe coatings, shell roofs, storage tanks, repair and rehabilitation of waterfront structures, irrigation channel linings, and refractory coatings. Polymeric fiber shotcrete is being used for intermediate covers of waste disposal sites.

12.1 Constituent Materials

The constituent materials for fiber-reinforced shotcrete are about the same as the ones for normal fiber-reinforced concrete presented in Chapter 5. Below are some of the special considerations for shotcreting applications.

12.1.1 Cement

Normal Type I cement can be used for most applications. In some instances where early strength is needed, such as tunnels, Type III cement can be used. The early strength development can help to resist the effects of blasting. Sulfate-resisting cements have been used in areas where sulfates are present in either the soil or the ground water.

12.1.2 Aggregates

The primary concern here is gradation. In most of the early applications, the matrix consisted of cement and sand with no coarse aggregates. Since mortar mixes require higher cement content, they were found to sustain higher creep and shrinkage strains, resulting in cracking. Hence, coarse aggregate addition has become the norm for recent applications.

Mixes containing coarse aggregates that are less than 10 mm (0.4 in.) are more common for fiber-reinforced shotcrete. A typical gradation chart used for mix design is shown in Figure 12-1 [12.3, 12.4]. In a typical mix 5 to 10 mm (0.2 to 0.4 in.) aggregates make up 20% to 35% of the total combined fine and coarse aggregates. For overhead applications,

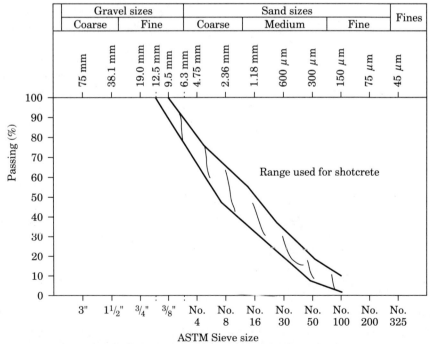

Figure 12-1 Gradation range used for shotcrete [12.3, 12.4].

the mixes are proportioned toward the finer side of the envelope. Mixes used for vertical applications such as walls fall in the midsection of the gradation curve, whereas coarser particles are used for applications on flat slopes.

Even though up to 19 mm (0.75 in.) maximum size aggregates have been used for plain concrete shotcreting, they are not recommended for fiber shotcrete. Fibers are more effective for shotcrete in matrices containing smaller aggregate particles. Note that fibers are placed essentially parallel to the surface being shotcreted because of the way they are carried during pneumatic projection.

12.1.3 Additives and admixtures

A variety of additives and admixtures are used especially for shotcreting applications. These added ingredients are needed for better adhesion and lower rebound, the material that falls down during the shotcreting process (Section 12.5). Naturally, minimum rebound is one of the primary requirements for mix design and placement techniques. Condensed silica fume is a popular additive used in conjunction with the addition of fibers. Other additives are accelerators, water-reducing admixtures, and polymer latexes.

Silica fume makes the mix cohesive, resulting in lower rebound, and makes it possible to build more thickness without chunks of material falling down. The addition of silica fume also provides an increase in strength, and a reduction in permeability. In most if not all cases, silica fume is used in conjunction with high-range water-reducing admixtures. Otherwise, a large amount of water is needed to satisfy the water demand produced by fine particles of silica fume.

A high-range water-reducing admixture (super plasticizer) is often used in shotcreting processes because it facilitates the placement of concrete with low water-cement ratios and results not only in higher strengths but also in higher earlier strengths and lower creep and shrinkage.

Accelerators are often used for shotcrete to shorten set time. The reduced set time results in less sagging or sloughing when thick sections are placed in a single pass. Admixtures containing calcium chlorides, fluorides, sulfites, sulfates, and nitrates are not recommended for steel fiber shotcrete because these chemicals are suspected to promote corrosion in steel fibers.

Polymer latexes are added in special situations where improved adhesion, permeability reduction, resistance to chloride attack, or improved freeze-thaw durability is needed. The latex modification also provides increased strength and higher impact resistance. Polymer latexes are normally used only in dry mix shotcrete. In wet mix, use of latex becomes difficult because of the increased adhesion and time limits on polymerization.

12.1.4   Fibers

Most of the fibers described in Chapter 5 can be used for shotcreting. However, in most applications only steel fibers are used. Polymeric fibers are gaining more acceptance. As mentioned earlier, glass fibers are used to fabricate different classes of products manufactured by spray-up processes.

For steel fibers the most common size range is from 0.5 to 1.5 in. × 0.01 to 0.1 in. (12 to 38 mm × 0.25 to 2.5 mm). The first number is the length and the second number is the diameter of the fiber. This fiber size range is easily handled and shot through a 2 in. (50 mm) diameter hose. Shorter fibers are easier to mix and shoot and they rebound less, but the resulting composite properties—especially toughness and postcrack resistance—are inferior compared with composites containing longer fibers. On the other hand, longer fibers create more plugging and higher rebound rates. Fibers that are about 1 in. (25 mm) long seem to be the optimum for both workability and hardened composite properties. Carbon steel fibers can be used for applications at ambient temperatures and at high temperatures up to 1500°F (816°C). For still higher temperature applications such as refractory linings where only one side of the composite is heated, stainless steel fibers are preferred [12.5].

Of the polymeric fibers, polypropylene in the fibrillated form has been investigated under both laboratory and field conditions. The development of polymeric fiber shotcrete is still in its infancy.

12.2   Mixture Proportions

12.2.1   Steel fiber shotcrete

As mentioned in Section 12.1.2, in earlier applications only a mortar mix was used. The sand-to-cement ratio was about 2.4 : 1 by weight, resulting in 940 lb of cement per cubic yard (558 kg/m^3). When 0.4 in. (10 mm) maximum size coarse aggregates are added, the cement content can be reduced. Typical ranges of various constituent materials are shown in Tables 12.1 and 12.2. The mixture proportions shown in Table 12.1 were used for tunnel linings. The mixture provides 28 day compressive strength of about 35 MPa (5000 psi). Typically in-place composite has a higher cement content than the initial mixture because more aggregate particles than cement particles are lost in the rebound.

The fiber volume fraction ranges from 0.5% to 2% (Table 12.1). Here again, the composite in place has fewer fibers than the original mixture because more fibers than matrix are lost in the rebound.

TABLE 12.1 Typical Steel Fiber–Reinforced Shotcrete Mixtures [12.6]

Material	Fine aggregate mixture (lb/yd³ [kg/m³])	0.4 in. (10 mm) aggregate mixture (lb/yd³ [kg/m³])
Cement	753–940 (447–558)	750 (445)
Blended sand		
¼ in. (6 mm) maximum size	2830–2500 (1679–1483)	1485–1175 (881–697)
0.4 in. (10 mm) aggregate	—	1180–1475 (700–875)
Steel fiber	66–265 (39–157)	66–250 (39–148)
Accelerator	(0.5 to 2%) varies	varies
Water-cement ratio		
(by weight)	0.40–0.45	0.40–0.45

12.2.2 Polymeric fiber shotcrete

In most instances polymeric fibers are simply added to the concrete mix. The primary objective of adding polymeric fibers is to reduce plastic shrinkage cracking; hence, the fiber volume fraction in most applications is about 0.1%. The contribution of fibers is assumed to be the same as in cast concrete (Chapter 11).

Higher volume fractions of polypropylene fibers have also been experimented. In this case the addition of fibers was found to improve toughness characteristics and to reduce plastic shrinkage cracking [12.7]. It was found that fibers can easily be added to the ready-mix truck up to a volume fraction of 0.66 percent (approximately 10.0 lb/yd³ or 6 kg/m³) [12.7]. The fibers were in the fibrillated form, and fiber lengths ranged from 0.75 to 2.25 in. (19 to 57 mm).

The matrix for polymeric fiber shotcrete can be proportioned using the same procedures for plain shotcrete. For higher fiber volume fractions, a larger dosage of high-range water-reducing admixtures may

TABLE 12.2 A Typical Shotcrete Mixture Proportion: Dry Bagged Materials [12.6]

Mix ingredients	Nominal volume at nozzle lb/yd³	Nominal volume at nozzle kg/m³
Type I, normal portland cement	775	460
10 mm (⅜ in.) coarse aggregate	860	510
5 mm (No. 4) fine aggregate	2057	1220
Accelerator	30	18
Steel fiber (hooked end)	100	60

be needed to maintain workability. Entrapped air content should be monitored especially at higher fiber volume fractions to make sure that excess air is not present.

12.3   Batching and Mixing

In dry mix process of shotcreting the ingredients, including the fiber, are premixed in the dry form; then they are premoistened and fed into the shotcreting equipment. Final make-up water is added at the nozzle. In the wet process, the ingredients are mixed as for cast-in-place concrete and fed to the shotcreting equipment. The following sections provide a brief description of the methods used and precautions to be taken.

12.3.1   Dry process

Batching for the dry process is done by either mixing the ingredients on site or using prebagged materials. Prebagged mixes are available in small paper bags and bulk synthetic cloth bags [12.6]. Paper bags normally contain either 50 or 88 lb (23 or 40 kg) of premixed materials. Bulk bags have capacities in the range of 2500 to 4000 lb (1100 to 1800 kg). The large bags are often used for bulk applications such as tunnel linings. Smaller bags are primarily used for repair work, where space and movements are restricted. The dry materials are premoistened up to a moisture content of 6 percent by weight before they are discharged into the shotcrete pot. Prebagged materials provide good control of the time between moisturizing and placement, thus providing good control of any hydration of cement that might occur before shooting.

The ingredients can also be mixed in a ready-mix truck or mixing plant. Different techniques have been employed to dispense fibers, including conveyor belts, shakers, and vibrating screens. In all cases care should be taken to avoid the entrance of fiber clumps to the shotcreting bin.

A turbine mixer (a stationary, cylindrical, flat-bottomed pan with revolving mixing arms) has also been successfully used to mix the dry ingredients [12.6]. The fibers were added through a 2.5 in. (63 mm) mesh over preplaced sand in the pan and mixed. The sand-fiber mix was transferred to a transit mixer and the cement was added at the job site. The cement-sand-fiber mix was then fed to a hopper equipped with a screen. The screen intercepted any fiber clumps formed during the transit and mixing of cement.

The following precautions should be taken when adding fibers to obtain a uniform mixture:

- Fibers that show a tendency to clump should be added through a screen and should not be allowed to accumulate in the mixer. The mechanisms carrying the fibers, such as belts or screw conveyors, should move fast enough to avoid formation of fiber clumps.

- Care should be taken to avoid bending of fibers. Bent fibers result in poor compaction and inferior mechanical properties. Bending of fibers was observed when a paddle mixer with small counterrotating paddle wheels was used.

- The materials should be fed to the shotcrete hopper through a screen to intercept fiber clumps. A good electrical ground to the gun and nozzle was found to reduce fiber clumping and plugging dramatically.

12.3.2 Wet process

In wet process shotcrete, procedures for batching and mixing are essentially similar to those used for cast-in-place fiber-reinforced concrete. Hence, the procedures presented in Chapter 6 also apply to shotcrete mixes.

Precautions should be taken to avoid fiber balling during the mixing process. Fibers tend to ball when they are added in bulk to the mixer. Performance of fibers depends on the fiber geometry, fiber volume fraction, and aspect ratio. Addition of fibers to fine aggregate on a conveyor belt was found to be successful.

When fibers are added directly to the transit mixer, fibers should land on the mixture and not on the mixing vanes, where they can form clumps. The drum should also rotate fast enough to avoid piling up of fibers. Collated fibers, fibers with aspect ratios less than 40, and some large-diameter fibers can be directly added in the final stage of mixing without causing problems.

As for dry process, it is recommended that the shotcrete be fed into the hopper through a mesh to intercept any fiber clumps formed during the batching and mixing processes.

Since shotcrete mixes have higher cement contents, the elapsed time for delivery and discharge of wet-mix should be limited to 20 minutes if the concrete is hauled in nonagitating trucks. When agitating trucks are used, a 40 minute time lapse is allowable. Retempering after 30 minutes from the initial mixing operation is not advisable.

Even though wet-process was introduced later, especially for steel fiber-reinforced shotcrete, it is gaining popularity. Norway has a very high success rate in using wet-process [12.8]. Wet-process is also being used more and more in Sweden and Finland. It was found that a robot rig can be efficiently used for wet-process shotcreting, improving both quality and productivity. It is also possible to obtain higher strengths in wet-process compared with dry-process.

12.4   Installation

Fiber-reinforced shotcrete can be installed (placed in position) with the same equipment normally used for plain concrete shotcrete. A wide range of equipment, from the original single- or dual-chamber feedwheel type to the more recent revolving-barrel and segmented-rotor types, is available in the market. Machines used for wet-process include the pressurized-chamber type, squeeze-pump type, and positive-displacement pumps.

Specialized equipment for feeding fibers separately and adding them at the nozzle has also been developed [12.9]. With the new equipment it may be possible to incorporate a higher volume fraction of fibers with larger aspect ratios. The fiber feeder equipment consists of an inclined rotating drum, Figure 12-2 [12.9]. The fiber mass fed into the drums is broken down to loose fibers by special spikes and fed into the ejector. The fibers are then carried to the nozzle using air pressure. The flow rate of fiber can be controlled by adjusting the size of the aperture through which the fibers are fed to the ejector. This feeder was found to be satisfactory for various types of fibers including smooth, deformed, and hooked-end fibers. Nozzle arrangements for this feeder are shown in Figures 12-3 and 12-4 for wet and dry processes, respectively.

In the wet process the fibers are fed into the center portion of the concrete and pushed through by forced air (Figure 12-3). In the dry process, fibers are mixed with cement and sand before the water is added (Figure 12-4). In this process large amounts of air are needed to carry the dry concrete mix. Some of the excess air is removed before the fiber is added, as shown in Figure 12-4. The fiber feeder of Figure 12-2 can also be integrated with the equipment used to place shotcrete, that is, the conventional gun and nozzle.

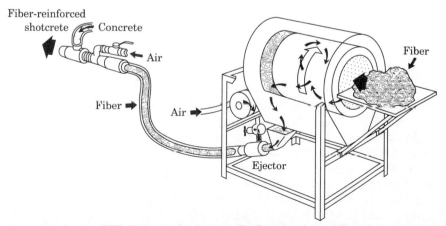

Figure 12-2   System BESAB for the production of high-strength steel fiber shotcrete [12.9].

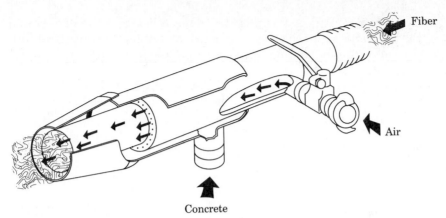

Concrete

Figure 12-3 Typical wet spray nozzle for system [12.9].

Below are some of the special precautions needed when placing fiber shotcrete. The provisions are especially applicable for steel fiber mixes.

The flow path should not have abrupt turns. The use of 90° elbows should be avoided. If the line size has to be reduced, a long tapered reducer should be used. Larger hose sizes—greater than or equal to 2 in. (50 mm) in diameter—are preferable. It is advisable to have the hose diameter at least twice the length of the fibers, even though in some instances 1.5 in. (38 mm) hoses have been successfully used to pump refractory shotcrete containing 1 in. (25 mm) long fibers.

Elastomeric wear linings at elbows should be removed. Adding vibrators or revolving wiper arms to the hopper screen can improve the material delivery. In some instances a motor with a higher capacity may be needed to handle the fiber mix.

Special precautions should be used when dry mix shotcrete is sprayed in multilayer applications to avoid laminations. In an experimental project, laminations were found in cored specimens sprayed

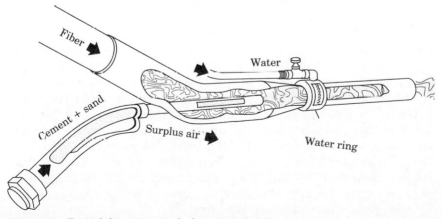

Figure 12-4 Typical dry spray nozzle for system [12.9].

in two layers [12.10]. A few minutes elapsed between the application of the first and second layers. Laminations produced premature failures in all modes of loading, namely, compression, flexure, and impact [12.10].

In shotcrete containing steel fibers, some fibers may be exposed at the surface or protrude. These exposed fibers can corrode and stain the surface even though corrosion essentially stops at the surface. In applications where this is a problem, a plain concrete mix can be used to provide a 0.5 in. (12 mm) thick surface coat.

12.5   Rebound

Rebound is a characteristic phenomenon of shotcreting. The pneumatic application of the composite causes part of the aggregates and fibers, which are traveling at high velocities, to bounce (rebound) off the surface. Rebound is significant while the initial layer is shot at the fresh rock surface, but typically diminishes after an initial layer of fresh composite is in place. The rebound material cannot be reused.

12.5.1   Factors affecting rebound of fibers

A systematic study, using high-speed photography, was conducted to observe the rebound of fibers [12.11]. The shotcrete airstream was photographed to observe the location and flow of fibers. The photographs showed that many fibers were at the outer portion of the airstream. Fibers near the periphery were found to be blown away radially from near the point of intended impact shortly before or after they hit.

Some fibers were blown up into the air and floated down. The residual air currents at the contact surface also contributed to the rebound of fibers. If lower air pressure was used, the rebound of the fibers was found to be less.

12.5.2   Comparison of plain and fiber shotcrete

For plain shotcrete, the rebound is 15% to 30% for sloping and vertical walls and 25% to 50% for overhead work. Various proportions of rebound have been reported for fiber shotcrete. Some studies have found that fiber shotcrete has less rebound than plain shotcrete [12.10–12.13].

Results in Reference [12.13] showed that plain matrix had a rebound of 31% compared with 10 percent for fiber matrix. The study was conducted in a tunnel with both vertical and overhead surfaces. The concrete was placed to an average thickness of 6 in. (150 mm). The mix contained 700 lb (317 kg) cement, and 2700 lb (1225 kg) sand per cubic yard (0.6 m³). The fiber mix had 150 lb (68 kg) of 0.5 in. (12 mm) × 0.01 in. (0.25 mm) steel fibers.

However, in Reference [12.11] the average rebound was 17.7% for fiber mix and 18.3% for plain mix. The small difference was attributed to change in mixture composition of the matrix rather than the presence of fibers.

Almost all investigators have reported that rebound of fibers is higher than rebound of matrix, resulting in a lower fiber content in the composite in place. Fiber retention was found to be as low as 40% [12.12] for overhead applications. The retention rate improved to about 65% for vertical surfaces [12.11, 12.12]. In another study where 2500 lb (1134 kg) of material was shotcreted to a 3 in. (75 mm) placement, a 22% overall rebound was recorded [12.14]. Fiber content in the original mixture (before gunning) was 3.3%, whereas the fiber content in the rebounded material was 4.6%.

12.5.3 Techniques to reduce rebound

The rebound process can be separated into two phases. In the first phase, when the material is projected onto fresh rock surface, anything that improves the adhesion will reduce rebound. Techniques that can be used include higher cement content; addition of more fines such as fine sand, fly ash, or silica fume; smaller maximum-size aggregate; finer gradation and proper wetness of aggregates so that particles are well coated with cement.

In the second phase, when material is projected against the soft initial layer, the most effective way to reduce rebound is to use the wettest consistency that is stable. Too much water can result in drop off. The mix should be made as soft as possible without causing any drop off.

In both phases, lower air pressure, lower air velocity, less air, more fines, shorter or thicker fibers, predamping to get the right moisture content, and shotcreting at the wettest stable consistency lead to lower rebound.

12.6 Physical Properties

12.6.1 Test methods

The test methods used for cast-in-place fiber-reinforced concrete can also be used for fiber shotcrete. The test procedures are described in chapters dealing with mechanical properties. In certain instances a variation may be allowed for shotcrete. For example, thinner beams than $4 \times 4 \times 14$ in. ($100 \times 100 \times 350$ mm) can be used to obtain flexural properties if the thickness of the field-applied shotcrete is less than 4 in. (100 mm). The details can be found in appropriate ASTM standards.

12.6.2 Compressive strength

The compressive strength of shotcrete is primarily governed by the compressive strength of the matrix. The inclusion of fibers, especially at higher volume percentages, may affect the compaction, resulting in lower compressive strength. Entrapped air also becomes a problem at higher fiber concentrations. For the mix proportions shown in Table 12.1, the 28 day compressive strength varied from 4200 to 7500 psi (29 to 52 MPa). Fiber addition was reported both to increase and decrease the compressive strength of the matrix. Four specific investigations, presented below, typify variations that can be expected.

12.6.2.1 Study by Ramakrishnan et al. [12.10]. Three types of fibers were considered in this investigation. The mix proportion for the matrix, fiber volume fractions, and the geometric details of the fibers are presented in Tables 12.3 and 12.4. The cement content was kept constant and the sand content was slightly reduced for fiber mixes. Low-carbon-steel coil fibers, fibers with hooked ends, and fibers from tire cord (brass-coated) were used at volume fractions of 0.6, 1.0, and 1.3 percent.

ASTM Type III cement and natural sand were used for all the mixes. Dry mix process was used to shotcrete test panels that were $22 \times 26 \times 1.5$ in. ($559 \times 660 \times 38$ mm) and $24 \times 26 \times 3$ in. ($610 \times 660 \times 75$ mm). The panels were kept covered under burlap for 24 hours and cured until the day of testing at ages of 2, 7, 14, and 28 days.

The compression test was conducted using cubes cut from test panels. Their dimensions were either $1.5 \times 1.5 \times 1.5$ in. or $3 \times 3 \times 3$ in. ($38 \times 38 \times 38$ mm or $75 \times 75 \times 75$ mm). The cubes were loaded perpendicular to the direction in which the concrete was shot. The test results are shown in Tables 12.5 and 12.6 for 1.5 in. and 3 in. (38 and 75 mm) specimens, respectively.

TABLE 12.3 Basic Mix Proportions [12.10]

Fiber volume fraction	Fibers (lb/yd³ [kg/m³])	Cement (lb/yd³ [kg/m³])	Fine aggregate (lb/yd³ [kg/m³])
0	0	745	2730
		(442)	(1620)
0.6%	78.3	745	2695
	(46.5)	(442)	(1598)
1.0%	132.3	745	2678
	(78.5)	(442)	(1589)
1.3%	172.8	745	2662
	(102.5)	(442)	(1579)

Notes: 1. All weights are saturated surface dry
2. Assumed water-cement ratio = 0.46
3. Assumed entrapped air = 5%

TABLE 12.4   Mix Designations, Number of Panels, and Fiber Types [12.10]

Mix designation	Fiber content (lb/yd³ [kg/m³])		Number of panels made	
			1.5 in. (38 mm)†	3 in. (75 mm)†
N₁	Plain Mix number 1		2	2
N₂	Plain Mix number 2		1	1
ZP 30/50–0.6%	78.3	(46.5)	2	2
ZP 30/50–1.0%	132.3	(78.5)	2	2
ZP 30/50–1.3%	172.8	(102.5)	2	2
ZP 30/40–0.6%	78.3	(46.5)	1	1
ZP 30/40–1.0%	132.3	(78.5)	1	1
FC–0.6%	78.3	(46.5)	2	2
FC–1.0%	132.3	(78.5)	2	2
TC–0.6%	78.3	(46.5)	-	2
TC–1.0%	132.3	(78.5)	2	2

*ZP = Hooked-end steel fibers
FC = Steel coil fibers
TC = Tire cord steel fibers
†Specimen thickness

From Tables 12.5 and 12.6 one can see that variation of compressive strength within the two plain mixes is high. Since water is controlled by the nozzle operator, the variation in water content can be considerable, resulting in variation in strengths. Thus, fiber addition can be assumed to have little effect on compressive strength. The trends in strength gain with time seem to be the same for fiber and plain mortar mixes.

12.6.2.2   Study by Morgan [12.15].   Mix proportions for this study are shown in Table 12.7. Mixes with and without coarse aggregates were investigated. The hooked-end steel fibers used were 30 mm (1.2 in.) long and 0.4 mm (0.016 in.) in diameter. The test panels were shotcreted using dry mix process. Panel dimensions were 25 × 25 × 5 in. (635 × 635 × 125 mm). The specimens were moist cured until the day of testing.

Compression strength tests were conducted using 4 × 4 × 4 in. (100 × 100 × 100 mm) cubes. The results presented in Table 12.8 indicate that fiber inclusion slightly increases the compressive strength at both 7 and 28 days. Mix C, containing coarse aggregate, had higher compressive strength than Mix E, containing no coarse aggregate. This again can be the result of more water being used for the all-sand mix.

12.6.2.3   Study by Opsahl, Buhre, and Hornfeldt [12.8].   This report in which wet-process was used for shotcreting, indicates a consistent increase of compressive strength with the addition of fibers. The addition

TABLE 12.5 Compressive and Flexural Strengths of Thinner Specimens [12.10]

Mix designation	Comp. strength			Cracking flex. strength			Ultimate flex. strength		
	2 day (psi)	7 day (psi)	28 day (psi)	2 day (psi)	7 day (psi)	28 day (psi)	2 day (psi)	7 day (psi)	28 day (psi)
N_1	3869	5586	5865	671	879	858	671	879	858
N_2	—*	6299	7494	—	838	908	—	838	908
ZP 30/50–0.6%	4248	5593	7643	666	703	832	677	703	832
ZP 30/50–1.0%	3700	5767	7023	736	977	1017	757	1166	1077
ZP 30/50–1.3%	5018	5316	5674	927	848	938	1180	1040	1087
ZP 30/40–0.6%	—	4837	5273	—	794	859	—	794	859
ZP 30/40–1.0%	—	4013	3010	—	773	876	—	826	923
FC–0.6%	6202	6367	8853	799	811	1086	799	811	1086
FC–1.0%	4697	6959	6957	770	1061	1091	770	1061	1091
TC–1.0%	3297	5210	5813	799	1089	1328	779	1089	1378

Specimens: cube size: ($1\frac{1}{2} \times 1\frac{1}{2} \times 1\frac{1}{2}$) in.; beam size: ($1\frac{1}{2} \times 1\frac{1}{2} \times 6\frac{1}{2}$ in.)
1 in. = 25.4 mm; 1 psi = 6.895 kPa
Values reported are average of three tests.
*—not tested

TABLE 12.6 Compressive and Flexural Strength of Thicker Specimens [12.10]

Mix designation	Comp. strength			Cracking flex. strength			Ultimate flex. strength		
	7-day (psi)	14-day (psi)	28-day (psi)	7-day (psi)	14-day (psi)	28-day (psi)	7-day (psi)	14-day (psi)	28-day (psi)
N$_1$	6491	–*	8283	514	–	657	514	–	657
N$_2$	7103	–	8032	585	–	732	585	–	732
ZP 30/50–0.6%	6672	–	8974	601	–	769	603	–	828
ZP 30/50–1.0%	5148	6334	6466	661	682	728	663	703	801
ZP 30/50–1.3%	4235	4536	6008	632	706	679	744	800	827
ZP 30/40–0.6%	3734	5562	5475	528	637	617	528	637	617
ZP 30/40–1.0%	4027	4777	5308	481	502	593	525	543	621
FC–0.6%	7379	7541	8192	670	696	842	670	696	842
FC–1.0%	6905	6938	7560	731	739	872	731	739	838
TC–0.6%	7720	8203	9905	676	752	809	676	752	809
TC–1.0%	7191	7397	8867	702	783	763	644	795	766

Specimens: cube size: (3 × 3 × 3 in.); beam size: (3 × 3 × 12 in.)
1 in. = 25.4 mm; 1 psi = 6.895 kPa
Values reported are average of three tests.
*—not tested

TABLE 12.7 Shotcrete Mix Proportions [12.15]

Mix number	A		B		C		D		E	
Description	Plain		0.5 percent fiber		1.0 percent fiber		1.5 percent fiber		1.0 percent fiber (sanded)	
	kg/m³	lb/yd³	kg/m³	lb/yd³	kg/m³	lb/yd³	kg/m³	lb/yd³	kg/m³	lb/yd³
Cement Type 10	448	755	448	755	448	755	448	755	448	755
⅜ in. aggregate	702	1183	702	1183	702	1183	702	1183	–*	—
Concrete sand	880	1485	880	1485	880	1485	880	1485	1582	2668
Superfine blend sand	180	302	180	302	180	302	180	302	180	302
Hooked-end steel fiber	—	—	39	66	78	132	117	198	78	132

*—no coarse aggregate

TABLE 12.8   Compressive Strength [12.15]

Mix Designation	Description	Compressive strength			
		7 days		28 days	
		MPa	psi	MPa	psi
A	Plain (control)	30.3	4395	35.3	5125
B	0.5 percent fiber	32.1	4655	50.0	7245
C	1.0 percent fiber	31.5	4570	47.3	6865
D	1.5 percent fiber	36.9	5350	57.9	8400
E*	Sanded, 1.0 percent fiber	23.7	3435	38.8	5630

Sanded means the mix contains only sand (Table12.7).

of silica fume was found to increase the compressive strength of both plain matrix and fiber-reinforced matrix. The increase due to fiber addition was higher for matrices containing silica fume than for matrices with no silica fume, but the increase is not very significant.

12.6.2.4 Study by Zellers [12.7]. The addition of polypropylene fibers was found to decrease the compressive strength, particularly for volume fractions exceeding 0.4 percent. Longer fibers reduced the strength more than shorter fibers. The investigation was carried out using 0.75 to 2.25 in. (19 to 57 mm) long fibrillated fibers.

12.6.2.5 Summary. All the investigations support the initial statement that the addition of fibers does not affect compressive strength considerably. For design purposes, the compressive strength of fiber shotcrete can be considered to be the same as that of the matrix unless a high volume fraction of fibers or special manufacturing techniques are used. Higher fiber volume fractions can result in substantially lower compressive strength.

12.6.3 Flexural strength

Fibers tend to orient parallel to the surface being shot. Hence, structural elements fabricated using shotcreting technique tend to have fibers aligned along the length (surface). This aspect is highly beneficial for improvement of both flexural strength and flexural ductility.

For flexural strength two values are significant. The first, known as first-crack strength, is primarily controlled by the matrix. The second, known as the ultimate flexural strength or the modulus of rupture, is determined by the maximum load that can be attained. For plain concrete matrices first-crack strength and modulus of rupture are the same. Fiber-reinforced composites could have moduli of rupture higher than first-crack strengths if sufficient volume fractions of fibers are incorporated.

Typical 28 day flexural strengths, determined by using $4 \times 4 \times 14$ in. ($100 \times 100 \times 350$ mm) beams tested using third-point loads, vary from 800 to 1500 psi (5.5 to 10.3 MPa) [12.10, 12.15].

Flexural strengths for shotcrete with mix proportions shown in Table 12.3 are presented in Tables 12.5 and 12.6 [12.10]. From them one can see that fiber composites have higher first-crack and ultimate strengths in all but one case. The increases in flexural strength are significant, especially in comparison with compressive strengths.

Table 12.9 summarizes the flexural strength results reported by Morgan [12.15]. Mix proportions used are shown in Table 12.7. Specimens tested were $4 \times 4 \times 14$ in. ($100 \times 100 \times 350$ mm) prisms. Again, fiber composites had higher strengths in all cases.

Specimens cast using wet-process registered even greater increases in flexural strength [12.8]. Specimens containing silica fume (10% by weight of cement) and 2% volume fibers had a flexural strength of 2000 psi (14 MPa). Strengths as high as 2900 psi (20 MPa) have been reported in the literature [12.2].

In composites reinforced with polypropylene fibers the increase in flexural strength is marginal [12.7]. The fibers provide excellent ductility. However, the postcrack strength is typically less than first-crack strength. Reports on polymeric fiber-reinforced composite are extremely limited, and definitive conclusions cannot be drawn regarding the behavior of these fibers.

12.6.4 Flexural toughness and load-deflection behavior

Flexural ductility (toughness) is an important parameter for structural design. One of the primary reasons for using fibers in concrete is their ability to improve ductility and energy absorption. All the in-

TABLE 12.9 Modulus of Rupture (Flexural Strength) [12.15]

Mix Designation	Description	7 days MPa	7 days psi	28 days MPa	28 days psi
A	Plain (control)	4.47	647	5.55	805
B	0.5 percent fiber	4.96	718	5.21*	754*
C	1.0 percent fiber	4.82	699	5.94	861
D	1.5 percent fiber	4.81	697	7.42	1076
E†	Sanded, 1.0 percent fiber	4.54	659	5.43	786

*Specimens contained sand lens
† *Sanded* means the mix contains only sand

vestigations on fiber shotcrete to date indicate that fiber contribution is substantial in post-crack resistance.

Typical load-deflection curves for four different fiber types and for plain matrix are shown in Figure 12-5 [12.10]. Basic mix proportions are listed in Table 12.3. From this figure one can see that fibers with hooked ends (B and C) and tire cord fibers (E) perform better than steel coil (straight) fibers (D). Figure 12-6, which is a plot of the effect of fiber volume fraction, indicates that a fiber content of about 1 percent is needed to provide good postcracking resistance [12.4]. This percentage is valid only for hooked-end fibers that are 30 mm (1.18 in.) long. The optimum volume fraction for postcrack ductility depends on fiber properties, volume fraction, and matrix composition. For example, the matrix containing silica fume provides higher density and better bond; the same fiber volume can yield different results compared with matrix containing no silica fume.

Composites containing silica fume typically show less postcrack ductility than plain cement matrix [12.16]. Part of the reduction in ductility can be attributed to improved bonding and simultaneous fracture of a large number of fibers.

Toughness is measured using the load-deflection curves of beams. As explained in Chapter 4, toughness can be quantified using the area under the curve (Japanese standard) or standardized with respect to the area under the curve up to first crack (ASTM). Most investigators have reported the toughness index of fiber shotcrete using the

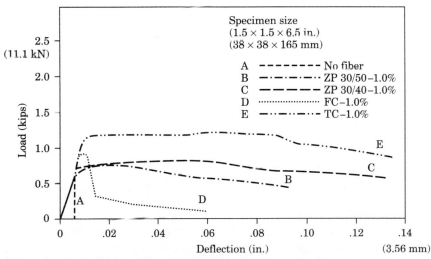

Figure 12-5 Load-deflection curves for different fiber types [12.10].

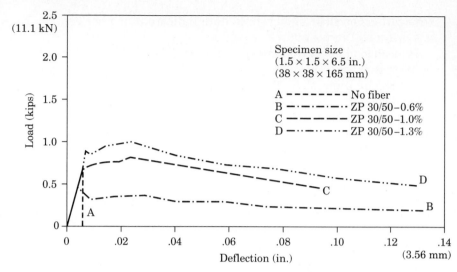

Figure 12-6   Load-deflection curves for hooked-end fibers at three volume fractions [12.10].

old ACI recommendation. In this procedure, the toughness index TI is computed as a ratio, defined by the following equation:

$$TI = \frac{\text{Area under the load-deflection curve up to a deflection of 0.075 in. (1.9 mm)}}{\text{Area under the load-deflection curve up to the first crack}}$$

The TI value for plain concrete is taken as 1.0.

The TI for steel fiber concrete is reportedly as high as 50 [12.2]. The TI values for mix proportions presented in Tables 12.3 and 12.7 are shown in Tables 12.10 and 12.11 respectively [12.10, 12.15]. From these tables one can see that hooked-end fibers provide excellent improvement in toughness.

Addition of polypropylene fibers also increases ductility. Longer fibers and higher volume fractions provide higher toughness indices [12.7]. The results reported for fibrillated polypropylene fibers show that I_5, I_{10}, and I_{20} (ASTM definitions, refer to Chapter 4) indices can be improved [12.7].

12.6.5   Tensile strength

Tensile strength tests are not usually conducted for shotcrete composites because the property is seldom used in design. Limited results indicate that tensile strength can be improved by about 10 percent by using fibers [12.8].

TABLE 12.10   Toughness Index [12.10]

Mixture designation	Fiber content (%)	Toughness index	
		1.5 in. (38 mm)*	3 in. (75 mm)*
N₁	0.0	1.0	1.0
FC	0.6	1.0	1.0
FC	1.0	5.0	4.0
TC	0.6	—†	9.8
TC	1.0	14.6	16.0
ZP 30/50	0.6	19.5	15.2
ZP 30/50	1.0	10.2	11.8
ZP 30/50	1.3	23.2	16.6
ZP 30/40	0.6	—	12.8
ZP 30/40	1.0	23.5	15.0

*Specimen cross section
†Not tested

12.6.6   Impact strength

Another important property to which fibers contribute is impact resistance. This can be measured using techniques explained in Chapter 9. Investigators who have evaluated fiber shotcrete have used the ACI procedure (Chapter 9), in which the number of blows required to crack and to shatter a 6 in. (150 mm) diameter disc is recorded. The blows are imparted on the disc using a 10 lb (4.54 kg) weight dropped from a height of 18 in. (450 mm).

Studies of impact resistance in References [12.10] and [12.15] are presented in Tables 12.12 and 12.13 (for corresponding mix proportions see Tables 12.3 and 12.7). In both cases fiber addition enhances performance 5 to 10 times. The improvement is about the same at ages ranging from 2 to 28 days of maturity. Again, hooked fibers were found to perform better than straight fibers. An increase in fiber volume fraction from 0.5% to 1.5% provides consistent improvement in impact resistance.

TABLE 12.11   Toughness Index at 28 Days [12.15]

Mixture designation	Description	Toughness index
A	Plain control	1.0
B	0.5 percent fiber	3.7
C	1.0 percent fiber	5.9
D	1.5 percent fiber	6.7
E	Sanded, 1.0 percent fiber*	10.8

*Sanded means the mix contains only sound.

12.6.7 Shrinkage

The magnitude of shrinkage strains is important in applications where the shotcrete is cast against an existing hardened surface. These include most of the repair jobs and some lining jobs. Typical curves for shrinkage strain versus time are shown in Figure 12-7 [12.10]. The addition of steel fibers reduces the shrinkage strains. Limited data available indicate that fiber shotcrete behavior is similar to that of the cast-in-place concrete (Chapter 10). Since shotcrete mixes contain higher cement content, higher magnitude of shrinkage strains should

TABLE 12.12 Impact Resistance of Shotcrete Reinforced with Different Fiber Types [12.10]

Specimen number	Mix†	N_1	ZP 30/50 0.6%	ZP 30/50 1.0%	ZP 30/50 1.3%	FC 0.6%	FC 1.0%	TC 1.0%
2-day‡ impact resistance								
1	F.CR* / FAIL.	5/8	14/94	34/120	187/712	14/16	13/50	45/99
2	F.CR / FAIL.	3/5	10/67	37/105	221/640	12/23	12/32	37/100
3	F.CR / FAIL.	7/10	10/89	50/420	100/256	=	14/35	29/80
Average	F.CR / FAIL.	5/8	11/83	40/215	169/536	14/19	13/39	37/93
14-day‡ impact resistance								
1	F.CR / FAIL.	4/8	13/96	45/190	104/396	7/23	64/162	80/202
2	F.CR / FAIL.	11/13	14/63	83/380	68/641	9/26	34/61	31/97
3	F.CR / FAIL.	3/6	23/108	60/280	150/611	9/22	143/164	24/142
Average	F.CR / FAIL.	6/9	17/89	63/283	107/549	8/24	80/129	45/147
28-day‡ impact resistance								
1	F.CR / FAIL.	12/14	13/100	79/215	176/412	=	46/70	41/162
2	F.CR / FAIL.	=	16/66	89/426	234/427	15/32	25/53	41/152
3	F.CR / FAIL.	9/13	31/124	77/456	208/531	17/40	56/75	62/151
Average	F.CR / FAIL.	10/13	20/97	82/366	206/457	16/36	42/66	48/155

*Number of blows to F.CR (first crack) and to FAIL (failure)
†For explanation of mixture designations see Table 12.4.
‡Number of days refers to elapsed time since concrete was made.
Specimens tested: 6 in. diameter × 1½ in. height (152 × 38 mm)

TABLE 12.13   Impact Resistance of Shotcrete Reinforced with Hooked-End Fibers at Different Volume Fractions [12.15]

Mixture designation*	Description	Impact resistance			
		7 days†		28 days†	
		Blows to first crack	Blows to failure	Blows to first crack	Blows to failure
A	Plain (control)	20	22	39	41
B	0.5 percent fiber	30	68	119	140
C	1.0 percent fiber	66	112	111	211
D	1.5 percent fiber	141	237	218	280
E	Sanded, 1.0 percent fiber	110	223	117	291

*For explanation of mixture designations see Table 12.7
†Number of days refers to elapsed time since concrete was made.

be expected for shotcrete compared with cast-in-place fiber-reinforced concrete.

12.6.8   Bond strength between shotcrete and rock surface

Bond strengths between fiber shotcrete and a granite rock surface have been obtained using drilled cores or embedded flat steel plates. For cores, a core drill was used to obtain a cylindrical specimen that was then pulled from the rock. The bond strength for fiber shotcrete was about 0.02 times the compressive strength (130 psi, 0.9 MPa) compared with 0.03 to 0.05 times the compressive strength (220 to 375 psi, 1.5 to 2.6 MPa) for plain shotcrete [12.14]. A bond strength of 145 psi (1 MPa) was reported for wet shotcrete [12.2].

12.6.9   Fatigue strength

Limited data available on fatigue strength of fiber shotcrete samples indicate that addition of fibers provides a noticeable improvement in fatigue strength [12.4]. Fiber addition also decreased the crack width under fatigue loading.

12.6.10   Permeability and porosity

Permeability and porosity provide an indication of the durability of a composite. These properties also affect alkali reactivity, leaching characteristics, resistance to chloride or sulfate attack, and corrosion of reinforcement steel. Shotcretes cast using dry process typically have higher permeability than cast-in-place concretes [12.17], whereas

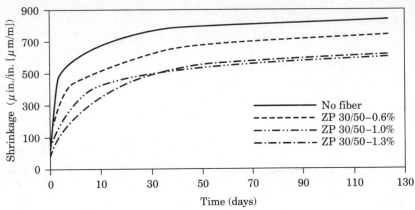

Figure 12-7 Shrinkage curves for different fiber contents. Time reckoned since end of wet curing at the age of 28 days [12.10].

wet-process shotcrete has a very low permeability [12.8]. Mixes with 1 percent fibers and compressive strength of 45 and 75 MPa (7 and 11 ksi) were found to have Darcy's permeability coefficient of 6×10^{-14} and 1×10^{-15} m/sec. Both mixes contained silica fume. Permeability was measured using drilled cores.

12.6.11   Freeze-thaw durability and air-void characteristics

The freeze-thaw durability factor of wet-process shotcrete measured using ASTM C666-77 was found to be 60% to 80% [12.8]. The compressive strength of the composite was in the range of 60 to 70 MPa (9 to 10 ksi) [12.8]. The following air void characteristics were reported for this mix based on two test samples [12.8]:

Air void content:	3.8% (no air-entraining admixture)
Specific surface:	33.6 mm²/mm³
Spacing factor:	0.15 mm
Air voids with diameters up to 100 μm:	0.48%
Air voids with diameters up to 200 μm:	1.21%
Air voids with diameters up to 300 μm:	1.68%

Resistance to deicing chemicals was found to be excellent for high-strength shotcrete and shotcrete containing air-entraining admixtures. Low-strength shotcrete with no air-entraining admixtures was found to have scaling problems [12.8].

12.6.12   Pull-out strength for measuring
quality of in-place concrete

Pull-out tests, used for measuring in-place strength of concrete, have been tried for estimating the compressive strengths of in-place shotcrete. The tests were conducted by pulling out embedded discs that were 1 in. (25 mm) in diameter and 1.25 in. (32 mm) deep [12.13]. A linear relationship between pullout strength and compressive strength is assumed for plain concrete. Data available for steel fiber shotcrete are not sufficient to verify that any relation exists between pull-out strength and compressive strength. Pull-out strengths were found to be about 1000 psi (6.9 MPa) and 1800 psi (12.4 MPa) for plain and fiber-reinforced shotcrete, respectively [12.13]. The plain shotcrete contained 750 lb (340 kg) of cement, 1825 lb (828 kg) of 0.4 in. (10 mm) aggregate, 1175 lb (533 kg) sand and 5 lb/yd^3 (3 kg/m^3) accelerating admixture. The aggregate and sand contents were changed to 1475 lb (669 kg) and 1300 lb (590 kg), respectively for fiber shotcrete. The fiber content was 250 lb/yd^3 (148 kg/m^3). The fiber length and diameter were 0.5 in. (12 mm) and 0.01 in. (.25 mm) respectively. Tests were conducted at the age of 14 days.

12.7   Applications

Fiber-reinforced shotcrete has been used for rock slope stabilization, tunnels, mines, dams, powerhouses, bridge arches, dome structures, rock caverns for oil storage, houses, boat hulls, stabilization of slopes to prevent landslides, repair of deteriorated concrete surfaces, water channels, and waste disposal site covers. The largest single volume use is in underground structures. The composite is used in most industrialized nations including Australia, Canada, England, Finland, Germany, Japan, Norway, Poland, Sweden, and the United States. A few specific applications are discussed in the following sections. These case studies are intended to provide information regarding site conditions, mix proportions, method of application, and special circumstances. Since the design of shotcrete systems is still in the developmental stage, these case studies can be used as a source of information by designers planning to use fiber shotcrete.

12.7.1   Slope stabilization

12.7.1.1   Snake River rock slope stabilization; near Little Goose Dam, Washington, USA [12.18].   This work, completed in 1974, involved the use of steel fiber shotcrete to stabilize a deteriorating section of rock slope near Little Goose Dam along the Snake River. A minimum thickness of 2.5 in. (63 mm) was used for the entire area, which was about

1550 ft (472 m) long and 15 to 45 ft (5 to 14 m) wide. The total area was about 6900 yd^2 (5800 m^2). The shotcreted slab was installed over rock bolts.

12.7.1.2 Joint Nordic Program (Nordforsk); Brofjorden, Sweden [12.19].
About 4500 m^2 (5400 yd^2) of rock surface was stabilized using 30 mm (1.2 in.) layer of steel fiber shotcrete placed over a 5 to 10 mm (0.2 to 0.4 in.) thick plain shotcrete applied to the rock. The steel fiber shotcrete was covered with another 5 to 10 mm (0.2 to 0.4 in.) thick layer of plain shotcrete.

12.7.2 Underground applications

12.7.2.1 Corps of Engineers tunnel adit; Rivie Dam, Idaho, USA [12.20].
The work, completed in 1972, involved lining a 40 ft (12 m) long exploratory tunnel adit on the right abutment of Rivie Dam. The lining thickness was 3 in. (75 mm). The lining successfully withstood the loads and shocks developed during the blasting operation with minor cracks.

12.7.2.2 Peace River Site C hydro Tunnels; British Columbia, Canada [12.21].
In this project, completed in 1982, a 2 in. (50 mm) thick steel fiber shotcrete was used to line several hundred feet of exploratory tunnels and test chambers. The following mix proportion was used for the shotcrete: Type 10 (Canadian) cement, 740 lb/yd^3 (439 kg/m^3); concrete sand, 1927 lb/yd^3 (1143 kg/m^3); 0.4 in. (10 mm) maximum size coarse aggregate, 610 lb/yd^3 (362 kg/m^3); fine blend sand, 376 lb/yd^3 (223 kg/m^3); and steel fibers, 100 lb/yd^3 (60 kg/m^3).

12.7.2.3 Atlanta subway tunnel lining; Georgia, USA [12.14].
A 200 ft (61 m) length of the subway tunnel was lined with 4 to 6 in. (100 to 150 mm) thick steel fiber-reinforced shotcrete by the dry-mix process. Field inspection review after 18 months showed satisfactory performance.

12.7.2.4 Bolidens Gruv AB—mines and ore shaft, Sweden [12.22].
Fiber shotcrete was used to line and stabilize a gravity ore-transfer shaft that was deteriorating from the impact of the ore. The thickness of shotcrete varied from 4 to 20 in. (100 to 500 mm).

12.7.2.5 Sydvatten AB; Malmö, Sweden [12.8].
Wet process shotcrete was used to provide support for a tunnel. In some stages a robot rig was used for shotcreting. The compressive strength of the composite was about 5000 psi (34 MPa).

12.7.2.6 British Rail arch and tunnel relining; England. Steel fiber–reinforced shotcrete was used for strengthening tunnel and brick arches under bridges. The steel fiber composite, which was 6 in. (150 mm) thick, was covered with a 0.5 in. (12 mm) coat of plain shotcrete.

12.7.2.7 British Columbia Railway; Canada [12.6]. Several kilometers of new tunnels constructed through the Rocky Mountains were lined with steel fiber shotcrete. A large part was shotcreted using bulk bin bag (mixed materials in bulk) materials. The pre-dry-bagging (mixed materials available in dry form) facilitated the continuation of the project even when outside temperatures dropped to as low as $-22°F$ $(-30°C)$.

12.7.2.8 Roadway tunnels; Japan. A number of tunnels have been lined with fiber shotcrete. In the Itaya tunnel, a 50-year-old deteriorated lining was repaired using 4 in. (100 mm) thick steel fiber shotcrete. The shotcrete was placed using wet-process and a squeeze-type pump.

12.7.3 Shell structures

Steel fiber–reinforced shotcrete has been successfully used to build dome-shaped structures. The form is usually a polyurethane foam sprayed onto the underside of an inflatable membrane. The shotcrete is applied both inside and outside of the foam to create a stiff fiber-composite shell [12.23, 12.24, 12.25]. The thickness of the foam was about 4 in. (100 mm) and the thickness of the fiber composite varied from 1.5 to 3 in. (38 to 75 mm). The resulting structure is both structurally and thermally efficient. These domes could be used for homes, offices, warehouses, and for storing a variety of products.

Fiber-reinforced shotcrete can also be used to fabricate cylindrical tanks using this process.

12.7.4 Repairs

Fiber-reinforced shotcrete has been used for a number of repair jobs. In Sweden, a lighthouse damaged by freezing and thawing and the interior of the chimney were repaired with steel fiber shotcrete.

In Australia fiber shotcrete was used to repair an eroded roof in a concrete bunker, reline a steel storage bin, and line curved sections of a stormwater drain.

Steel fiber shotcrete has also been used to repair marine structures in Canada, and resurface a rocket flame deflector at Cape Canaveral, Florida, USA. It has also been employed for coal mine strengthening and sealing of stoppings by the National Coal Board, England and for stabilization of the Tuve landslide in Sweden.

12.7.5   Miscellaneous applications

Fiber shotcrete has also been used for building boat hulls, with and without incorporating wire meshes. Recently, polymeric fiber-reinforced shotcrete has been used as a cover material for waste disposal sites in Canada.

12.8   Design Procedures

The basic design procedure for plain shotcrete can be used for fiber-reinforced shotcrete by incorporating the improved mechanical properties. At present most data available for fiber shotcrete are derived from studies on ground support, such as tunnel linings. The primary concern in these applications is to provide support for unstable or loose natural formations.

Both empirical rules and analytical models have been formulated to predict shotcrete-rock behavior for thin shotcrete linings. These are briefly discussed below.

12.8.1   Empirical design

Empirical rules for estimating plain shotcrete thickness were developed using the performance of shotcrete applied in various instances [12.26] and for various site conditions [12.27–12.32]. Some of these empirical rules take into account the quality of the rock on which shotcrete is placed.

Data on steel fiber-reinforced shotcrete are given in Reference [12.33]. Performance of steel fiber shotcrete and wire-mesh reinforced shotcrete was compared. Fiber shotcrete provided good residual capacity even at deformations of 2 in. (50 mm); and its behavior is comparable to wire-mesh reinforced shotcrete. Wire-mesh reinforcement is more expensive because it must be placed in position and anchored properly. Postcrack behavior of fiber shotcrete was also found to be similar to that of wire-mesh reinforced shotcrete [12.21].

12.8.2   Analytical models

Analytical models have been tried to simulate the behavior of shotcretes by modeling them as either flat rectangular or circular plates [12.26, 12.29, 12.34–12.36]. In general, a minimum thickness of 2 in. (50 mm) is suggested to avoid failures due to deterioration caused by shrinkage cracks and overloads occurring due to construction and water accumulation. The analytical methods need to be refined further for use in practical design.

Analytical models based on large-scale laboratory testing and field tests have also been proposed for fiber-reinforced shotcrete [12.37–

12.39]. A number of assumptions have to be made to convert the actual structure to a designable structural system.

12.8.3   Precautions

The following precautions should be considered in designing shotcrete support systems.

Shotcrete may be used as the sole support of underground excavations only in cases where a good shotcrete-rock bond can be obtained. The shotcrete should also be able to act as a structurally continuous lining. When a shotcrete layer is acting as an arch, the effects of thickness variation and variation in centerline geometry should be considered in the design. When extra supports are needed, these can be provided in the form of rock bolts, steel ribs, and so on.

The sealing action of shotcrete can build up hydraulic pressure and, hence, proper drainage should be provided.

Large openings such as those created by blasts should not be spanned by fiber shotcrete without continuous bar reinforcement. Failures were reported on openings larger than 20 ft. (6.1 m) covered by shotcrete.

In locations where rock surfaces are smooth, extra anchors should be provided.

For vertical side walls more than 10 ft (3 m) in height, anchors should be provided.

12.9   References

12.1   ACI Committee 506. "Guide to Shotcrete," ACI 506R. 90, American Concrete Institute, Detroit, 1990, 41pp.

12.2   ACI Committee 506. "State-of-the-Art Report on Fiber Reinforced Shotcrete," *Concrete International: Design and Construction,* Vol. 6, No. 1, 1984, pp. 15–27.

12.3   ACI Committee 506. "Specification for Materials, Proportioning and Application of Shotcrete," ACI 506.2-90, *American Concrete Institute,* Detroit, Michigan, 1990, 8 pp.

12.4   Ramakrishnan, V. "Steel Fiber Reinforced Shotcrete, a State-of-the-Art Report," *Steel Fiber Concrete,* S. P. Shah and A. Skarendahl (eds.), Elsevier, New York, 1986, pp. 7–24. (Also published by Swedish Cement and Concrete Research Institute, Stockholm, 1985.)

12.5   Glassgold, I. L. "Refractory Shotcrete: Current State-of-the-Art," *Concrete International: Design and Construction,* Vol. 3, January 1981, pp. 41–49.

12.6   Morgan, D. R.; and McAskill, N. "Rocky Mountain Tunnels Lined with Steel Fiber Reinforced Shotcrete," *Concrete International: Design and Construction,* Vol. 6, No. 12, 1984, pp. 33–38.

12.7   Zellers, R. C. "High Volume Applications of Collated Fibrillated Polypropylene Fiber," *Fiber Reinforced Cements and Concretes: Recent Developments,* Elsevier, New York, 1989, pp. 316–325.

12.8   Opsahl, O. A.; Buhre, K. E.; and Hornfeldt, R. "Why Wet Process Steel Fiber Reinforced Shotcretes," in *Steel Fiber Concrete,* S. P. Shah and A. Skarendahl (eds). Elsevier, New York, 1986, pp. 51–65. (Also published by Swedish Cement and Concrete Research Institute, Stockholm, 1985.)

12.9   Sandell, N. O.; Dir, M.; and Westerdahl, B. "System BESAB for High Strength Steel Fiber Reinforced Shotcrete," in *Steel Fiber Concrete,* S. P. Shah and A.

Skarendahl (eds.), Elsevier, New York, 1986, pp. 25–39. (Also published by Swedish Cement and Concrete Research Institute, Stockholm, 1985.)

12.10 Ramakrishnan, V.; Coyle, W. V.; Dhal, L. F.; and Schrader, E. K. "A Comparative Evaluation of Fiber Shotcrete," *Concrete International: Design and Construction,* Vol. 3, No. 1, 1981, pp. 56–69.

12.11 Parker, H. W.; Fernandez, G.; and Lorig, L. J. "Field-Oriented Investigation of Conventional and Experimental Shotcrete for Tunnels," Report No. FRA-OR&D 76-06, Federal Railroad Administration, Washington, D.C., August 1975, 628 pp.

12.12 Ryan, T. F. "Steel Fibers in Gunite, An Appraisal," *Tunnels and Tunneling (London),* July 1975, pp. 74–75.

12.13 Henager, C. H. "The Technology and Uses of Steel Fibrous Shotcrete: A State-of-the-Art Report," Battelle-Northwest, Richland, Washington, September 1977, 60 pp.

12.14 Rose, D. C., et al. "Applied Research for Tunnels," *The Atlanta Research Chamber,* Report No. UMTA-GA-06-0007-81-1, U.S. Department of Transportation, Washington, D.C., March 1981.

12.15 Morgan, D. R. "Steel Fiber Shotcrete—A Laboratory Study," *Concrete International: Design and Construction,* Vol. 3, No. 1, 1981, pp. 70–74.

12.16 Ostfjord, S. "Lack of Practicable Standards and Test-Methods Restrict the Development of Steel Fiber Shotcrete," in *Steel Fiber Concrete,* S. P. Shah and A. Skarendahl (eds.), Elsevier, New York, 1986, pp. 41–50. (Also published by Swedish Cement and Concrete Research Institute, Stockholm, 1985.)

12.17 Krantz, G. W. "Selected Pneumatic Gunite for Use in Underground Mining: A Comparative Engineering Analysis," Bureau of Mines Information Circular, IC8984, U.S. Department of Interior, 1984.

12.18 Kaden, R. A. "Slope Stabilized with Steel Fibrous Shotcrete," *Western Construction,* April 1974, pp. 30–33.

12.19 Malmberg, B.; and Ostfjord, S. "Field Test of Steel Fibre Reinforced Shotcrete at San-Raff, Brofjorden," Fiberbetong, Nordforsk Projekt Committee for FRC-Material Delvapporter, Cement Och Betonginstitutet, Stockholm, 1977, pp. 1–16.

12.20 Kaden, R. A. "Fiber Reinforced Shotcrete Rivie Dam and Little Goose (CPRR) Relocation," *Shotcrete for Ground Support,* American Concrete Institute/American Society of Civil Engineers, Detroit, Michigan, 1977, pp. 66–88.

12.21 "Peace River Development Site C Project, Shotcrete Testing," Hydroelectric Generation Projects Division, Geotechnical Department, British Columbia Hydro, Vancouver, Canada, January 1983.

12.22 Sandell, B. "Steel Fiber Reinforced Shotcrete (Steltfiberarmerad Sprutbeton)," Proceedings, Information-Dagen 1977, Cement-Och Betonginstitutet, Stockholm, 1977, pp. 50–75.

12.23 Wilkinson, B. M. "Foam Domes, High Performance Environmental Enclosures," *Concrete Construction,* Vol. 23, No. 7, July 1978, pp. 405–406.

12.24 _____Shotcrete and Foam Insulation Shaped over Inflated Balloon Form," *Concrete Construction,* Vol. 27, No. 6, June 1982, pp. 511–513.

12.25 Nelson, K. O.; and Henager, C. H. "Analysis of Shotcrete Domes Loaded by Dead-weight," Preprint No. 81-512, ASCE Convention, St. Louis, October 1981.

12.26 Mahar, J. W.; Parker, H. W.; and Wuellner, W. W. "Shotcrete Practice in Underground Construction," Report No. FRAOR&D 75-90, Federal Railroad Administration, Washington, D.C., August 1975, 482 pp.

12.27 Alberts, C. "Berforstakning genom Beton sputtering och Injecterine", Proceedings 1965 Rock Mechanics Symposium, Publication No. 142, Swedish Academy of Sciences, Stockholm, 1965.

12.28 Kohler, H. G. "Dry-Mix Coarse-Aggregate Shotcrete as Underground Support," *Shotcreting,* American Concrete Institute, Detroit, Michigan, SP-14, 1966, pp. 33–58.

12.29 Cecil, O. S. "Correlations of Rock Bolt-Shotcrete Support and Rock Quality Parameters in Scandinavian Tunnels," Ph.D. Thesis, University of Illinois, Urbana, 1970.

12.30 Hener, R. E. "Selection/Design of Shotcrete for Temporary Support," *Use of Shotcrete for Underground Structural Support,* SP-45, American Concrete Institute/American Society of Civil Engineers, Detroit, Michigan, 1974, pp. 160–174.

12.31 Deere, E. U.; Peck, R. B.; Monsees, N. E.; and Schmidt, B. "Design of Tunnel Liners and Support Systems," Contract No. 3-0152 (NTIS PB 183 799), Office of High Speed Ground Transportation, U.S. Department of Transportation, Washington, D.C., 1969, pp. 387–391.

12.32 Wickham, G. E.; Tiedemann, H. R.; and Skinner, E. H. "Ground Support Prediction Model RSR Concept," Proceedings, North American Rapid Excavation and Tunneling Conference, American Institute of Mining, Metallurgical, and Petroleum Engineers, New York, 1974, Vol. 1, pp. 691–707.

12.33 Morgan, D. R. "Report on Steel Fibre Shotcrete for Tunnel Support Lining," Hardy Associates, Ltd., Vancouver, Canada, March 1981.

12.34 Peng, S. S. *Coal Mine Ground Control,* John Wiley & Sons, New York, 1978, pp. 415–416.

12.35 Rabcewicz, L. "The New Austrian Tunneling Method, Parts I, II, III," *Water Power* (London), November, December 1964, and January 1965.

12.36 Rabcewicz, L. "Stability of Tunnels Under Rock Roads, Parts I, II, III," *Water Power* (London), June 1969, pp. 225–234; July 1969, pp. 266–273; and August 1969, pp. 297–302.

12.37 Fernandez-Delgado, G.; Mahar, J. W.; and Parker, H. W. "Structural Behavior of Thin Shotcrete Liners Obtained from Large Scale Tests," *Shotcrete for Ground Support,* SP-54, American Concrete Institute/American Society of Civil Engineers, Detroit, Michigan, 1977, pp. 399–442.

12.38 Holmgren, J. "Thin Shotcrete Layers Subjected to Punch Loads," *Shotcrete for Ground Support,* SP-54, American Concrete Institute/American Society of Civil Engineers, Detroit, Michigan, 1977, pp. 413–459.

12.39 Fernandez-Delgado, G.; et al. "Thin Shotcrete Linings in Loosening Rock," *The Atlanta Research Chamber,* Report No. UMTA-GA-06-0007-81-1, U.S. Department of Transportation, Washington, D.C., March 1981.

Glass Fiber–Reinforced Concrete

Glass fiber–reinforced concrete (GFRC) is a general term used to describe all types of cement composites reinforced with glass fibers. The matrix can be concrete, mortar, or cement paste with additives. Glass fiber–reinforced concrete is extensively used for architectural cladding, for which the fibers are predominantly made of alkali-resistant glass (AR-glass). The favored manufacturing process is the spray-up process, in which continuous strands of glass fiber are fed into a compressed air–powered spray gun and are chopped into predetermined lengths. The chopped glass strands and a cement mortar slurry are sprayed onto a mold surface. The typical fiber content is 5% by weight of the composite. The fiber lengths generally vary from 1.5 to 2.0 in. (38 to 50 mm).

Glass fiber–reinforced concrete can also be made using a premix process. In this process, the fiber length and content in the fresh matrix has to be limited because of workability problems. Polymer-modified matrix systems have also been used with conventional borosilicate glass (E-glass) fibers. New types of cements that provide better long-term durability are being developed.

The primary concern in the development of GFRC is the durability of the glass fibers embedded in the highly alkaline concrete matrix. Most of the research efforts are focused on the development of fiber and matrix compositions whose long-term durability and effectiveness are ensured.

This chapter presents the basic information regarding GFRC composition, manufacturing, composite behavior, and applications. Attention is focused on thin-sheet products that are widely used for various applications including cladding, ducts and shafts, and silos and storage tanks.

13.1   Development of GFRC

In the beginning, researchers used conventional borosilicate glass fibers (E-glass) and soda-lime-silica glass fibers (A-glass) to reinforce brittle cement matrices. These fibers had very high tensile strengths (450 to 500 ksi, 280 to 3500 MPa) and relatively high elastic moduli (9400 to 10,400 ksi 3.1 to 3.5 GPa). Hence the resulting fiber cement composite had a high modulus of rupture (3 to 4 times the matrix strength) and good ductility. However, the fibers were found to lose their reinforcing effectiveness rather quickly. Both E- and A-glass fibers, whose typical chemical compositions are shown in Table 13.1 were found to lose their strength when placed in a typical cement-based matrix of high alkalinity (pH $\geq$ 12.5) [13.1]. Since this kind of composite is not suitable for long-term use, researchers began to look for ways to improve the long-term durability of the fibers.

The researchers took the two logical developmental paths: one involved the development of fibers that are durable in an alkaline environment and the other focused on the development of a less-alkaline matrix. The first approach led to the development of alkaline-resistant glass (AR-glass) fibers. The second approach led to the development of a polymer-modified mortar matrix and the use of less-alkaline special cements. The following sections provide a brief description of these developments. It should be noted that a large number of researchers focused their attention in developing test methods for accelerated aging and improving the long-term durability of the fibers [13.2–13.32].

TABLE 13.1   Chemical Composition of Selected Glass Fibers

Component	A-Glass (%)	E-Glass (%)	Cem-FIL* AR-Glass (%)	NEG† AR-Glass (%)
SiO_2	73.0	54.0	62.0	61.0
Na_2O	13.0	—	14.8	15.0
CaO	8.0	22.0	5.6	—
MgO	4.0	0.5	—	—
K_2	0.5	0.8	—	2.0
Al_2O_3	1.0	15.0	0.8	—
Fe_2O_3	0.1	0.3	—	—
B_2O_3	—	7.0	—	—
ZrO_2	—	—	16.7	20.8
TiO_2	—	—	0.1	—
Li_2O	—	—	—	1.0

*Trade name for product by Pilkington Brothers Ltd, U.K. (Cem-FIL)
†Nippon Electric Glass (NEG)

13.1.1   Alkali-resistant glass fibers

It was found that if a minimum of 16% zirconia was incorporated into the glass composition, the durability of the fiber in an alkaline environment improved considerably. Typical composition of AR-glass is shown in Table 13.1. The physical properties of the modified glass did not change much, as shown in Table 13.2. Early commercial developments were undertaken by Pilkington Brothers Ltd. in the United Kingdom, Nippon Electric Glass Company Ltd. in Japan, and Owens-Corning Fiberglass in the United States.

At present, AR-glass fiber–reinforced composites are by far the ones most widely used for the manufacture of GFRC products. More details regarding the long-term durability of fibers are presented later in this chapter.

13.1.2   Modified-cement matrices

Efforts were also made to modify the cement matrix so that it would be less hostile to E-glass fibers. These attempts included the use of polymers, silica fume, and low-alkaline cements; they had varying levels of success [13.2–13.5]. For polymer-modified GFRC, a certain minimum amount of polymer is added to the matrix. The polymer is intended to coat the glass fibers and hence protect them from the alkaline environment of concrete. The use of silica fume is intended to provide a similar protective system. Specially formulated cements have been shown to provide the best solution to date [13.5]. Comparisons of the various systems are presented in Section 13.5.1.5.

13.2   Constituent Materials of GFRC: Special Requirements

The materials used for the manufacture of GFRC are cement, aggregates, water, admixtures, and fibers. The materials should gener-

TABLE 13.2   Properties of Selected Glass Fibers

Property	A-Glass	E-Glass	Cem-FIL AR-Glass	NEG AR-Glass
Specific gravity	2.46	2.54	2.70	2.74
Tensile strength (ksi)	450	500	360	355
Modulus of elasticity (ksi)	9400	10,400	11,600	11,400
Strain at break (%)	4.7	4.8	3.6	2.5

1 ksi = 6.895 MPa

ally conform to appropriate ASTM standards, and the fiber properties should be appropriate for the particular application. The following sections provide a brief review of the constituent materials used in common GFRC products.

13.2.1 Cement

The portland cement should conform to ASTM C150, *Specifications for Portland Cement.* Type I cement (Chapter 5) is recommended for general use. Type III cement may be used when early strength development, smaller cement particle size, or both are needed. Special cements could also be used for unique application requirements.

If GFRC is to be used for architectural cladding, cement from the same source (preferably the same batch) may be needed for matching color uniformity with the architectural face mix. This factor becomes more important if white cement and pigments are used to obtain the specified colors in the face mix.

13.2.2 Aggregates

The primary aggregate in GFRC is silica sand. Natural sands with rounded particles were found to be easier to pump; they cause less blockage than crushed sands. Washed-and-dried silica sand meeting the requirements of ASTM C144 "Specifications for Aggregate for Masonry Mortar" is recommended for the spray-up process. Typically, silica content should be in the range of 96% to 98%, and organic matter (including clay) should be less than 0.5%.

For GFRC made using the spray-up process, it is advisable to limit the maximum particle size to 1/16 in. (1.18 mm), or the sand should pass through ASTM U.S. No. 16 sieve. Excessive fines should be avoided because they increase water demand. Fines passing through U.S. No. 100 (150 mm) sieve should be limited to 2%.

If improved long-term strength and ductility are needed, only alkali-resistant (AR) glass fibers should be used [13.2]. It should be noted that even AR fibers lose their reinforcing effectiveness after long exposure times, but their durability is substantially better than unmodified E-glass GFRC. E-glass fibers should be used only in special modified-matrix formulations that provide better long-term durability. Typical fiber lengths for the spray-up process range from 1.5 to 2 in. (38–50 mm). For special situations, shorter fibers have been used.

13.2.3 Admixtures

A standard necessity for all GFRCs is a water-reducing admixture. It could be a normal water-reducing agent or a high-range water-reducing admixture (superplasticizer). These admixtures reduce water demand, resulting in a composite with higher strength and better shrinkage characteristics, which are necessary for sprayable slurries in the spray-up process. Sometimes, acrylic thermoplastic copolymer dispersions are also used to improve strength and to reduce permeability and shrinkage. The addition of polymer latex also reduces the water demand.

The other chemical admixtures used in GFRC include accelerators, retarders, and air-entraining admixtures (premix only). Accelerators are used to gain early strength and early form removal. The admixture is also helpful when the work is to be performed on existing vertical or steep walls. Accelerators containing calcium chloride were found to cause excessive shrinkage, especially in mixes with high cement content; hence they are not recommended. Retarders may be needed to reduce the heat of hydration or to prevent cold joints when spraying is done in hot climates. Air entrainment is typically used to improve the freeze-thaw durability of premixed GFRC components that are exposed to outdoor environments.

The mineral admixtures used in GFRC include fly ash, silica fume, and coloring pigments. Fly ash and silica fume improve workability and reduce shrinkage. Silica fume also makes the plastic mix more cohesive and provides early strength. White or color pigments are typically added to obtain the desired color for architectural cladding.

13.2.4 Coatings

Sometimes GFRC panels are coated with clear layer of silicone or siloxane to reduce efflorescence and moisture absorption. The panels can also be stained or painted using water-resistant materials. Latex masonry paints and methyl methacrylate–based paints offer a wide variety of colors. It is important that the surface coatings are vapor-permeable to allow the water vapors to escape from the panels.

13.2.5 Decorative face mixes

Decorative aggregates are sometimes used in bonded face mixes for GFRC architectural panels. Unreinforced face mixes are generally 0.125 to 0.25 in. (3 to 6 mm) thick and are bonded to the GFRC. The thickness of the face mix is generally controlled by the maximum size of aggregate. A face mix is placed in the mold and vibrated and screened. This is followed by the spraying of the GFRC. The decorative aggregates stick to the GFRC matrix resulting in an intergral panel.

13.3   Fabrication of GFRC

As mentioned at the beginning of the chapter, the two basic processes used for fabricating GFRCs are the spray-up process and the pre-mix process. Spray-up is more widely used than premix because of its ability for producing thin sections of various shapes and dimensions.

13.3.1   Forms

A wide variety of forming materials are used. The choice of the forming material depends on panel shape, required form stiffness, water resistance, number of uses, ease of fabrication, and cost. The materials currently used include fir plywood, birch-faced plywood, resin-coated plywood, GFRC, fiberglass-reinforced plastic-coated wood, reinforced concrete, and steel. Some of these forming materials have to be coated to make them water-resistant and removeable.

The forms should be designed to support their weight and the weight of the GFRC. Dimensional accuracy is an important factor, and hence forms should not deflect during the fabrication process. When designing the forms, removal of the finished product from the form should be considered. Chamfers, rounded corners, giving slope to vertical walls, built-in air connections for product release, and the placement of jacking points make stripping easier. Easier stripping not only induces less handling stress on the finished GFRC but also increases the form life, allowing more uses. Forms should also be coated with release agents that inhibit bonding and permit the release of the finished product. The chosen release agent should not react with any coating that may be applied to the finished panel.

Glass FRC panels generally shrink more than normal concrete. In some instances oversize forms are needed to compensate for this shrinkage.

13.3.2   Mixture proportions

The primary objective of mixture proportioning is to use the least amount of water to obtain a workable mix. In the spray-up process, the mix should be fluid enough to pass through the nozzle without clogging; the fibers should be adequately moistened. The fluidity can be checked using a flow test described in the test procedures used for quality assurance (Section 13.4.3). Mortar sand and mineral admixtures can be added to improve the shrinkage characteristics. The amount of sand to be added depends on the panel thickness and surface texture needed. Typical cement-to-sand ratios vary from 1 : 1 to 1 : 3, with 1 : 1 being the most commonly used ratio today.

The most commonly used fiber content for the spray-up process is 5% by weight of the total mix. As mentioned earlier, the lengths normally vary from 1.5 to 2.0 in. (38 to 50 mm).

If copolymers are used as curing admixtures, care should be taken to disperse them uniformly. A minimum polymer volume of 5% is needed to provide sufficient curing [13.6]. The pot life life of the polymer should be long enough for the fabrication process. The pot life of polymer is the time period for which it remains workable.

13.3.3 Fabrication using the spray-up process

The spray-up process is commonly preferred because it tends to place the fibers in a two-dimensional plane surface. The resulting thin composite is very effective in resisting bending forces in thin-sheet applications involving GFRCs. Spraying also facilitates the manufacture of thin sections with good fiber distribution. In addition, a section of virtually any shape can be sprayed, enabling the manufacture of architectural cladding of various shapes as well as surface textures.

In the spray-up process, cement/sand mortar and chopped glass fibers are simultaneously deposited from a spray gun onto a prepared mold surface, (Figure 13-1). Normally, the required thickness is obtained by spraying the composite in a number of layers. Each pass of the gun provides an approximate thickness of 0.125 in. (4 mm). For a typical 0.375 in. (10 mm) thick GFRC panel, two or three passes are needed. After spraying each layer, the wet composite is compacted by using a roller. Compaction by the roller not only removes the entrapped air but also presses the composite against the mold so that it conforms to the shape of the molding surface. Roller compaction is normally used with the manual handgun process.

Spraying can also be done using an automated process. In this process, compaction is achieved by a dewatering process. Dewatering is done by applying suction to the underside of a permeable mold. In an automated assembly line, panels move to the appropriate stations for spraying and dewatering. An added advantage of dewatering is that it lowers the water-cement ratio of the composite.

The GFRC panels are usually removed from their molds after 24 hours and are moisture-cured for at least seven days to obtain reasonable strength. Since the panels are thin, the chances for premature drying are high, and therefore proper curing is an absolute necessity for GFRC panels. The curing temperatures should range from 60° to 120°F (16°–50°C). However, if spray-up GFRCs are manufactured with a copolymer latex additive, the polymer addition can substitute for moisture curing. Only the copolymers proven by laboratory testing to eliminate the need for curing should be used.

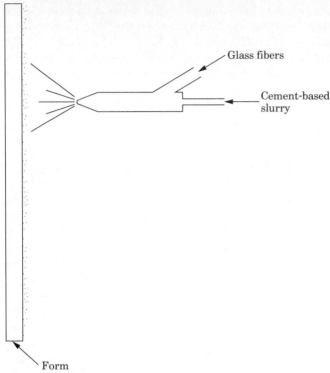

Glass fibers

Cement-based slurry

Form

Figure 13.1   Fabrication of a GFRC panel: hand spray method.

13.3.4   Fabrication using the premix process

In the premix process, cement, sand, admixtures, and fibers are mixed together. Flow and mixing aids such as high-range water-reducing agents are often used to facilitate mixing. Overmixing should be avoided because the fibers could be damaged. A volume of up to 5% glass fiber has been incorporated using the premixing process [13.7]. Typical fiber contents used in the field range from 2% to 3%.

The premixed composite can be molded to the desired shape by pouring it into a mold or a casting or by extruding, press molding, or slip forming. The curing scheme is the same as the one used for the spray-up process.

13.3.5   Surface finish

Surface finish is an important aspect for GFRC products because they are widely used as architectural cladding. A smooth, off-the-form finish may be the most efficient and economical approach. This type of finish normally leads to surface crazing. The hairline cracks may not have any significance in the area of strength and durability but play a major roll in aesthetics, especially when dirt appears in these cracks. It is also difficult to obtain color uniformity with off-the-form finish. The appearance of the panels could be improved by creating profiled surfaces such as fluted, sculptured, or rough board finishes. Coatings could also be used to improve the appearance.

Since GFRCs contain no coarse aggregate, a variety of surface textures and patterns can be obtained by using the appropriate molds. It is advisable to avoid sharp corners.

Typically, panels are produced with 0.125 to 0.25 in. (3 to 6 mm) thick bonded face mixes embedded with decorative aggregates. In most cases, the embedded aggregates are exposed by using retarders, sand blasting, acid etching, or honing and polishing. It is essential that differential shrinkage between the face mix and the GFRC be considered in the panel design. The compatibility of the two mixes for temperature and moisture-induced volume changes is critical to reduce the potential for differential shrinkage–related cracking.

13.4 Quality Control Tests

Some of the quality control tests needed for GFRC are the same as the ones used for normal concrete or FRC. They include checking the quality of the cement, aggregates, water, admixtures, and fibers. Some of the strength tests, such as the flexural strength test, are also the same even though special care is taken to provide proper support conditions since GFRC is normally thin, is often uneven, and tends to deflect more than normal FRC beams. The tests that are unique for GFRC are (1) equipment calibration tests (slurry flow rate and glass fiber flow rate tests), (2) a test to check the thickness during manufacture, (3) the slurry consistency slump test, (4) the wet density test, (5) the fiber content (wash-out test), (6) the anchor pull-off or shear tests, and (7) various types of aging tests. These tests are briefly described in the following sections. More details can be found in Reference [13.33].

13.4.1 Equipment calibration tests

Equipment calibration tests ensure that the fiber content is within the specified limits. Assuming all other factors are the same, density variation can be directly related to strength variation. Air entrapment is the common cause for reduction in density and strength.

13.4.1.1 Slurry flow rate test (bucket test). The slurry flow rate is determined by simply collecting and weighing the slurry that is discharged from the spray nozzle for a period of at least 30 s. An average of three

30 second collections is required. The average is multiplied by two, resulting in the slurry flow rate expressed in lb/min (kg/min).

13.4.1.2   Fiber flow rate test (bag test).   The fiber flow rate, which is also called a fiber roving chopping rate or chopper output rate, is determined by collecting the fibers from the chopping gun for 30 seconds and weighing the collected fibers. An average of three 30 second collections is required. The average is multiplied by two, resulting in the fiber flow rate in lb/min (kg/min).

13.4.2   Thickness test

This simple test ensures both the minimum thickness requirement and the tolerances for the specific variations across the panel. Pin gauges are used to measure thickness. It is recommended that a minimum of one thickness measurement be made per 5 ft^2 (0.5 m^2) of area [13.2]. Particular attention should be paid near edges, corners, and attachment inserts. When unreinforced architectural face mixes are first cast in the bottom of the mold, special attention should be paid to ensure constant mix thickness. This is important because subsequent GFRC thickness is checked by measuring the total thickness (face mix + GFRC) from the top of the last sprayed layer to the mold surface. If the face mix thickness is not held constant to a known thickness, then the quality control on GFRC thickness is lost.

13.4.3   Slurry consistency slump test

This test can be used to ascertain the suitability of the slurry for spraying. More importantly, it is used to maintain the consistency of the mix from batch to batch.

The test consists of filling a 3.187 in. (80.9 mm) tall 2.2 in. (57.2 mm) inside diameter Plexiglas tube with slurry and letting it spread onto a flat Plexiglas target plate that is marked with concentric circles. Three repetitions are recommended [13.2]. The tube is placed on the target plate, filled with cement slurry, screened and leveled with the edge of a spatula, and then lifted vertically off the plate allowing the slurry to flow onto the target plate. The diameter of the settled slurry is used as an indicator of the sprayability of the slurry. The slump values are expressed using a 0 to 8 scale. The value 0 means that the mix is stiff, and the diameter of the settled slurry is ≤ 2.5 in. (63.5 mm). A settled slurry diameter of 8.875 in. (225 mm) represents a very fluid slurry having a slump number of 8. The slump values vary with time since slurry stiffness varies with time. The time involved in the testing should therefore be correlated with the time needed for the spraying process.

13.4.4   Wet density test

The wet density or unit weight is determined by weighing the slurry that fills in a rigid 8 oz cup. An average of three measurements should be made. The procedure is similar to ASTM C138, except that a smaller sample is taken and the slurry is not consolidated before weighing.

13.4.5   Fiber content (wash-out test)

The uniformity of fiber distribution across an area and at various depths is checked using a test called the wash-out test. In this method, a sample of freshly formed GFRC board is cut from a small quality control board and is weighed. Then the slurry is washed out using water. The fibers are then dried and weighed. The ratio of the weight of the dried fibers to the weight of the total mix (the weight of the fresh board) provides the fiber content. Samples taken from various locations across the board area and at different depths should have fiber contents within acceptable limits.

13.4.6   Tests for hardened GFRC

Typically the strength of GFRC is established by a flexural test conducted according to ASTM C497, "Standard Test Method for Flexural Properties of Thin Section Glass-Fiber-Reinforced Concrete." This is a typical beam test with third-point loading. Since the specimens are thin, extra care should be taken at supports and loading points to ensure that a uniform line load is applied across the width of the specimen from the start of the test. This test is used to establish (1) proportional elastic limit (PEL), or matrix cracking strength, (2) modulus of rupture (MOR), or ultimate strength, and (3) flexural modulus of elasticity.

The other tests conducted on a hardened composite include density absorption and porosity according to ASTM C948 and the anchor pull-off and shear tests. Anchor pull-off and shear tests are used to establish the safe loads that can be transferred from the panel to the anchors. Determining whether these loads that can be safely transferred from the sheet to anchors is very important, because the sheets are attached to the buildings using these anchors. Live and dead loads that are applied to GFRC sheets are transferred to the structural frames through these anchors. The test methods are described in Reference [13.33].

Test methods are also available to check the compatibility of sealers within GFRC [13.2].

13.5   Physical Properties

The physical properties needed for the design of GFRC panels are tensile, flexural, and shear strengths. The properties that affect long-term performance are creep, thermal expansion, freeze-thaw resistance, and shrinkage and moisture movement. Depending on the type of application, proper design can require knowledge of other physical properties such as permeability, density, thermal conductivity, fire endurance, and acoustics. This section briefly describes various physical properties of GFRC.

The primary factors that affect the physical properties are the composition of the matrix and the lengths, orientation, and volume fraction of the fibers. The long-term properties are affected by the nature of fibers such as E-, A-, or AR-glass fibers. The addition of a polymer to the matrix may change composite behavior both in strength and in serviceability.

13.5.1   Behavior under tensile and flexural loads: short- and long-term behaviors

Properties often used for design are tensile and flexural strengths. For GFRCs, strengths (unaged strength) and long-term strengths are not the same. Hence strengths are normally designated as 28 day strength and aged strength.

The load-deflection curves shown in Figure 13-2 clearly indicate the difference in behavior for 28 day and aged beam specimens. The specimens were 2 in. (50 mm) wide and 0.375 in. (10 mm) thick [13.8]. They were tested on a 10 in. (254 mm) span using third-point loading. The mix composition and the details of the aging process are presented in this section. The following important points can be noted from Figure 13-2.

- The aging process of typical AR-GFRC composites will reduce the flexural strength to about half of the (unaged) 28 day flexural strength.

- The post-cracking ductility of the composite is completely lost after aging.

- The proportional elastic limit (PEL, the point at which the load-deflection curve deviates from a straight line) remains about the same or shows improvement. Additional strength beyond the PEL is lost during the aging process.

Specimens subjected to tensile loading also show loss of strength and ductility.

The following sections cover typical behavior of a GFRC at 28 days and after long-term exposure, accelerated aging to simulate long-

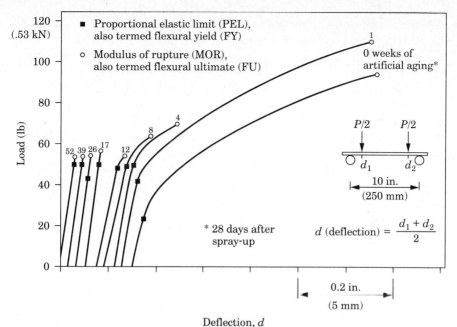

Figure 13.2 Typical load-deflection curves: AR-GFRC in bending [13.8].

term exposure, and developments to improve the durability of the fiber.

13.5.1.1 Typical load-deformation response of the unaged GFRC. The term unaged is used to designate the specimen after 28 days of curing. At this stage, the glass fibers contribute to the increase in flexural strength and ductility through an effective fiber pull-out mechanism.

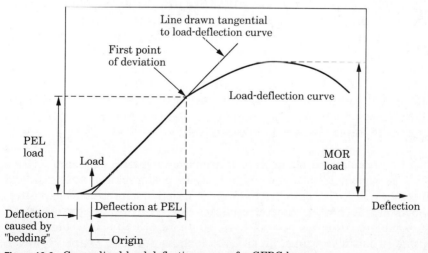

Figure 13-3 Generalized load-deflection curve of a GFRC beam.

A typical load-deflection curve of a prism in bending is shown in Figure 13-3. The curve consists of an initial linear elastic part and a nonlinear postcracking part. The initial elastic part primarily represents the behavior of the matrix. The fibers contribute little to an increase in the elastic modulus of the composite. Recent studies show that fibers do increase or extend the linear portion of the curve by increasing the strain capacity of the matrix. The nonlinear (postcracking) increase in strength is mainly provided by the debonding of fibers and fiber pull-out. While debonding, the fibers provide sufficient resistance to increase the composite strength beyond the pure matrix strength. Recent studies have shown that even though the matrix is brittle and has little or no postcracking strength, the fibers embedded in the matrix provide substantial resistance after cracking. The load-deflection curve starts to descend after the contribution of fiber resistance peaks.

The flexural strength measured using the load at the end of the linear portion of the curve is normally called the proportional elastic limit (PEL). The strength obtained using the maximum load is called the modulus of rupture (MOR). In both cases, the stresses are computed by dividing the maximum moment by the section modulus ($bd^2/6$), even though, because of excessive deflection and cracking of the specimen, the actual section modulus at maximum load is not ($bd^2/6$).

The ductility of the specimen can be expressed using the toughness indices I_5, I_{10}, and I_{20} defined in ASTM C1018 (Chapter 4 of this book), or it can be expressed using the area under the load-deflection curve per unit cross section of the beam. The PEL is used for designing GFRC panels for long-term loading and for handling stresses.

A typical load-elongation response of a tension specimen is shown in Figure 13-4. This curve essentially looks like the load-deflection curve of the bending specimen except that the elongation values at failure are much smaller and the elastic limit is a little smaller. The elastic limit is called the bend-over point (BOP), and the peak (maximum) strength is called ultimate tensile strength (UTS).

Typical PEL values in bending range from 900 psi to 1500 psi (6–10 MPa); MOR values range from 2500 to 4000 psi (16–30 MPa); and BOP and UTS values in tension range from 400 to 800 psi (5–7 MPa) and 1000 to 1600 psi (7–11 MPa) respectively.

13.5.1.2 Accelerated aging tests to simulate long-term exposure.

Glass FRC is known to lose nearly all of its postcracking strength and ductility (i.e., to become brittle) over a period of time that is exposure-dependent. In hotter, moister climates, the natural aging process can be complete after as few as 8 to 10 years [13.33]. In cooler, dryer climates, natural aging can take as long as 30 to 40 years. However, in terms of panel design, it must be assumed that all postcracking

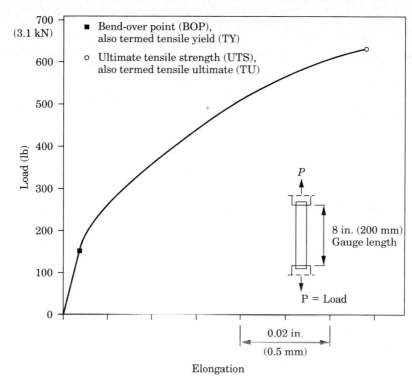

Elongation

Figure 13-4 Typical load-elongation curve at 28 days: E-polymer GFRC [13.8].

strength and ductility will be lost within the anticipated service life of the panel. Panel design must be based on the PEL of the composite.

In order to evaluate the long-term performance of a GFRC, accelerated aging tests were developed. In the popular accelerated aging test, the GFRC specimens are stored in lime-saturated water maintained at elevated temperature levels. Elevated temperature levels accelerate the aging process. The use of lime-saturated water prevents the leaching away of naturally occurring lime. For example, one day immersion of the sample in lime-saturated water maintained at 122°F (50°C) has been shown to be equivalent to 101 days of natural weathering exposure in the United Kingdom (mean annual temperature 50.7°F, 10.4°C) [13.9]. Typically, test specimens stored in water baths are tested either in tension or flexure at various time intervals (up to 52 weeks) to determine changes in strength and ductility.

Information about accelerated and natural aging processes is obtained by comparing the MOR values at various ages. For example, Figure 13-5 shows the MOR values at various stages of aging for both accelerated and natural weathering conditions. When the accelerated aging curves at various temperatures are horizontally displaced to the

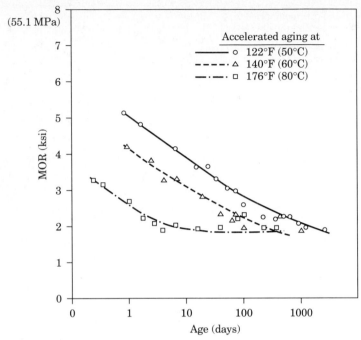

Figure 13-5 Modulus of rupture versus Accelerated aging time [13.9, 13.32].

right, they tend to agree with natural weathering with good accuracy, Figure 13-6 [13.9, 13.32]. It was therefore concluded that the results obtained at various accelerated aging temperatures correlate well with natural weathering results.

The explanation presented above deals only with strength. The loss of ductility could follow a different mechanism and hence accelerated aging may not reflect the behavior under natural weathering [13.10].

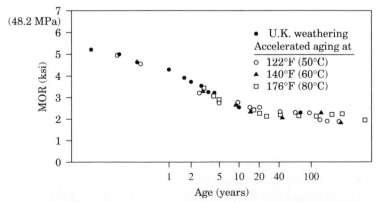

Figure 13-6 Comparison of accelerated and natural aging [13.13, 13.32].

13.5.1.3 Long-term behavior: accelerated aging. This section presents the loss in strength and ductility that occurs because of aging, based on accelerated aging tests. Results based on actual exposures, up to ten years, are presented in the next section. The results presented here were taken from Reference [13.8] and deal with a GFRC made with AR-glass fibers and a polymer-modified GFRC made with E-glass fibers. Both flexural and tension tests were conducted up to an accelerated age of 52 weeks. The results could be considered a typical case study.

Materials. The materials used consisted of Type I cement, washed silica with a maximum particle size of 0.02 in. (0.5 mm), normal or high-range water-reducing admixture, polymer latex (48% solids), and AR-glass fibers containing 16% zirconia or E-glass (borosilicate) fibers. The fibers were chopped to a length of about 1.5 in. (38 mm).

Mixture Proportions. The mixture proportions used for the regular and polymer-modified matrices are shown in Table 13.3a, b, and c respectively. The cement-to-sand ratio was approximately 1 to 0.5 for the AR-glass GFRC and for E-glass P-GFRC Composition 1, and was 1 to 0.20 for the other E-glass P-GFRC. The water-cement ratio varied from 0.22 to 0.35. The fiber content was 5% and 5.9% by total weight of mix. The polymer content was 10% and 15% of total volume of mix.

Fabrication. Test specimens were cut from $36 \times 48 \times 0.375$ in. ($910 \times 1220 \times 10$ mm) boards fabricated using the hand spray method. First, a mist coat was applied; it was followed by three layers of slurry-fiber mix to obtain a total thickness of 0.375 in. (10 mm). The composite was rolled between sprayings.

For AR-glass panels, the composite was covered with a polyethylene sheet after spraying and was stored overnight at 73°F (23°C). The slab was removed from the form on the next day and was cured for six days under 100% relative humidity and 73°F (23°C). The slab was then stored at a relative humidity of 50% for the remaining 20 days.

The slabs manufactured with E-glass P-GFRC were left under 50% relative humidity immediately after spraying. Note that a dry environment is preferable for polymer-modified GFRC compositions.

At the end of the 27th day, flexural and tension specimens were sawed to dimensions of $12 \times 2 \times 0.375$ in. ($305 \times 50 \times 10$ mm) and $12 \times 1 \times 0.375$ in. ($305 \times 25 \times 10$ mm) respectively. After cutting, the specimens were placed in water for one day prior to testing or aging.

TABLE 13.3a Mixture Proportions for AR-GFRC [13.8]

Specific gravity	Ingredients	Weight (lb)	Percent by weight	Volume (ft³)	Percent by volume
3.15	Cement	94.0	51.3	0.478	35.5
2.64	Sand	47.0	25.7	0.235	21.1
1.00	Water*	33.0	18.0	0.529	39.3
2.78	Glass fibers	39.2	35.0	0.053	4.1
Totals		183.2	100.0	1.345	100.0

*Also 1.3 ml water reducer per pound of cement was added. (1 lb = 0.45 kg, 1 ft³ = 0.028 m³)

TABLE 13.3b Mixture Proportions for E-Glass P-GFRC, Composition 1 [13.8]

Specific gravity	Ingredients	Weight (lb)	Percent by weight	Volume (ft³)	Percent by volume
3.15	Cement	94.0	60.6	0.478	41.2
2.65	Sand	18.8	12.1	0.114	9.8
1.00	Water*	20.9	13.5	0.335	28.9
1.12	Polymer solids†	12.1	7.9	0.174	15.0
2.55	E-Glass fibers	9.2	5.9	0.058	5.0
Totals		155.1	100.0	1.159	100.0

*Total water = Batch water plus water contained in polymer latex compound.
†Also 1.3 ml water reducer per pound of cement was added.
(1 lb = 0.45 kg, 1 ft³ = 0.028m³)

TABLE 13.3c Mixture Proportions for E-Glass P-GFRC, Composition 2 [13.8]

Specific gravity	Ingredients	Weight (lb)	Percent by weight	Volume (ft³)	Percent by volume
3.15	Cement	94.0	51.6	0.478	36.3
2.65	Sand	47.0	25.8	0.284	21.6
1.00	Water*	22.8	12.5	0.365	27.7
1.12	Polymer Solids†	9.2	5.1	0.132	10.0
2.55	E-Glass	9.1	5.0	0.057	4.3
Totals		182.1	100.0	1.316	100.0

*Total water = Batch water plus water contained in polymer latex compound.
†Also 1.3 ml water reducer per pound of cement was added.
(1 lb = 0.45 kg, 1 ft³ = 0.028 m³)

Accelerated Aging. The unaged specimens were tested at the 28th day after a one day soaking in water. The specimens subjected to accelerated aging were stored in lime-saturated water maintained at 122°F (50°C). Tests were conducted at accelerated ages ranging from 0 to 52 weeks.

Test Procedure. Flexural specimens were tested in third-point bending with a 10 in. (254 mm) span (Figure 13-7). The load was applied at a crosshead speed of 0.09 in./min (2.3 mm/min). Deflections under the loading points and corresponding loads were recorded using a x-y plotter.

The tension tests were run at an elongation rate of 0.5%/min (Figure 13-8). Elongation over a gauge length of 8 in. (203 mm) was measured to obtain the load-elongation curve.

Test Results. Typical load-deflection and load-elongation curves at various ages are shown in Figures 13-2 and 13-9 respectively. The strength results for flexure are presented in Tables 13-4 through 13.6. Flexural strength was calculated assuming linear elastic behavior. Deflection when the beam breaks into two pieces was taken as deflection at failure (total deflection). The other terms used in the tables are described in the previous sections.

The tensile strength results are presented in Tables 13.7 and 13.8. The definitions of the terms are the same as those for deflection.

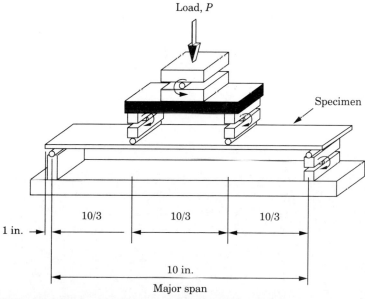

Figure 13-7 Flexural test setup (1 in = 25.4 mm).

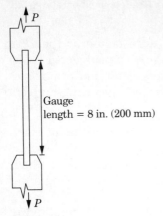

Figure 13-8
Uniaxial
tensile test
setup [13.8].

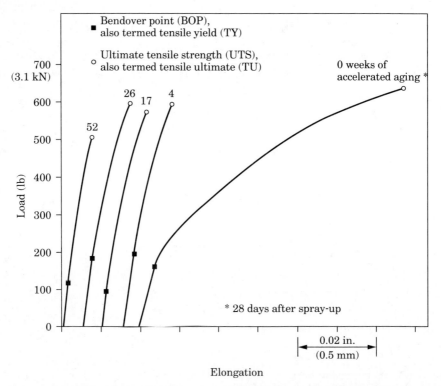

Figure 13-9 Typical load-elongation curves, aged and unaged specimens: E-polymer GFRC [13.8].

TABLE 13.4 Summary of Experimental Results for AR-GFRC in Bending [13.8]

Property	Accelerated aging period (weeks)								
	0	1	4	8	12	17	26	39	52
Proportional elastic limit (PEL), psi	1040	1405	1660	1730	1700	1640	1640	1735	1690
Relative value of PEL	100.0	135.1	159.6	166.3	163.5	157.7	157.7	166.8	162.5
Modulus of rupture (MOR), psi	3500	3760	2330	2390	2060	1865	1945	1900	1840
Relative value of MOR, %	100.0	107.4	66.6	68.3	58.9	53.3	55.6	54.3	52.6
Modulus of elasticity, ksi	2800	2700	2230	2340	2910	3400	3940	3420	3180
Toughness/cross-sectional area, lb/in.	63.04	44.14	10.93	6.57	3.40	1.07	1.37	1.65	1.15
First-crack deflection, in.	0.019	0.025	0.036	0.037	0.021	0.021	0.021	0.025	0.026
Peak deflection, in.	0.505	0.347	0.129	0.091	0.055	0.026	0.026	0.028	0.029
Total deflection, in.	1.038	0.669	0.276	0.205	0.129	0.086	0.031	0.038	0.033
Relative value of toal deflection, %	100.0	64.53	26.68	19.75	12.43	8.29	3.02	3.69	3.22

1 psi = 6.895 kPa, 1ksi = 6.895 MPa, 1 in. = 25.4 mm, 1 lb/in. = 0.18 N/mm

TABLE 13.5 Summary of Experimental Results for P-GFRC, Composition 1, in Bending [13.8]

Property	Accelerated aging period (weeks)					
	0	1	4	17	26	52
Proportional elastic limit (PEL), psi	1900	1765	1960	2140	1975	1770
Relative value of PEL,%	100.0	92.9	103.1	112.6	103.8	93.1
Modulus of rupture (MOR), psi	4115	2950	2600	2845	2995	2625
Relative value of MOR, %	100.0	71.77	63.2	69.1	72.8	63.7
Modulus of elasticity, ksi	1365	1645	1865	2225	2300	2560
Toughness/cross-sectional area, lb/in.	53.49	11.75	4.66	4.86	5.42	3.71
First-crack deflection, in.	0.071	0.054	0.054	0.048	0.042	0.034
Peak deflection, in.	0.461	0.140	0.082	0.069	0.072	0.057
Total deflection, in.	0.531	0.149	0.082	0.078	0.078	0.062
Relative value of total deflection, %	100.0	28.11	15.46	14.67	14.61	11.80

1 psi = 6.895 kPa, 1 ksi = 6.895 MPa, 1 in. = 25.4 mm, 1 lb/in. = 0.18 N/mm

TABLE 13.6 Summary of Experimental Results for E-Glass P-GFRC, Composition 2, in Bending [13.8]

Property	Accelerated aging period (weeks)					
	0	1	4	17	26	52
Proportional elastic limit (PEL), psi	1700	1700	1660	2025	1725	1910
Relative value of PEL,%	100.0	100.3	97.8	119.2	101.5	112.4
Modulus of rupture (MOR), psi	3680	2495	2365	2540	2560	2540
Relative value of MOR, %	100.0	67.8	64.3	69.1	69.6	69.0
Modulus of elasticity, ksi	1820	1975	2090	2530	2720	3420
Toughness/cross-sectional area, lb/in.	47.87	7.96	4.21	3.64	3.39	2.75
First-crack deflection, in.	0.050	0.043	0.405	0.044	0.032	0.029
Peak deflection, in.	0.432	0.102	0.067	0.061	0.056	0.043
Total deflection, in.	0.532	0.112	0.083	0.071	0.060	0.049
Relative value of total deflection, %	100.0	20.99	15.69	13.34	11.25	9.29

1 psi = 6.895 kPa, 1 ksi = 6.895 MPa, 1 in. = 25.4 mm, 1 lb/in. = 0.18 N/mm

TABLE 13.7 Summary of Experimental Results for E-Glass
P-GFRC, Composition 1, in Tension [13.8]

Property	Accelerated aging period (weeks)				
	0	4	17	26	52
Bend-over point (BOP), psi	445	560	315	510	335
Relative value of BOP, %	100.0	125.6	70.2	114.4	75.3
Ultimate tensile strength (UTS), psi	1740	1650	1510	1640	1355
Relative value of UTS, %	100.0	94.7	86.9	94.1	78.0
Modulus of elasticity, ksi	1365	1865	2225	2300	2560
First-crack elongation, in.	0.003	0.002	0.001	0.002	0.001
Peak elongation, in.	0.066	0.013	0.011	0.012	0.007
Total elongation, in.	0.068	0.013	0.012	0.011	0.007
Relative value of total elongation, %	100.00	18.92	18.04	16.28	9.68

1 psi = 6.895 kPa, 1 ksi = 6.895 MPa, 1 in. = 25.4 mm

Discussion of Results. A study of Tables 13.4 through 13.8 and of Figures 13-2 and 13-9 leads to the primary conclusion that postcracking ductility (deformation beyond the PEL and the BOP for bending and tension, respectively) is essentially lost by the aging process. This aspect can be clearly seen in Figure 13-10, which shows the variation of toughness (measured as the area under the load-deflection curve) with respect to aging. The flexural toughness of GFRC specimens made

TABLE 13.8 Summary of Experimental Results for E-Glass
P-GFRC, Composition 2, in Tension [13.8]

Property	Accelerated aging period (weeks)				
	0	4	17	26	52
Bend-over point (BOP), psi	485	530	515	330	365
Relative value of BOP, %	100.0	109.3	106.2	68.1	73.9
Ultimate tensile strength (UTS), psi	1080	1050	1270	1335	1430
Relative value of UTS, %	100.0	97.0	117.4	123.5	132.2
Modulus of elasticity, ksi	1820	2090	2530	2720	3420
First-crack elongation, in.	0.002	0.002	0.002	0.001	0.001
Peak elongation, in.	0.003	0.008	0.008	0.007	0.006
Total elongation, in.	0.032	0.014	0.010	0.008	0.007
Relative value of total elongation, %	100.00	42.63	29.47	26.02	20.69

1 psi = 6.895 kPa, 1 ksi = 6.895 MPa, 1 in. = 25.4 mm

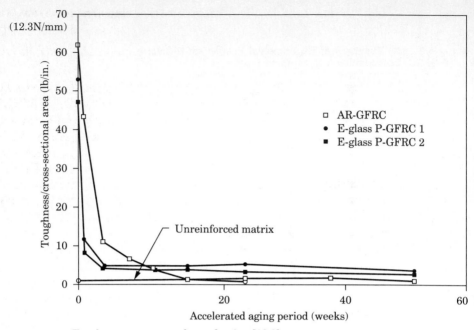

Figure 13-10 Toughness versus accelerated aging [13.8].

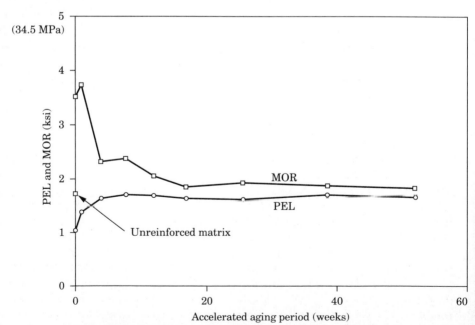

Figure 13-11 PEL and MOR versus accelerated aging: AR-GFRC [13.8].

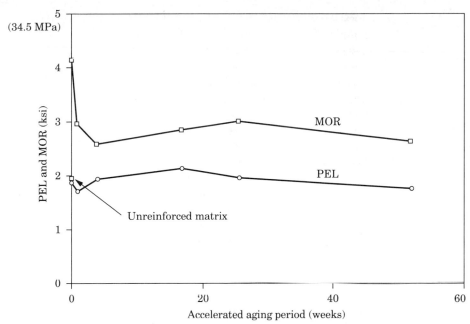

Figure 13-12 PEL and MOR versus accelerated aging: E-glass P-GFRC [13.8].

using AR-GFRC and E glass P-GFRC is about the same as the unreinforced matrix after about three weeks of accelerated aging at 122°F
(50°C).

In terms of flexural strength, the aging process essentially eliminated the strength contribution of fibers. Figure 13-11 shows that
the MOR essentially reached PEL values after about seven weeks of
accelerated aging for AR-GFRC. A polymer-modified matrix provides
slightly better results (Figure 13-12); the polymer addition resulted
in a composite that gave a nonlinear P-Δ response. Evaluating the
curves with the strict rules of first deviation from linearity results in
underestimating the PEL point. However, an examination of fracture
surface showed no contribution by fibers.

The flexural strength is higher than the tensile strength because the
increase in flexural strength is also based on the postpeak (descending)
part of the tensile σ-ϵ curve. Since the ductility is lost in tension, the
strength loss in flexure is greater [13.8].

The strength results concerning tension and flexure, coupled with
the above discussion, support the hypothesis that the alkali-resistant
glass fibers do not disintegrate with aging, but instead get embedded more firmly in the matrix. The improved bond results in fracture (instead of pull-out) of fibers and hence the composite failure is
brittle. Pull-out tests on glass fiber confirm this conclusion [13.36].

Pull-out load versus slip response was observed from matrices that were unaged and aged under an accelerated-aging environment. Aged specimens had higher bond strength and bond fracture energy than the unaged specimen. The fact that a lower cement content (P-GFRC, Composition 2) provided an increase in tensile strength with accelerated aging further supports the above hypothesis (Figure 13-13).

The other explanation for the loss of strength and ductility is that the fibers simply deteriorate in the highly alkaline environment (pH $\geq$ 12.5) of concrete. This seems to be true for the E-glass fibers embedded in the normal cement matrix. However, AR-glass in a normal cementitious matrix appears to become embrittled and loses its ability to pull out, thus decreasing ductility of the composite. Studies of the fracture surfaces also indicate the presence of undeteriorated AR-glass fibers in aged composites. However, studies of the fracture surfaces of E-glass P-GFRC show definite signs of glass deterioration and etching by alkalis in the cement matrix [13.25]. The behavior under accelerated aging conditions seems to be similar under natural aging conditions presented in the next section.

13.5.1.4 Long-term behavior: natural aging.

As mentioned earlier, results under natural aging conditions are available for periods up to ten years. Some of the experimental data are shown in Figures 13-14 through 13-16 [13.11, 13.12, 13.32]. All three figures show the results

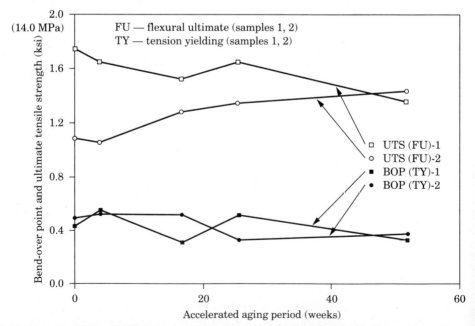

Figure 13-13 BOP and UTS versus accelerated aging: E-glass-GFRC [13.8].

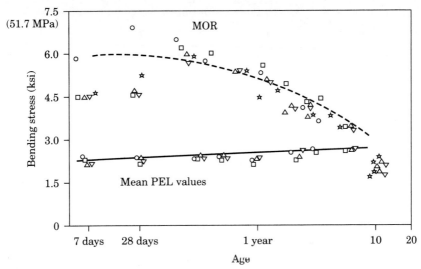

Note: Results are from five similar specimens.

Figure 13-14 PEL and MOR versus Natural Aging: AR-GFRC [13.12, 13.32].

for AR-GFRC. Figure 13-14 shows the variation of the modulus of rupture (MOR) and the proportional elastic limit (PEL) for specimens exposed to natural weathering conditions in the United Kingdom [13.13, 13.14]. It can be seen that in ten years, the composites' strength capacity (MOR) reaches its elastic limit (PEL), and all reinforcing qualities are essentially lost.

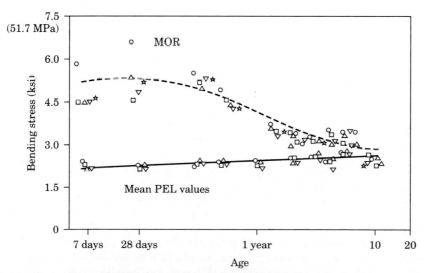

Note: Results are from five similar specimens.

Figure 13-15 PEL and MOR versus aging in water at 64° to 68°F (18°–20°C), AR-GFRC [13.12, 13.32].

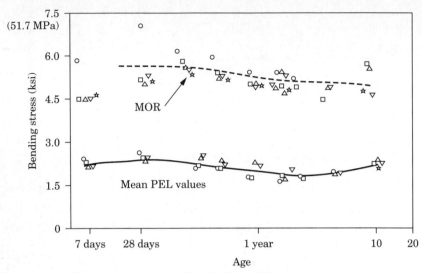

Note: Results are from five similar specimens.

Figure 13-16 PEL and MOR versus aging in air at 68°F (20°C) and 40% Relative Humidity: AR-GFRC [13.12, 13.32].

The results shown in Figure 13-15 are for specimens stored in water maintained at 64° to 68°F (18°–20°C), whereas results shown in Figure 13-16 are for specimens stored at 68°F (20°C) temperature and 40% relative humidity. The specimens stored in water are shown to lose strength faster than specimens exposed to natural weathering conditions. Specimens exposed to 40% relative humidity maintain a good portion of their strength. The loss of strength in moist conditions results can he attributed to better hydration cement.

A comparison of specimens exposed to natural weathering conditions and accelerated aging seems to provide the following approximate correlations [13.9].

Accel. aging temperature (° C)	No. of days outdoors in the U. K. eqeuivalent to one day of accel. aging
80	1672
70	693
60	272
50	101

The mean temperature in the United Kingdom is 10.4°C. As this temperature rises, the natural weathering period for complete embrittlement decreases. For example, accelerated aging at 50°C is equivalent to 101 days in the United Kingdom. This day-period increases to 160 days for conditions in Montreal, whereas it decreases to 82, 55, and 18 days for New York, Tokyo, and Miami where the mean temperatures are 12°, 15°, and 24°C respectively. Relative humidity also plays an important role in the aging process.

13.5.1.5 Techniques to improve durability. As mentioned earlier, the techniques investigated for improving durability can be broadly grouped as (1) the improvement of fibers by changing their chemical composition or surface coating and (2) the modification of the matrix. The addition of a minimum of 16% zirconia to the chemical composition of the fiber substantially improves its durability. As described elsewhere, this modified fiber is called alkaline-resistant (AR) glass fiber. Adding more then 16% zirconia was found to be more expensive because of the higher energy levels needed in the manufacturing process.

To further improve the durability of AR-glass fibers, various types of coatings have been tried. Cem-FIL2 and NEG AR glass H200 are two commercially available coated fibers. The coated fibers were found to maintain an initial GFRC strength level for much longer periods of aging than uncoated fibers [13.13]. It is believed that the coating reduces the affinity of glass fibers for calcium hydroxide, which may be the cause of both deterioration and embrittlement. Alkali-resistant organic materials were also used for coatings. The coated fibers provide longer strength retention both under tensile and flexural loads, as shown in Figures 13-17 and 13-18 [13.14, 13.32].

Microsilica directly applied to AR-glass rovings (by hand-dipping the glass strands directly into a solution of pure microsilica) was found to significantly increase the durability of the GFRC composite [13.4]. But its practicality with a hand-sprayed GFRC system was not investigated.

The primary focus on matrix modification was placed on reducing the alkalinity and specifically eliminating calcium hydroxide. Researchers have tried using pozzolans, polymers, and special cements with varying amounts of success.

The use of blast furnace slag to partially replace cement was found to reduce the initial strength without substantially improving durability. After two years of aging, GFRC specimens containing 50% cement and 50% blast furnace slag had strengths almost equivalent to those of plain matrix specimens [13.15]. This observation was confirmed by another research team [13.16], and improvements in strength retention in the presence of fly ash and blast furnace slag are reported in Reference [13.17]. The investigation of supersulfated cements led to the conclusion that some strength retention is possible depending on the type of cement and on the curing conditions [13.18].

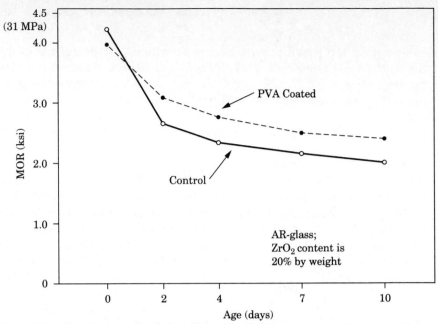

Figure 13-17 Tensile strength of glass fiber strands with various coatings embedded in portland cement paste and stored at 176°F (80°C) [13.14, 13.32].

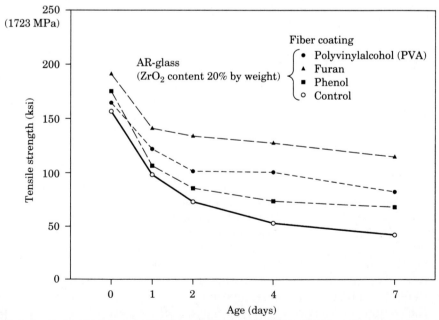

Figure 13-18 The MOR of the GFRC made with coated AR-glass fibers and stored in water at 176°F (80°C) [13.14, 13.32].

A number of investigators have studied polymer addition to the matrix for improving the durability of GFRCs [13.19–13.26, 13.31]. A minimum of 5% polymer solids by weight of composite appears to give some long-term strength improvement. The addition of polymer also improves workability, and it reduces permeability and water absorption. It was found that moisture curing is not required if a minimum of 5% polymer latex was added to the matrix.

The use of low-alkalinity cement also shows good promise for strength retention [13.27–13.31]. Adding large percentages of silica to the cement matrix was found to improve long-term durability [13.4, 13.14]. But adding large quantities of silica fume is not cost-effective.

Thus far, the most promising low-alkaline cement composition that has been developed appears to be calcium silicates C_4A_3s − Cs slag-type low-alkaline cement. Typical results obtained using this cement are presented in Figures 13-19 though 13-25 [13.5, 13.14, 13.32]. From Figure 13-19 it can be seen that an AR-glass matrix made using the new cement provides an excellent combination. Figures 13-20 and 13-21 present the comparison of MOR and PEL values for bending specimens. Figures 13-22 and 13-23 present the flexural stress (load)-deflection responses of beams subjected to accelerated aging at 80°C up to 10 days. Figures 13-24 and 13-25 present the responses for tension

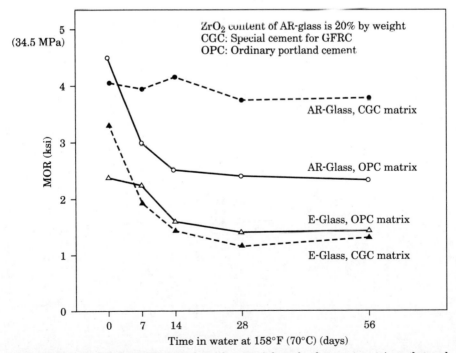

Figure 13-19 The MOR of GFRC made with a special portland cement matrix and stored in water at 158°F (70°C) [13.14, 13.32].

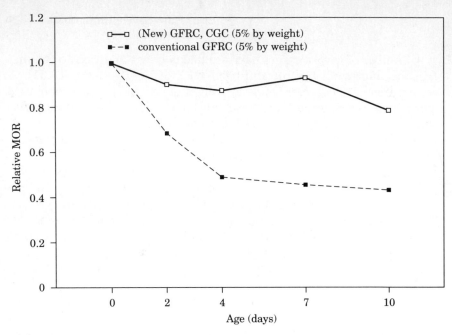

Figure 13-20 The LOP and MOR of GFRC made with a special portland cement matrix, stored in water at 80°C (176°F) [13.5].

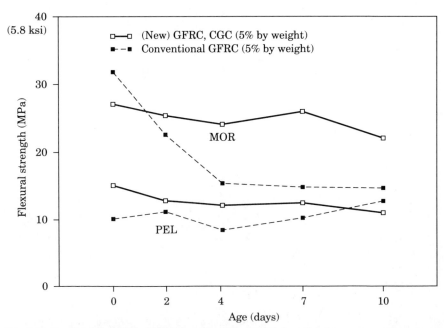

Figure 13-21 Relative MOR of GFRC made with a special portland cement matrix, stored in water at 80°C (176°F) [13.5].

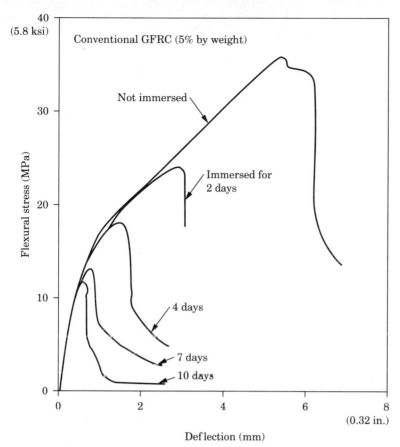

Figure 13-22 Flexural stress versus deflection of GFRC made with portland cement, stored in water at 80°C (176°F) [13.5].

specimens. The new cement type appears to retain both the strength and deformation capacities of the initial GFRC. Note that E-glass fibers are not effective with this new cement. Hence, only AR-glass fibers should be used.

13.5.1.6 Summary of properties in tension and flexure. In summary, it can be said that GFRCs provide excellent strength and ductility improvements at early ages. However, aging reduces the early strength level to approximately that of the matrix. More importantly, the ductility provided by the fibers is totally lost over a period of time, and the composite becomes brittle. The period of time is dependent on climate (moisture and temperature).

13.5.2 Compressive strength

Compressive strength primarily depends on the matrix strength. Compressive strength along the plane of the board is normally different

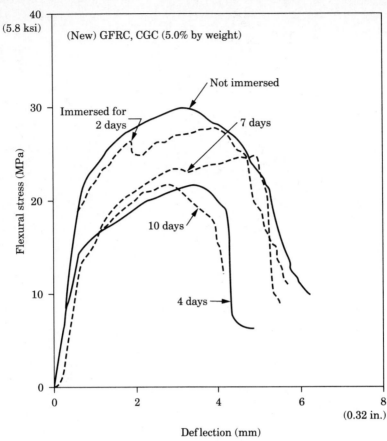

Figure 13-23 Flexural stress versus deflection of GFRC made with special cement, stored in water at 80°C [13.5].

from the strength across the thickness because of the fiber orientation. Since the matrix is strong in compression, the compressive strength of a GFRC panel rarely controls panel design.

13.5.3 Shear strength

Since most GFRC panels are made using the spray-up technique, there is a possibility for a weak link between the layers. Care should be taken to avoid cold joints between the layers. Even if cold joints are not present, there is a possibility for shear failure between the layers, since each layer is reinforced with random fibers in two directions. However, such failures do not normally occur in the field because loading configurations do not generally induce stresses between layers.

Panels will also be subjected to shear forces across the thickness and along the lengths and widths. The shear strengths along the length and width are called in-plane shear strengths. It was found that tensile

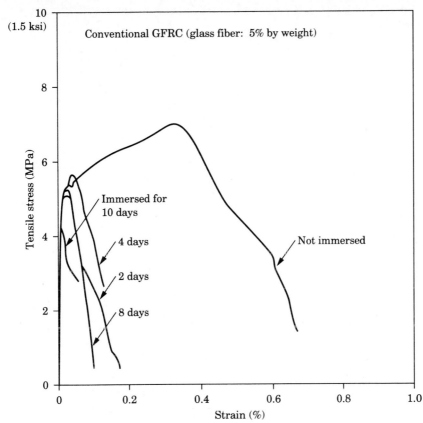

Figure 13-24 Tensile stress versus strain of GFRC made with portland cement, stored in water at 80°C (176°F) [13.5].

strength and in-plane shear strength are about the same [13.2]. Shear stresses encountered across the thickness are very small and hence are not considered in panel design.

13.5.4 Elastic modulus

The elastic modulus is needed in design for the computation of panel stiffness. Since for most panels the predominant loading mode is flexure, modulus of elasticity values are usually calculated from the load-deflection curves in bending. The modulus of elasticity is measured using the initial linear portion of the load-deflection curve. Typical modulus of elasticity values range from 1.5 to 2.9×10^6 psi (10,000 to 20,000 MPa). The matrix composition, including the admixtures, the water-cement ratio, and the type and amount of cement, controls the modulus of elasticity. Fibers increase the modulus of elasticity, but the magnitude of increase is not significant.

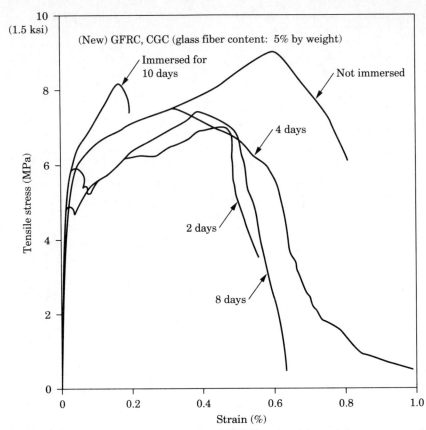

Figure 13-25 Tensile stress versus strain of GFRC made with special cement, stored in water at 80°C (176°F) [13.5].

13.5.5   Creep and shrinkage

Since the primary component of a GFRC is the cement matrix, GFRC panels exhibit creep behavior under sustained loads. The Creep behavior of a GFRC is very much similar to normal concrete components. The creep rate decreases with time on a logarithmic basis. Creep strains are proportional to initial elastic stress (strain), time at loading, time under loading, and relative humidity. Saturated GFRC specimens were found to exhibit more creep deformations for a short initial period.

Creep curves obtained using beams are shown in Figure 13-26 [13.2]. The trend of the creep curves is similar to the one reported for plain concrete.

In the area of shrinkage, GFRC panels are more sensitive than normal concrete elements. Since the matrix has a much higher cement content than that normally used in concrete, the shrinkage strains are normally higher for GFRC. The addition of sand substantially reduces

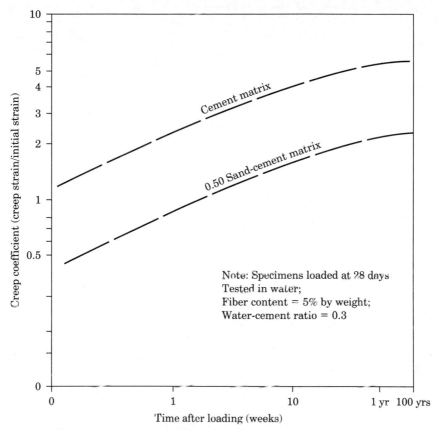

Figure 13-26   Flexural creep of GFRC [13.2].

the shrinkage, as shown in Figure 13-27 [13.2]. The fact that the panels are thin contributes to dimensional changes when they are exposed to wet and dry conditions. When the panel is placed in a wet condition, it expands a little, resulting in so-called reversible shrinkage (Figure 13-27). The magnitude of shrinkage depends on the cement-sand ratio, the water-cement ratio, curing conditions, exposure conditions, and the age of the composite.

Dimensional changes occur because of shrinkage or drying and wetting. External restraints provided by the structural frame does not allow the expansion or contraction of panels. This results in internal stresses in the panels. The internal stresses could produce cracking in the panels. Fibers generally increase the strain capacity at which cracks occur, and they aid in crack distribution, resulting in a reduction of maximum crack width. The shrinkage cracks tend to follow the paths where fibers are not present. Hence a uniform distribution of fibers results in a narrower maximum crack width.

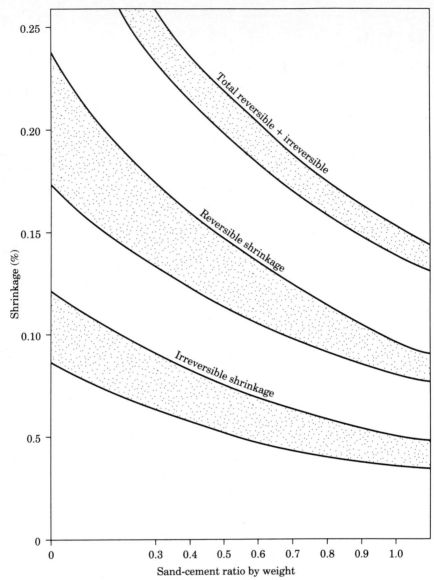

Figure 13-27 Influence of sand on shrinkage of GFRC [13.2].

Typically, dimensional change between saturated and oven-dry conditions is about 0.15% [13.2]. The observed movement for large building panels is about 0.10%, for a matrix composition with cement-sand ratio of 1.0 and water-cement ratio of 0.35. More details on shrinkage can be found in Reference [13.33].

For applications, it is advisable to measure dimensional changes for the actual panels to be placed in the building. These measurements can be used to properly design the fixtures and to avoid stresses caused by poor fit.

13.5.6   Freeze-Thaw Resistance

Glass FRC panels are exposed to some freezing and thawing cycles in the field, and hence freeze-thaw resistance is important. There are two tests used to evaluate the freeze-thaw resistance. The most severe test is the one specified in ASTM C666, Procedure B. In this test the specimens are subjected to freezing at 0°F (−18°C) in air and thawing at 40°F (4°C) while immersed in water. The test is normally done up to 300 cycles. Each cycle takes about 4.5 h. The other test is British Standard 4624, for which artificially aged specimens are subjected to 50 cycles of freezing and thawing in air. Freezing is accomplished by exposing the specimens to −4°F (−20°C) for 16 h, and thawing is accomplished at 68°F (20°C) for 8 h. After each cycle, the specimen has to be soaked in water for 48 h.

Tests done using the British Standard indicated that a GFRC can withstand 50 cycles of freezing and thawing without losing its mechanical strength. The test specimens were cut from sprayed boards containing 5% AR-glass fibers; cement-sand ratio was 5 : 1. The specimens were subjected to accelerated aging by storing them in water maintained at 122°F (50°C) for 90 days. The performance of the samples after aging and 50 cycles of freezing and thawing was evaluated using the modulus of elasticity, ultimate flexural and yield strengths, and the area under the load-deflection curve. None of the above properties were affected by the freeze-thaw portion of the test. Visible inspection also showed no deterioration from freezing-thawing [13.2].

Tests conducted using ASTM C666, Procedure B, showed a slight decrease in the bending strength after freeze-thaw cycling. Plain matrix and GFRC specimens were subjected to 300 cycles of freezing and thawing. Bending strengths were obtained after 100, 200, and 300 cycles. Tests were done using both aged and unaged specimens. The GFRC specimens were aged for 0, 8, and 26 weeks. The aging was done by immersing the specimens in water maintained at 122°F (50°C). The results obtained are shown in Figures 13-28 and 13-29 [13.2].

From the figures it can be seen that the plain matrix specimens disintegrated before 200 freeze-thaw cycles. The GFRC specimens had a slight decrease in both matrix strength and modulus of rupture. Unaged specimens showed more decrease in strength in the 0-to-100 cycle range (Figure 13-29). Flaking was also observed on the surfaces. Flaking on the troweled face was more severe than on the face against the formwork. The results are consistent with the results of normal concrete. Rich non–air entrained mixes always show more

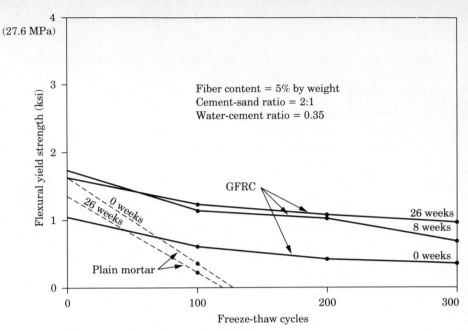

Figure 13-28 LOP versus freeze-thaw cycles [13.2].

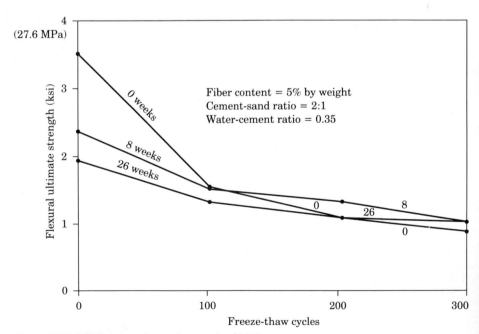

Figure 13-29 MOR versus freeze-thaw cycles [13.2].

deterioration if there is no mechanism for the redistribution of the thermal stresses. Even though 26 weeks of accelerated aging embrittles the GFRC, fibers are still effective in preserving the matrix through
freeze-thaw cycles. This can be seen by observing the constant sloping in the 0, 8, and 26 week curves.

13.5.7 Moisture absorption and density

Moisture absorption indirectly affects the freeze-thaw resistance of a GFRC. More moisture absorption results in higher expansion under freezing and creates more damage. Typical moisture absorption values range from 11% to 16% by weight. For normal applications with relative humidity less than 60%, moisture absorption is less than 10% by weight [13.2].

The densities of GFRC range from 120 to 140 lb/ft^3 (1900 to 2250 kg/m^3) compared to about 145 lb/ft^3 (2330 kg/m^3) for normal-weight concrete. Denser specimens typically absorb less water.

13.5.8 Fire resistance

Since the major use of GFRCs is for architectural building panels, fire resistance becomes an important factor in design. Typical ASTM standards used for fire rating are ASTM E136, "Test Method for Behavior of Materials in Vertical Tube Furnace at 750°C" and ASTM E119, "Methods of Fire Tests of Building Construction and Materials." The British Standard for a similar test is BS476, Parts 4, 5, and 6, "Fire Tests on Building Materials and Structures."

Glass FRC panels made with cement, sand, and glass fibers are noncombustible. When polymers are used as curing agents and to improve durability, the panels do not pass the combustibility test. Most tests for fire endurance are done using GFRC panel assemblies that contain steel studs and cavity insulations. Specific recommendations are given in Reference [13.2] for improving the fire endurance of GFRC panels.

13.5.9 Other physical properties

The following physical properties are also important for certain applications.

Water vapor permeability should range from 5 to 11 per in. (7 to 16×10^{-9} g/Pa·s·m). Air permeability should range from 0.2 to 4.6 per in. (0.3 to 6.7×10^{-9} g/Pa · s · m) depending on the relative humidity. Air permeability increases with a decrease in relative humidity.

Thermal conductivity should range from 3.5 to 7.0 Btu/in./h/ft^2/°F (0.5 to 1.0 W/m/°C). Thermal expansion range from 6 to 9×10^{-6} in./in./°F (11 to 16×10^{-6} mm/mm/°C). An increase in sand content results in a decrease in thermal expansion.

13.6 Design Procedures

This section provides a broad outline of design stresses and loads to be considered. For a comprehensive design, a more detailed manual should be consulted.

The long-term flexural strength of a GFRC is the main design parameter. Since it is not established that the increase in flexural strength provided by fibers can be retained for long-term loading, the allowable stresses are conservatively limited to the matrix cracking strength. Even though the PEL increases with time, the design stresses are conservatively based on 28 day strengths.

The Prestressed Concrete Institute's (USA) committee on GFRC panels recommends the following allowable flexural stress f_u for design [13.2].

$$f_u = \phi s f_u'$$ (13.1)

where ϕ = strength reduction factor
    s = shape factor
    f_u' = aged flexural strength

The recommended strength reduction factor ϕ is 0.67. This strength reduction factor is used to account for errors, material and dimensional variations, and assumptions involved in test methods and design procedures. The ductility of the member under normal load conditions and the importance of the panel to the structural integrity of the building are also considered in arriving at this constant.

Flexural strength is determined by test using solid prismatic beams. Hence, if the panel consists of solid section, the shape factor s is 1.0. If flanged or box sections are used, the shape factor can be taken as 0.5 to allow for stress concentrations. Values greater than 0.5 can be used in design if substantiated by tests and analysis.

The assumed ultimate strength f_u' is limited to a maximum of 1300 psi (9 MPa), the PEL at 28 days, or 1/3 of the MOR at 28 days, whichever is smallest. Hence f_u' is taken as the lesser of

$$f_{yr}(1 - tV_y)/0.9$$

or

$$1/3 f_{ur}(1 - tV_u)/0.9$$

or

$$1300\text{psi}   (9 \text{ MPa})$$

where $\quad f_{yr}$ = 28 day PEL (proportional elastic limit)

$\qquad\quad f_{ur}$ = 28 day MOR (modulus of rupture)

$\qquad\qquad t$ = statistical constant for student distribution

$\quad V_y, V_u$ = coefficient of variations for PEL and MOR

The factors $(1 - tV_y)/0.9$ and $(1 - tV_u)/0.9$ are introduced to obtain statistically predictable strength values. The strength values (f_{yr} and f_{ur}) are to be determined using 20 consecutive tests of six specimens each. The factors 0.9 and t are used to control the number of low tests. A value of 2.539 for t provides for no more than 1 in 100 samples will fall below 90% of average strengths.

Shear strength rarely controls the design. Principal tensile stresses in the panel should be limited to $0.4\phi f'_u$. Shear introduced locally by connections should not exceed interlaminar shear strength.

Deflection should generally be limited to span/360. More deflection could be allowed if it does not affect the adjacent members.

13.6.1 Design loads and load combinations

The primary loads to be considered for design are dead loads, live loads, and wind loads, and earthquake loads. In addition, loads that develop owing due to thermal and moisture changes should also be considered.

Movements caused by thermal or moisture changes and gradients can induce buckling, excessive loads on connections and fasteners, or excessive stresses within the skin. For example, temperature or moisture change or a combination of them could result in the expansion or contraction of the panel, inducing stresses wherever it is restrained. If a bonded architectural facing is used, the facing might expand at a different rate than the panel itself, inducing stresses on the panel and the connections.

The minimum load factors and the load combination recommended by the Prestressed Concrete Institute are the following [13.2].

$$0.75 \, [1.4D + 1.7 \, (\text{greater of } L, W, \text{or } 1.1E) + 1.6 \, (\text{greater of } M \text{ or } T)]$$

where $\quad D$ = dead load

$\qquad\quad E$ = earthquake load

$\qquad\quad L$ = live load

$\qquad\quad M$ = self-straining forces and effects arising from contraction or expansion caused by moisture changes

$\qquad\quad T$ = Self-straining forces and effects arising from contraction or expansion caused by temperature changes

$\qquad\quad W$ = wind load

13.7 Supporting Systems

During the early development of GFRC, three basic panel types evolved: the sandwich panel, the integral rib panel, and the steel stud flex-anchor panel. Of these three, the steel stud flex-anchor panel became commonplace because of its adaptability for construction; hence, only it is discussed in this section. A steel stud framing system involves the fabrication of the supporting steel frame and the attachment of the GFRC skin to the frame using either flex-anchors or gravity anchors, as described in the following sections. More details can be found in references [13.34] and [13.35].

13.7.1 Design and fabrication of steel stud frame

The primary purpose of the steel frame is to transfer the weight and the loads from the GFRC skin to the main structural frame. In most cases, the stud frames should have an in-plane rigidity comparable to the GFRC skin, such that each stud can support its tributary portion of the loads transferred from the skin. The frame should also be rigid enough in the direction perpendicular to its plane to withstand forces developed by any minor bowing of the skin. Care should be taken to reduce the development of bowing, especially in systems carrying bonded architectural face mixes. Most of the bowing tendencies can be avoided if the GFRC skin and the bonded face mix are dimensionally compatible.

Studs in the frame should be dimensioned to avoid buckling. If long studs are used, bridging and diagonal bracing may be required. A GFRC skin should not be expected to prevent stud buckling.

The load transfer takes place from GFRC skin to studs through the anchors. The studs transfer the load to the structure through horizontal tracks and vertical stud connections. Hence the GFRC skin has to be analyzed as a simple or continuous beam or as a two-way slab system spanning between anchors. The steel stud frame should be designed to transfer all the loads coming from the anchor points to horizontal tracks and building connection locations. The design of the frame depends on the configuration of the stud and cross-bracing. Generally, the stud frame is attached to the structural frame using two load-bearing connections to create a simply supported system. Note that a simply supported system reduces the loads developed by dimensional changes. Additional nonload-bearing connections may be provided to improve the lateral stability.

The steel stud frames can be fabricated using the guidelines provided in *Lightweight Steel Framing Systems Manual*, prepared by the Metal Lath / Steel Framing Association. Typical stud spacing ranges

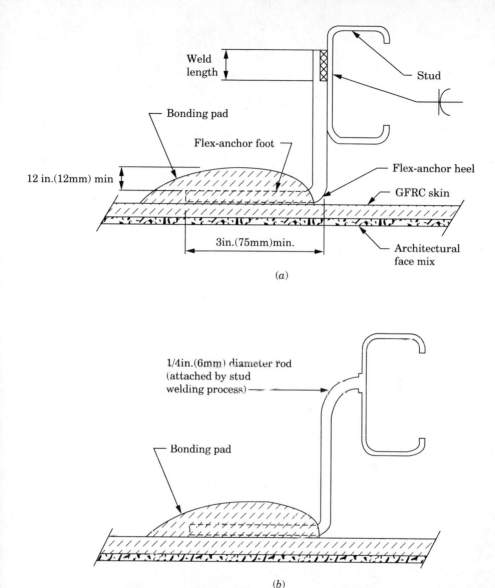

Weld length

Bonding pad

Flex-anchor foot

Stud

Flex-anchor heel

GFRC skin

12 in.(12mm) min

3in.(75mm)min.

Architectural face mix

(a)

1/4in.(6mm) diameter rod (attached by stud welding process)

Bonding pad

(b)

Figure 13-30   Flex-anchors [13.2, 13.32].

from 16 to 24 in. (400 to 600 mm). The prefabricated frame is often moved quite a few times. Hence welded rather than simple screw connections are recommended, especially to maintain the dimensional accuracy. A minimum of 16 gauge material is recommended if welding is used for the connections. Surface coatings required for the frame depend on exposure conditions. The frame and welds should be protected from corrosion.

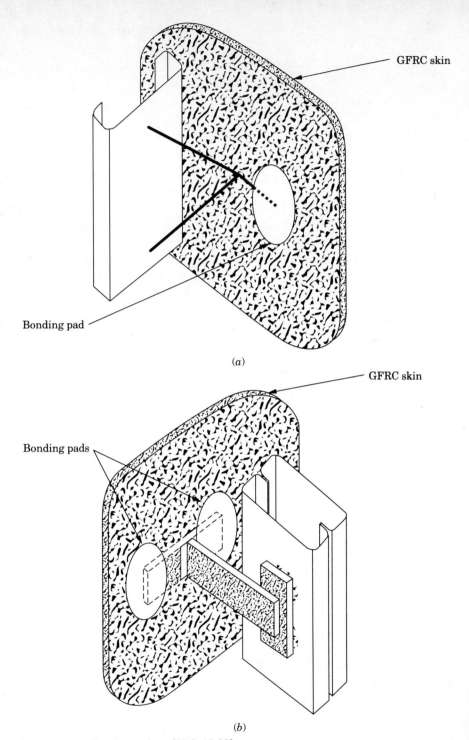

Figure 13-31 Gravity anchors [13.2, 13.32].

Gravity or flex-anchors are attached to the frame by welding (Figure 13.30). Typical anchor spacing ranges from 16 to 36 in. (400 to 900 mm). The anchors should also be protected against corrosion. They could be zinc- or cadmium-plated or painted with a zinc-rich coating. They could also be manufactured using stainless steel.

13.7.2 Flex-anchor connections

In the flex-anchor connection, the weight of the GFRC skin is transferred to the steel studs by the bending strength of the anchor (Figure 13-30). The L-shape anchors should have sufficient strength and rigidity to transfer the load from the skin to the steel stud without undergoing excessive deformation. At the same time, they should be flexible enough not to develop excessive loads when the skin undergoes dimensional changes caused by moisture or temperature change.

Flex-anchors are typically made with rods 0.25 in. (6 mm) in diameter. The rod can be placed at the top with groove welds (Figure 13-30a) or can be bent at the top and directly welded to the stud (Figure 13-30b). The second arrangement is more flexible. In some instances square bars are also used with filled welds at the top. The GFRC skin is attached to the anchors using bonding pads, as explained in Section 13.7.4.

A plastic sleeve may be used at the foot of the anchor to facilitate movement. The toes are normally placed toward the center of the panel so that drying shrinkage will pull the anchor away from the stud instead of pushing the stud.

Unsupported edges of panels could warp and bow, creating gaps and unevenness at the junction of the panels. Hence it is advisable to place the anchors as close to the edge as possible.

13.7.3 Gravity anchor connections

The GFRC skin can be also attached to the steel studs using gravity anchors (Figure 13-31). If the stud frame is rigid, then the skin can be attached to it by a series of trussed gravity anchors (Figure 13-31a). The anchors, which are made of smaller-diameter bars for transverse loads, can be used at every stud or at alternate studs.

If the stud frame is not rigid, the panel can be attached to the stud frame where it is connected to the main structural frame. In this way, the weight of the panel is directly transferred to the structural frame. Figure 13-31b shows a typical gravity anchor made of plates. The plates can be made deep so that they can carry heavier weights and at the same time be flexible in the lateral direction.

13.7.4 Anchor details for seismic areas

In seismic areas, resistance to sideways movement is needed during earthquakes. This resistance has to be provided without totally restraining the system. In a flat plate tee system, as shown in Figure 13-31*b*, one of the anchors can be strengthened to carry the load devel-

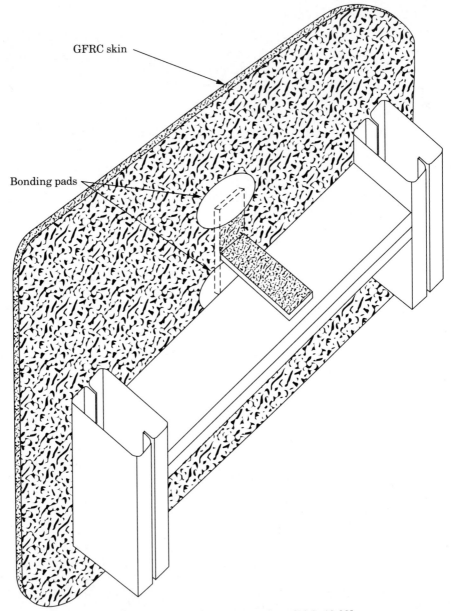

GFRC skin

Bonding pads

Figure 13-32 Special gravity anchors for seismic regions [13.2, 13.32].

oped by seismic motion. Seismic load can be carried using a horizontally oriented flat plate, as shown in Figure 13-32. The anchor should be designed to carry any rotational force developed because the anchor system is not colinear with the center of mass of the skin.

If flex-anchors are used, the stiffness of the steel stud along the weak axis should be considered in the design. Gravity anchors can be combined with flex-anchors to obtain a balanced design.

If gravity anchors are provided at the midheight of the panels, over-turning moments can be reduced. But this setup results in tensile stresses at the bottom half of the panel, which are undesirable.

13.7.5 Attaching a GFRC panel to a steel stud frame

The steel stud frame is attached to the GFRC skin using a bonding pad (Figures 13-30, 13-31). Typically, the frame is positioned against the GFRC skin and is held in position by jigs until the bonding pad develops sufficient strength. The bonding pad is cast using one of the three processes: (i) the green sheet overlay process, (ii) the hand-pack method, or (iii) the direct spray-up process. In all three methods, the pad is manually kneaded to the GFRC skin by operators. The time delay between the final roller compaction of GFRC skin and the casting of the pad should be minimized in order to provide a good bonding. If there is excess time delay, the pad could delaminate from the skin.

Sometimes the anchor "telegraphs" through the skin and shows on the outside face. To avoid this, and for production convenience, the anchors are sometimes positioned such that they are 1/8 to 3/8 in. (3 to 10 mm) away from the surface of the GFRC skin. The anchor pads should be at least 0.5 in. (13 mm) thick over the top of the anchor foot with a bonding area sufficient to transfer the loads from the skin to the frame. Typically, bonding areas vary from 18 to 32 in.^{2} (120 to 210 cm^2). The actual size should be determined using strengths obtained on actual axial and shear pull-off tests using fully aged specimens.

It is recommended that at least seven tests be conducted using actual field conditions and that the maximum pull-off load be taken as the average of the five middle values (eliminating the maximum and minimum value) [13.2]. Aging is known to reduce the anchor pull-off strengths by as much as 50%. A safety factor of 4 to 5 is recommended to obtain the pull-off strengths at service load conditions.

13.8 Applications

The most common application of a GFRC is in the area of cladding and architectural panels for buildings. The composite is also being used as a surface bonding material for existing structures to provide a durable and continuous exterior skin. In new construction, concrete-block walls are sometimes bonded with 3/16 in. (5 mm) thick GFRC skins on either side to create sandwich walls. Surface bonding is also used as a sealant in mine shafts.

Table 13.9, taken from Reference [13.32], lists a wide variety of applications in which GFRC has already been used or is contemplated for use.

TABLE 13.9   Applications of GFRC

General Area	Specific Examples
Agriculture	Livestock husbandry products
	Water troughs
	Feeding troughs
	Sheep dips
	Sheds
	Irrigation channels
	Reservoir linings
Architectural cladding	Interior panels
	Single skin
	Double skin (thermally insulated)
	Paint, tile, aggregate facings
	Exterior panels
	Single skin
	Double skin (thermally insulated)
	Profiled single skin
	Paint, tile, aggregate facings
	Doors and door frames
	Architectural surface finish
	Windows, subframes, sills
	Raised-access floor panels
	Internal fixtures
	Prefabricated bathroom units
	Lavatory units
	Bench tops
	Shelving
	Shells
Asbestos replacement	Simple sheet cladding
	Flat
	Profiled
	Promenade and plain roof tiles
	Fire-resistant pads
	General molded shapes and forms
	Pipes
Ducts and shafts	Track-side ducting for cables and switchgear
	Internal service ducts
Fire protective systems	Fire doors
	Internal fire walls
	Calcium silicate insulation sheets

(continued)

TABLE 13.9—cont'd

General Area	Specific Examples
General building (excluding wall systems, cladding)	Roofing systems (tiles, shingles, lintels) Cellar grills and floor gratings Decorative grills and sun shades Hollow nonstructural columns or pillars Impact-resistant industrial floors Brick facade siding panels Cellular concrete slabs
Low-cost housing, schools, factory buildings	Single- and double-skin cladding on timber frame construction Prefabricated floor and roof units
Marine applications	Hollow buoys Pontoons Marina walkways Workboats, dinghies

Metal replacement	Sheet piling units for canal, lake, or ocean revetments
	Covers
	    Manholes
	    Motors
	    Gasoline storage tanks at
	        service stations
	    Grating covers for guttering
	Hoods
	Stair treads

Miscellaneous	Sun-collector castings Artificial rocks for zoos or park settings
Pavements	Overlays (to control reflection cracking)
Permanent and temporary formwork	Bridge decking Parapets Abutments Waffle forms Columns and beams
Site-applied surface bonding	Bonding of dry-block walls Single-skin surface bonding to metal lath substrates Low-cost shelters (stacked, unmortared mud brick)
Small buildings and enclosures	Sheds Garages Acoustic enclosures Kiosks Telephone booths

(continued)

TABLE 13.9—cont'd

General Area	Specific Examples
Small containers	Telecommunication junction boxes Storage tanks, silos Stopclock and meter encasements and covers Manhole encasements and covers Utility boxes
Street furniture components	Seats and benches Planters Litter bins Signs Noise barriers Bus shelters Revetment facing panels

Water applications	Low-pressure pipes
	  Drainage
	  Sewerage
	Sewer linings
	Water channels (culverts)
	Canal linings
	Field drainage components
	  Inspection chambers
	  Hydrant chambers
	  Headwall liners
	  Pipe drain inlets
	  Drainage covers, traps
	  Guttering
	Tanks
	  Swimming pools, ponds
	  Fish farming
	  Sewage treatment
	  Septic tanks
	  Storage tanks

13.9   References

13.1 Larner, L. J.; Speakman, K.; and Majumdar, A. J. "Chemical Interactions between Glass Fibers and Cement," *Journal of NonCrystalline Solids,* Vol. 20, 1976, pp. 43–74.

13.2 Prestressed Concrete Institute (PCI) *Recommended Practice for Glass Fiber Reinforced Concrete Panels,* PCI, Chicago, Illinois, October 1987, 87 pp.

13.3 Bijen, J. "E-Glass Fiber-Reinforced Polymer-Modified Cement," Proceedings, International Congress on Glass Fiber Reinforced Cement, London, October 1979, pp. 62–67.

13.4 Bentur, A.; and Diamond, S. "Effects of Direct Incorporation of Microsilica into GFRC Composites on Retention of Mechanical Properties After Aging," Proceedings, Durability of Glass Fiber Reinforced Concrete Symposium, PCI, Chicago, Illinois, 1986, pp. 337–356.

13.5 Akihama, S.; Suenga, T.; Tanaka, M.; and Hayashi, M. "Properties of GFRC with Low Alkaline Cement," *Fiber Reinforced Concrete Properties and Applications,* SP-105, American Concrete Institute, Detriot Michigan, 1987, pp. 189–210.

13.6   Daniel, J. I.; and Pecoraro, N. E. "Effect of Forton Polymer on Curing Require-ments of AR-Glass Fiber Reinforced Cement Composites," Report by CTL to For-ton Inc., Skokie, Illinois, 1982.

13.7   *Design Guide: Glass Fibre Reinforced Cement,* Pilkington Brothers Ltd., Steel House Press, Liverpool, England, 1979.

13.8   Shah, S. P.; Daniel, J. I.; and Ludiraja, D. "Toughness of Glass Fiber Rein-forced Concrete Panels Subjected to Accelerated Aging," *PCI Journal,* September-October 1987, pp. 82–99.

13.9   Litherland, K. L.; Oakley, D. R.; and Proctor, B. A. "The Use of Accelerated Aging Procedures to Predict the Long-Term Strength of GFRC Composites," *Journal of Cement and Concrete Research,* Vol. 11, 1981, pp. 455–466.

13.10   Litherland, K. L. "Test Methods of Evaluating Long-Term Behavior of GFRC," Proceedings, Durability of GFRC Syposium, PCI, Chicago, Illinois, November, 1985.

13.11   Proctor, B. A.; Oakley, D. R.; and Litherland, K. L. "Developments in the As-sessment and Performance over Ten Years," *Composites,* April 1982.

13.12   "Properties of GRC: Ten-Year Results," *BRE Information Paper,* IP36/79, Building Research Station, U.K., Department of Environment, Garston, Hertford, Novem-ber 1979, 4 pp.

13.13   Proctor, B. A. "Past Development and Future Prospect for GRC Materials," Pro-ceedings of the International Congress on Glass Fiber Reinforced Cement, Paris, November 1981.

13.14   Hayashi, M.; Sato, S.; and Fujii, II. "Some Ways to Improve Durability of GRC," Proceedings, Durability of Glass Fiber Reinforced Concrete Symposium, PCI, Chicago, Illinois, 1986, pp. 270–284.

13.15   Mills, R. H. "Preferential Precipitation of Calcium Hydroxide on Alkali-Resistant Glass Fibers," *Cement and Concrete Research,* Vol. 11, 1981, pp. 689–697.

13.16   Singh, B.; Majumdar, A. J.; and Ali, N. A. "Properties of GRC Containing PFA," *International Journal of Cement Composites and Lightweight Concrete,* Vol. 6, No. 2, 1986, pp. 67–74

13.17   Kohno, K.; and Horri, K. "Use of By-Products for Glass Fiber Reinforced Con-crete," *Transactions,* Japan Concrete Institute, Tokyo, Vol. 7, 1985, pp. 25–32.

13.18   Singh, B.; and Majumdar, A. J. "GRC Made from Supersulphated Cement: 10-Year Results," *Composites* (Guildford, U.K.), Vol. 18, No. 4, September 1987, pp. 329–333.

13.19   Bijen, J. "Glass Fiber Reinforced Cement: Improvements by Polymer Addition," Proceedings, Symposium L, Advances in Cement-Matrix Composites, Boston, November 1980, Materials Research Society, University Park, Maryland, pp. 239–249.

13.20   Bijen, J.; and Jacobs, M. "Properties of Glass Fibre Reinforced Polymer Modified Cement," *Journal of Materials and Structures,* Vol. 15, No. 89, 1982, pp. 445–452.

13.21   Jacobs, M. J. N. "Forton PGRC—A Many-Sided Construction Material," Proceed-ings, GRCA Conference: GFRC in the 80s, Paris, 1981, pp. 31–49.

13.22   Jacobs, M. J. N.; and Bijen, J. "Durability of Forton Polymer Modified GFRC," Proceedings, Durability of Glass Fiber Reinforced Concrete Symposium, Novem-ber 1985, Prestressed Concrete Institute, Chicago, Illinois, 1986, pp. 293–304.

13.23   Ball, H. P., Jr. "The Effect of Forton Compound on GFRC Curing Requirements," Fourth Biennial Congress of the Glass Fiber Reinforced Cement Association, Stratford-upon-Avon, October 1983, pp. 56–65.

13.24   Daniel, J. I.; and Shultz, D. M. "Long Term Strength Durability of Glass Fiber Reinforced Concrete," R/D Serial No. 1780, Portland Cement Association, Skokie, Illinois, April 1985, 48 pp.

13.25   Daniel, J. I.; and Schultz, D. M. "Durability of Glass Fiber Reinforced Concrete Systems," Proceedings, Durability of Glass Fiber Reinforced Concrete Sympo-sium, November 1985, Prestressed Concrete Institute, Chicago, Illinois, 1986, pp. 174–198.

13.26   Majumdar, A. J.; Singh, B.; and West, J. M. "Properties of GRC Modified by Styrene-Butadiene Rubber Latex," *Composites* (Guildford, U.K.), Vol. 18, No. 1, 1987, pp. 65–65.

13.27 Cao, Y. K. "The Durability of Low Alkalinity Sulfoaluminate Cement Reinforced with Alkali Resistant Glass Fibers," Research Institute of Building Materials, Beijing.

13.28 Lu, H. T.; Song, Y.; and Wang, Y. M. "Study of the Glass-Fibre Reinforced Low Alkalinity Cement," *Silicate Journal,* Vol. 9, No. 4, December 1981.

13.29 Xue, J. G.; Zhang, P. X.; Xu, W. X.; and Lu, B. S. "Investigation of the Corrosion Mechanism of Low Alkalinity Cement to Glass Fibers," Cement Research Institute, Research Institute of Building Materials, People's Republic of China, *Ceramic Institute* Vol. 8, No. 3 1982, pp. 104–107.

13.30 Majumdar, A. J.; Singh, B.; and Ali, M. A. "Properties of High Aluminate Cement Reinforced with Alkali Resistant Glass Fibers," *Journal of Materials Science* (London), Vol. 16, 1981, p. 2597.

13.31 Daniel, J. I.; and Pecorato, M. E. "Effect of Forton Polymer on Curing Requirements of AR-Glass Fiber Reinforced Cement Composites," Construction Technology Laboratories, Skokie, Illinois, October 1982, 48 pp.

13.32 Daniel, J. I. "Glass Fiber Reinforced Concrete," Fiber Reinforced Concrete, Report no. 2493D and 2614D, Construction Technology Laboratories, Inc., Skokie, Illinois, 1988, pp. 5.1–5.30.

13.33 Daniel, J. I.; Roller, J. J.; Weinmann, J. L.; Oesterle, R. G.; and Schultz, D. M. "Quality Control and Quality Assurance for the Manufacture and Installation of GFRC Facades," Seventh Biennial Congress of the GRCA, September 26–28, 1989, Maastricht, The Netherlands, pp. 243–266.

13.34 Oesterle, R. G.; Schultz, D. M.; and Glikin, J. D. "Design Considerations for GFRC Facades," *Thin Section Fiber Reinforced Concrete and Ferrocement.* SP-124, American Concrete Institute, Detroit, MI, 1990, pp. 157–182.

13.35 Hanson, N. W.; Roller, J. J.; Daniel, J. I.; and Weinmann, T. L. "Manufacture and Installation of GFRC Facades," *Thin Section Fiber Reinforced Concrete and Ferrocement.* SP-124, American Concrete Institute, Detroit, Michigan, 1990, pp. 183–213.

13.36 Shah, S. P.; Li, Z.; and Mobasher, B. "Effect of Aging on Interfacial Properties of Glass Fiber Reinforced Concrete," *Fiber Reinforced Cementitous Materials*, Materials Research Society, Vol. 211, 1991, pp. 113–118.

14

Thin-Sheet Products

The manufacturing technique used for thin-sheet products is quite different from the techniques used for FRC with coarse aggregates. In most cases the fiber volume fraction is much higher. The industries dealing with thin-sheet products are growing considerably because of the replacement of asbestos fibers.

The fibers can be metal (low-carbon or stainless steel), mineral (glass or asbestos), synthetic (carbon or polymer), or natural organic (cellulose, sisal, etc.). In some applications, more than one fiber type is being used. The properties of glass fiber–reinforced composites are presented in Chapter 13; the properties of composites containing large volume fractions of steel fibers (typically used in thin sections) are presented in Chapter 15.

This chapter is dedicated to the composites reinforced with carbon, polymeric, and naturally occurring fibers. Even though asbestos fibers were widely used until recently, their current use is being greatly curtailed because of adverse health effects. Hence the use of asbestos fibers is not discussed in detail. Most of the development in thin-sheet cement products can be attributed to the excellent properties that can be achieved by incorporating asbestos fibers in a cement matrix. However, even after intense research for more than a decade, asbestos fibers could not be replaced with other types of fibers without sacrificing some of the excellent properties. Asbestos fibers have a tensile strength of 36,000 MPa (5225 ksi), an elastic modulus of 150 GPa (21,770 ksi), and a very large length-diameter ratio. They could be incorporated in a cement matrix up to 15 percent by weight. The composites are also heat-resistant.

The favored method for manufacturing asbestos cement is the Hatschek process. In this process, a sheet is formed around a rotating drum sieve that is immersed in a dilute cement–asbestos fiber slurry. The resulting sheet is used to manufacture a variety of components including flat sheets, corrugated sheets, and pipes. The fibers are fine enough to provide adequate retention of cement particles. About 12 percent of fibers can improve the flexural strength of a plain cement matrix by as much as five times. The elastic modulus can be doubled. The only drawback is that these fibers provide only a limited increase in ductility. The use of asbestos fibers is being discontinued because it causes asbestosis, which is a severe lung disease.

The fibers that were evaluated for replacing asbestos include steel, glass, carbon, aramid, polypropylene, polyethylene, polymide, polyester, rayon, polyvinylalcohol, cellulose, and HM-polyacrylonitril. A number of researchers around the world have concluded that asbestos cannot be replaced by a single fiber for all types of applications. The following are some of the problems encountered in using other types of fibers.

- Special manufacturing techniques are needed for steel fibers. Moreover, they are expensive and heavy.

- Glass fibers are too coarse and brittle for the Hatschek process of manufacturing. Alkali resistance can also be a problem, as explained in Chapter 13.

- Polymeric fibers and polymeric pulp have typically lower moduli of elasticity. Almost all the polymeric fibers have problems in the area of thermomechanical properties. Some of them have adhesion problems and chemical instability in the cement environment.

Attempts have been made to improve both the fiber and the matrix properties. Fibers have been coated with various compounds to improve their adhesion to cement and increase alkaline resistance. The elastic modulus has been increased for some polymeric fibers by modifying the processing techniques. For example, an elastic modulus of 37 GPa (~550 ksi) has been reported for polyvinylalcohol (PVA) fibers. Matrix modifications have been tried to reduce the alkalinity of the matrix and to improve the density; techniques tried include

- Using inert or pozzolanic fillers such as fly ash and silica fume
- Changing the chemical composition of the cement
- Improving the density of the composite by compression or dewatering
- Changing the curing procedure
- Treatment of the products after curing

Based on the results available so far, polymeric and cellulose fibers show good promise. However, many improvements are still needed, especially in the area of thermomechanical properties.

14.1 Thin Sheets Reinforced with Polymeric Fibers, Fabrics, or Meshes

Polypropylene fibers and fabrics have been evaluated by a number of researchers for use in thin sheets [14.1–14.3]. Other polymeric fibers investigated include polyvinylalcohol [14.4, 14.5], polyethylene [14.6], and polyester [14.5]. The following sections provide a brief discussion of the engineering properties of the composites containing different types of polymeric reinforcement. The matrices consist of either cement paste or cement mortar.

14.1.1 Polypropylene Fibers

The mechanical properties of 0.5 in. (12 mm) thick plates made using polypropylene fibers are reported in Reference [14.1]. The fibers consisted of the following: (1) 400 D (Denier [D] = the mass in grams of 900 m of yarn) of single filament fiber, 500 D one-ply twisted ribbon yarn with number of twists per inch varying from 1.5 to 6 and (2) a 2 × 500 D two-ply twisted ribbon yarn with 3 to 6 twists per inch and draw ratio (stretch ratio) varying from 6:1 to 12:1. The two-ply twisted fibers were found to perform better than the other fibers. These fibers had a tensile strength of 85 ksi (586 MPa), an elastic modulus of 980 ksi (6.7 GPa), and an average elongation at failure of 21 percent. The fibers were 1 in. (25 mm) long.

Specimens were cast using both a premixing of the fibers in the matrix and an infiltration of a fiber bed with a cement mortar slurry. The volume fraction of fibers varied from 2% to 6%. The matrix com-

TABLE 14.1 Typical Mortar Mix Proportions Used for Polypropylene-Reinforced Sheets [14.1]

	Mix proportions by weight		
	Mix 1	Mix 2	Mix 3§
Cement (C)*	1	1	1
Sand (S/C)	1†	1†	0.4‡
Water (W/C)	0.50	0.40	0.32
Superplasticizer/C (Melament)	0.023¶	0.023	0.023

* Cement was mostly ASTM Type III.
† Graded Ottawa silica sand, ASTM C-109.
‡ Fine Ottawa silica sand, all passing sieve no. 40.
§ All components were cooled (to about 40°F) prior to mixing.
¶ Used up to above amount only if necessary.

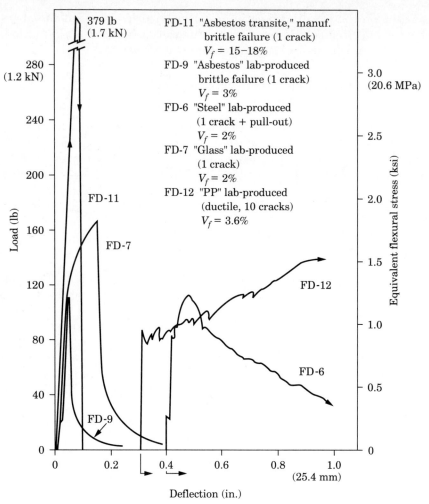

Figure 14-1 Comparison of the load-deflection responses of various fiber-reinforced cement composites [14.1].

position details are shown in Table 14.1. Specimens were cast using Plexiglas molds. They were cured in a room maintained at 100% relative humidity for one week to one month.

Prisms had dimensions of $12 \times 3 \times 0.5$ in. ($300 \times 75 \times 12$ mm). The prisms were tested in bending using a third-point load over a simply supported span of 10 in. (250 mm). The load-deflection responses were recorded by using an x-y plotter.

The load-deflection responses are shown in Figures 14-1 through 14-3. Figures 14-1 and 14-2 present the comparison between polypropylene (PP), asbestos, steel, and glass fibers. Polypropylene fibers compare well with steel, glass, and laboratory-produced asbestos fibers.

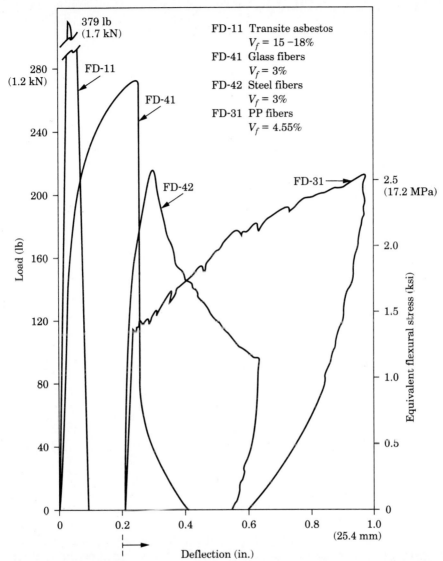

Figure 14-2 Comparative performances of steel, glass, polypropylene, and asbestos fiber-reinforced sheets [14.1].

Polypropylene fibers also provide better ductility than the other fibers. Figure 14-3 provides a comparison of the premixing and preplacing manufacturing techniques. Premixing provides better results. However, the preplaced fiber technique could provide a better composite if entrapped air is removed.

The influences of fiber volume, aspect (length-diameter) ratio, fineness of fibers, and Young's modulus of fibers on the mechanical proper-

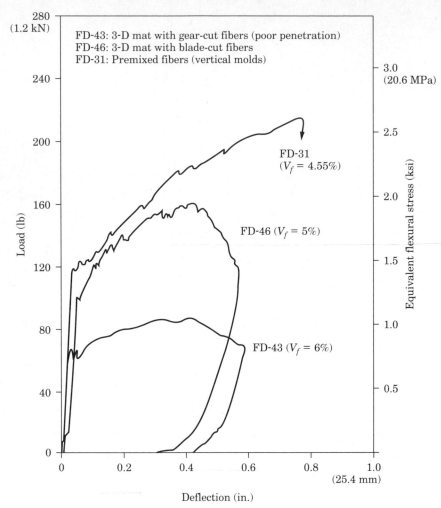

Figure 14-3 Comparison of polypropylene 3-D mats and premixed fibers [14.1].

ties of polypropylene fiber–reinforced composite are reported in Reference [14.7]. The specimens were prepared by mixing fibers and cement in a dilute cement slurry and forcing the mixture into thin discs under pressure. Excess water was removed using suction. High volume fractions of thin fibers could be incorporated using this technique.

The 120 mm (4.8 in.) diameter discs were tested in bending using simple supports, as shown in Figure 14-4. Load-deflection curves of the composite are shown in Figures 14-5 through 14-8. Figure 14-5 shows the influence of fiber diameter on the modulus of rupture and the load-deflection behavior. Since the lengths of the fibers were constant except for the 250 μm fiber (Figure 14-5), the aspect ratios were higher for the thinner fibers. The figure also shows the effect of fiber modulus for

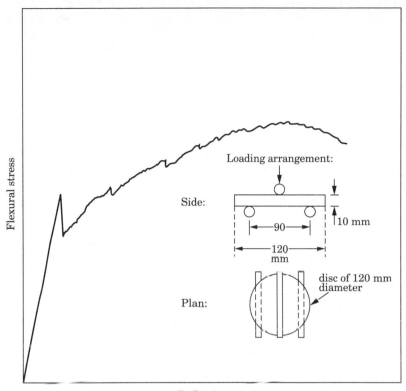

Figure 14-4 Flexural stress versus deflection of polypropylene-reinforced thin cement section [14.7] (1 in = 25.4 mm).

the 250 μm fiber. Both higher modulus and higher compaction pressure produce better postpeak response. Higher compaction pressure also results in higher first-crack strength. Lower fiber diameter provides much better postcrack response. The first-crack strength is not influenced by the fiber diameter. The improvement provided by the smaller fiber diameters seems to level off at about 20 μm. The leveling off could be due to mixing problems encountered when using fibers with large aspect ratios.

The effect of fiber length is shown in Figure 14-6. The curves were obtained using 49 μm diameter fibers. Here again, longer fibers with higher aspect ratios provide better postpeak response. The first-crack strength is slightly decreased for longer fibers. The improvement in performance levels off around an aspect ratio of 400.

Figure 14-7 shows the effect of volume fraction. There is a considerable improvement when the volume fraction is increased from 2.2% to 4.2%. Volume fractions higher than 4.2% do not produce a considerable increase in flexural strength. The postcrack load-deformation

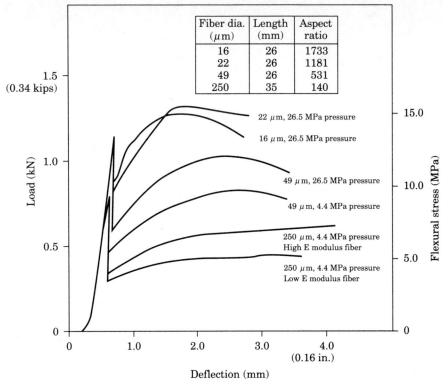

Figure 14-5 Effect of fiber diameter and preparation pressure on load-deflection behavior [14.7] (1 ksi = 6.895 MPa, 1 in. = 25.4 mm).

behavior is similar for fiber volume fractions greater than 4.2%. The optimum fiber volume fraction is about 5%.

The effect of fiber modulus is shown in Figure 14-8. As expected, a higher modulus provides better postcrack performance. The shapes of the curves are similar for moduli varying from 2.92 to 6.80 GPa (424 to 987 ksi), whereas the fiber with a modulus of 9.72 GPa (1410 ksi) provides a much steeper postcrack curve.

Overall, higher compaction pressure and higher fiber moduli provide better results. Finer fibers provide better results if they can be uniformly dispersed in the matrix.

14.1.2 Polypropylene Film

Open networks of polypropylene films have been used to reinforce cement matrices by a number of investigators [14.2, 14.8–14.12]. The use of continuous open networks allows for the incorporation of a higher fiber volume fraction. In addition, the continuous network provides better overall bonding and hence better ductility. In most instances, the composite develops a large number of fine cracks before

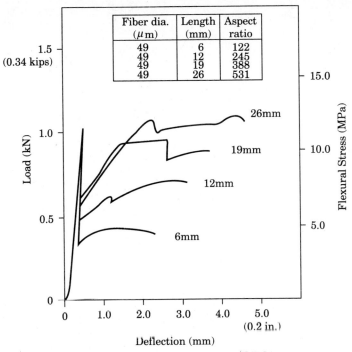

Figure 14-6 Effect of fiber length on load-deflection behavior [14.7] (1 in. = 25.4 mm, 1 ksi = 6.89 MPa)

failure. The results of a typical investigation on the behavior of thin sheets are discussed in the following paragraphs.

Cement sheets reinforced with polypropylene films were manufactured in the laboratory and tested to compare their behavior with asbestos cement sheets [14.2]. The asbestos sheets were 6–9 mm (0.2 to 0.4 in.) thick with corrugations at a pitch of 146 mm (5.7 in.). The overall depth was 57 mm (2.2 in.).

TABLE 14.2 Corrugated Sheet Dimensions and Section Properties [14.2]

	Asbestos cement	Polypropylene-reinforced cement	
		Sheet A	Sheet B
Overall length (mm)	1820	1660	1810
Overall breadth (mm)	1080	920	1010
Overall depth (mm)	57	56	55
Average thickness (mm)			
crest	9.8	9.1	7.2
side	6.1	8.0	6.3
trough	9.0	8.0	6.2
Second moment of area (mm⁴/m width)	2643×10^3	2885×10^3	2182×10^3
Section modulus of extreme tensile fiber (mm³/m width)	90×10^3	99×10^3	77×10^3

1 in. = 25.4 mm

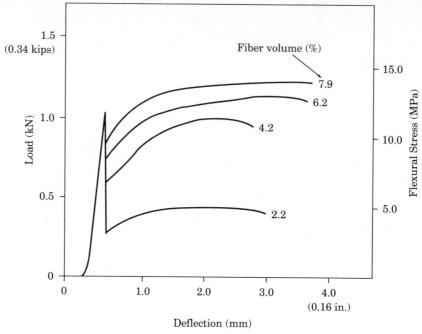

Figure 14-7 Effect of fiber volume fraction on load-deflection behavior [14.7] (1 in. = 25.4 mm, 1 ksi = 6.89 MPa).

The polypropylene-cement composite was reinforced with four layers of film networks. The tensile strengths and Young's moduli were 300 to 400 MPa (44 to 58 ksi) and 2 to 3 GPa (290 to 435 ksi) respectively. The reinforcement volume fraction was 5% in the longitudinal direction and 3% in the transverse direction. The matrix consisted of 1 part cement, 0.25 part pulverized fuel ash, 0.19 part sand, 0.34 part water, and 0.017 part high-range water-reducing admixture. The composite was manufactured as a flat sheet by placing film networks and working the fresh matrix into layers. The sheets were corrugated by pressing

TABLE 14.3   Direct Tensile Test Results: Average of Eight Specimens [14.2]

	Flat sheet A		Flat sheet B	
	Longitudinal	Lateral	Longitudinal	Lateral
σ_{co} (MN/m^2)	6.1	5.1	5.8	4.1
δ_{co} (MN/m^2)	6.8	5.6	6.3	4.8
E_c (GN/m^2)	24.5	22.4	24.5*	22.4
t_μ (microstrain)	279	250	258	212

σ_{co} = initial composite cracking stress (stress at 0.1% strain)
δ_{co} = mean composite cracking stress = stress at strain $\epsilon_{mc/2}$ where ϵ_{mc} is the strain at the end of multiple cracking
E_c = modulus of elasticity of uncracked composite
$t_\mu = \sigma_{co}/E_c$ = mean cracking strain
*The results from Sheet A were used. Sheet B results were unreliable because of warping.

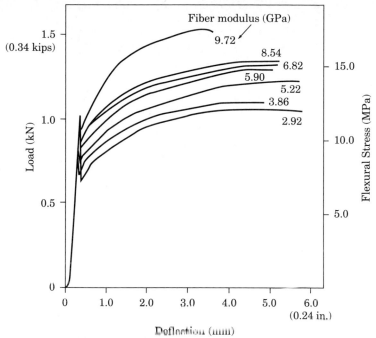

Figure 14-8 Effect of Young's modulus of fiber on load-deflection behavior [14.7] (1 ksi = 6.89 MPa).

the flat sheets into shape; they were covered with polyethylene sheets for 24 hours and cured for seven days in water.

The dimensions of the corrugated sheets tested are shown in Table 14.2. Sheet A was almost similar to the asbestos sheets; sheet B was thinner. The flat sheets were 6 mm (0.25 in.) thick. Prism specimens that were 300 mm (12 in.) long and 25 mm (1 in.) wide were cut from 600 mm (24 in.) square sheets for testing. The corrugated sheets were tested over a span of 1380 mm (54 in.), which is the recommended

TABLE 14.4 Important Values from Load-Deflection Curves for Polypropylene-Reinforced Corrugated Cement Sheeting [14.2]

	Sheet A	Sheet B
Equivalent maximum u.d. (uniformly disributed) load (kN/m²)	6.1	4.3
Equivalent u.d. load at LOP (kN/m²)*	3.5	2.8
Deflection at LOP (mm)	3.0	3.8
Equivalent u.d. load at a deflection of span/250 (kN/m²)	4.1	3.2
Approximate maximum average deflection that was recorded (mm)	20.2	14.8
Average residual deflection at zero load (mm)	7.4	5.1

1 in. = 25.4 mm, 1 lb = 4.4 N
*LOP—limit of proportionality

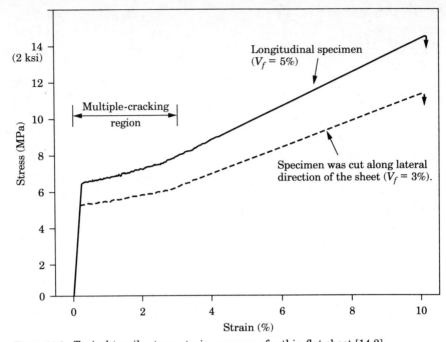

Figure 14-9 Typical tensile stress-strain response of a thin flat sheet [14.2].

purlin spacing for roof applications. The flat sheets were tested in direct tension at a strain rate of 10 percent per minute.

The results of tension tests are shown in Table 14.3. Typically, the sheets are stronger and stiffer in the longitudinal direction. The stress-strain response is shown in Figure 14-9. The curves have an initial elastic modulus that is approximately equal to the matrix modulus.

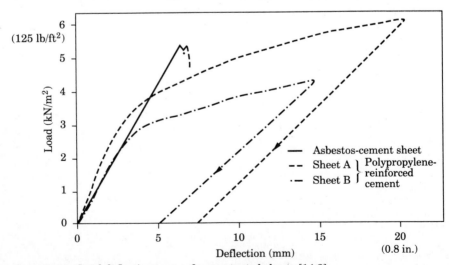

Figure 14-10 Load-deflection curves for corrugated sheets [14.2].

After cracking, the fibers play a major role in controlling the deformation. The post-crack response is almost linear, indicating the stable multiple cracking of the composite. The slope of the postcrack stress-strain curve is a function of fiber modulus and fiber volume fraction.

Table 14.4 and Figure 14-10 present the results of the bending tests. The asbestos-cement sheet had a higher first-crack strength but was very brittle. The asbestos sustained a bending stress of 14.5 MPa (2.1 ksi). Sheets reinforced with polypropylene films had an initial stiffness equivalent to that of the asbestos sheet. After the matrix cracked, the fibers were able to sustain higher loads.

The polypropylene-reinforced sheets were also subjected to a sustained load of 0.75 kN/m^2 (15.7 lb/ft^2) over a span of 1380 mm (54 in.). The deflections were found to be less than span/350 after two years of sustained load.

14.1.3 Woven Polypropylene Fabrics

Earlier studies indicated that films are better than woven fabrics [14.8]. However, the investigations carried out later show that woven fabrics provide better results [14.3, 14.13]. Woven fabrics are easy to

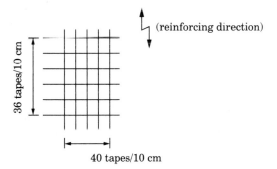

Fabric structure

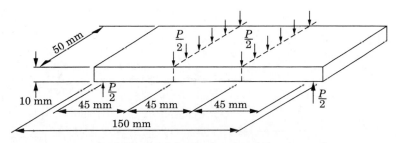

Test specimen and rig dimensions

Figure 14-11 Dimensions and loading and fabric structure of test specimens [14.11]. (1 in. = 25.4 mm)

handle and easy to place, using both manual and automated manufacturing techniques. Since woven fabrics are thicker, the number of
layers required is much smaller than with films. For example, 6 mm
(0.25 in.) thick sheets need only six layers of woven fabric compared
to 48 layers of opened networks [14.11].

Tests were conducted using both smaller-size plates [14.13] and
large plates [14.3]. Results reported in Reference [14.13] deal with
the flexural behavior of 150 × 50 × 10 mm (6 × 2 × 0.4 in.) prisms,
loaded using third-point bending, (Figure 14-11). The sheets were
manufactured on a form with a plan dimension of 600 × 900 mm
(24 × 36 in.). The matrix was cement mortar with a water-cement

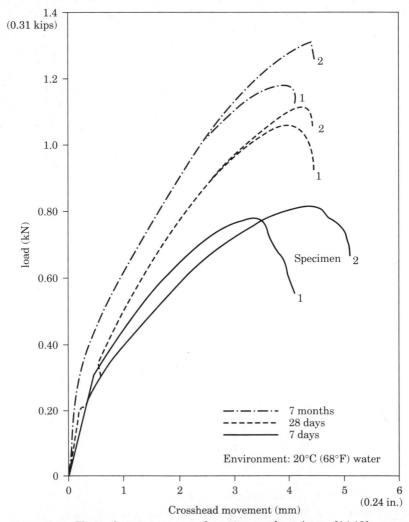

Figure 14-12 Flexural response curves for water-cured specimens [14.13].

ratio of 0.5. The paste and the fabric were placed in alternating layers. Ten layers of reinforcement were needed to obtain a 10 mm (0.4 in.) thick sheet. Limited tests were also conducted using 6 mm (0.25 in.) thick sheets. Three types of curing regimes were used. The samples covered with damp burlap sacking were left on the form for 48 hours. After this 2 day period, one set of samples was cured in water at 20°C (68°F), the second set was merely stored in the laboratory environment, and the third set was placed on the roof of the laboratory. The effects of aging were studied using specimens exposed to the aforementioned environments for 7 months.

Figure 14-12 shows the load-deflection response of bending specimens after 7 days, 28 days, and 7 months. The specimens were con-

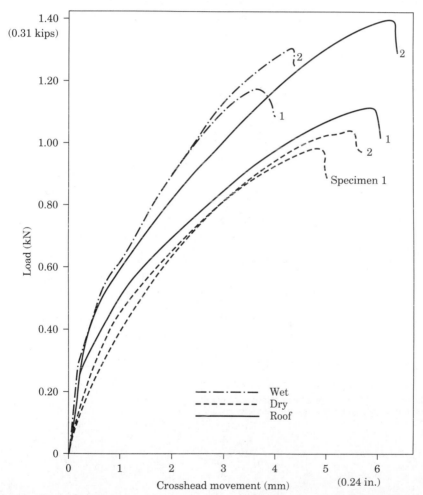

Figure 14-13 Flexural response curves at 7 months after curing in three different environments [14.13].

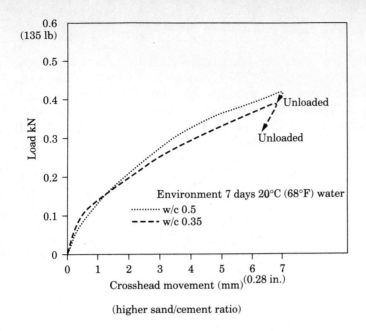

(higher sand/cement ratio)

(lower sand/cement ratio)

Figure 14-14 Flexural response curves for 6 mm thick specimens with different water-cement and sand-cement ratios [14.13].

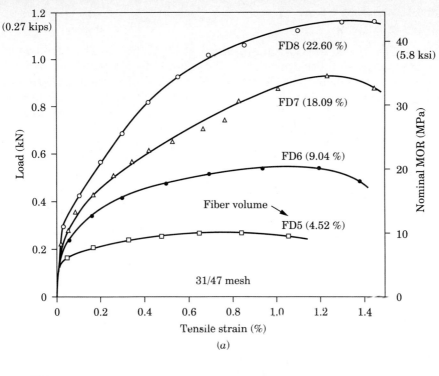

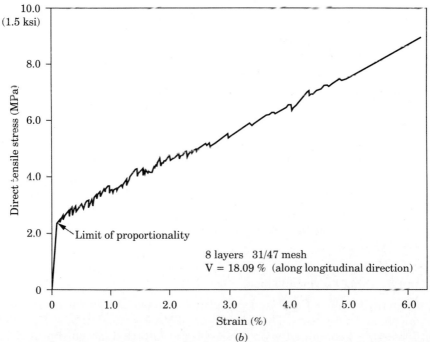

Figure 14-15 Load-tensile strain for various fiber contents [14.3].

tinuously stored in water maintained at 20°C (68°F). The increase in age resulted in higher strengths. This should be expected because the matrix strengthens more with age. The shape of the curves does not seem to change in character. The composite has good postcrack behavior at all three ages. Both initial and postcrack stiffness increase with age. This increase in stiffness could be due to better bonding between fibers and the matrix at latter stages of maturity.

The effects of curing regimes are shown in Figure 14-13. Water curing provided the best results. The specimens left in the dry environment had the lowest strengths. Specimens cured outdoors had better strength than the specimens stored inside the laboratory, even though they were subjected to some freezing and thawing cycles.

Figure 14-14 shows the effect of water-cement and sand-cement ratios. Lower water-cement ratios and higher cement content provide higher strengths at an early age. The difference decreases with age.

The effect of fiber volume fraction can be seen in Figure 14-15 [14.3]. The tests were conducted using $500 \times 100 \times 10$ mm ($20 \times 4 \times 0.4$ in.) specimens. The matrix consisted of cement and fly ash. As expected, a higher volume fraction resulted in better postcrack behavior.

Tests conducted using corrugated sheets of $1000 \times 600 \times 10$ mm ($40 \times 24 \times 0.4$ in.) loaded over an effective span of 900 mm (36 in.) indicate that the fabrics provide consistent postcrack resistance [14.3]. For fiber volume fractions ranging from 1.2% to 10%, the limits of proportionality vary from 7.5 to 13.5 MPa (1.1 to 2 ksi). The flexural strengths were 7.5 to 54 MPa (1.1 to 8ksi). The crack patterns and postcrack behavior of corrugated sheets are similar to those of plain sheets.

14.1.4. Comparison of Filament, Film, and Woven Fabrics

Table 14.5 provides a comparison of strength values for the three types of polypropylene reinforcement. Woven fabric seems to provide higher strengths than the other two types of reinforcement. Postcrack performance is similar even though fabrics provide stiffer response after the first crack.

14.1.5 Other Polymeric Fibers

Polyvinylalcohol (PVA) and polyethylene have also been investigated for use in thin sheets. The advantage of PVA is its high modulus. This fiber is also combined with other fibers. Polyethylene pulp provides a modulus of rupture as great as 20 MPa (~3 ksi) [14.6]. Both these fibers provide good postcrack performance [14.4, 14.6].

TABLE 14.5 Ultimate Load Capacities for Various Fiber Types [14.3]

Reference	Matrix	Reinforcement	Curing conditions	Ultimate moment (N · m)
	Lightweight matrix	8 layers 24/31 polypropylene mesh, $v = 5.78\%$ (WC4)	28 days in fog room: 20°C, RH 100%	41.0
	Lightweight matrix	8 layers 16/16 polypropylene mesh, $v = 1.64\%$ (WC8)	28 days in fog room: 20°C, RH 100%	29.0
[14.3]	Cement paste	4 layers 31/47 polypropylene mesh, $v = 9.04\%$ (WD6)	28 days in fog room: 20°C, RH 100%	65.6
	Cement paste	10 layers 31/47 polypropylene mesh, $v = 22.6\%$ (WD8)	28 days in fog room: 20°C, RH 100%	144.0
		10 layers 36/40 polypropylene mesh, $v = 11\%$ (d = 10 mm)	28 days in water at 20°C	24.8
			7 months in water at 20°C	29.3
[14.11]	Cement mortar	10 layers 36/40 polypropylene mesh, $v = 11\%$ (d = 10 mm)		
		6 layers 36/40 polypropylene mesh, $v = 1$: (d = 6 mm)	5 months in water at 20°C	15.0
		4% volume chopped filaments of polypropylene (22 μm diameter)	28 days under water	18.8
[14.7]	Cement paste	7.9% volume chopped filaments of polypropylene (49 μm diameter)	28 days under water	19.4
[14.12]	Lightweight matrix	7.5% volume of fibrillated filament	28 days in wet room at 20.5°C, RH 90%, and oven dried at at 50°C for 48 hours	8.1
	Lightweight matrix	7.5% volume of monofilament	28 days in wet room at 20.5°C, RH 90%, and oven dried at at 50°C for 48 hours	7.4
[14.13]	Cement mortar	6% volume of opened polypropylene film networks		40.5

1 in. = 25.4 mm, 1lb = 4.4 N, °F = (°C × $\frac{9}{5}$ + 32)

14.2   Thin Sheets Reinforced with Carbon Fibers

Carbon fibers had been used for reinforcing plastics for a long time. Investigations that attempted to use carbon fibers in cement matrix date back to the 1970s [14.14, 14.15]. The earlier investigations dealt with high-modulus, high-strength carbon fibers called PAN fibers. These fibers, made by baking acrylic fibers, are very expensive, and hence the composite was not promoted for field use. In the 1980s, low-cost carbon fibers were produced using pitch, contained in petroleum and coal. Even though the pitch-based fibers had lower moduli of elasticity and strengths less than PAN fibers, their moduli and strengths were still quite high and hence were suitable for use in cement matrix. The tensile strengths and moduli of elasticity of pitch-based fibers vary from 400 to 780 MPa (~60 to 110 ksi) and 27 to 55 GPa (~4000 to 8000 ksi), respectively. The fibers can be obtained in various lengths and diameters. Typical lengths vary from 2 to 12 mm (0.08 to 0.5 in.), and the diameters vary from 10 to 18 μm (~0.4 to 0.7 $\times$ 10^{-3} in.). These fibers have relatively high aspect (length-diameter) ratios. The specific gravity is about 1.6. Carbon fibers are also available in the form of mats and long continuous filaments.

A number of investigators have evaluated the various engineering properties of carbon fiber–reinforced cement composites using pitch-based fibers [14.16–14.20]. The current use of carbon fiber–cement composites include cladding, free-access floor panels, lightweight decorating frames, shell structures, and protective coatings for structural elements exposed to harsh environments [14.18]. The following sections provide a brief discussion on the engineering properties of carbon fiber–cement composites.

14.2.1   Materials and Mixture Proportions

The materials used for carbon fiber-cement composites typically consist of portland cement, fine aggregate, and admixtures. The fine aggregate could be a normal or lightweight sand. Since the addition of fibers decreases the workability of the mix, high-range water-reducing admixtures are often used to improve the workability. The addition of silica fume was found to improve the dispersion of fibers. Furthermore, silica fume improves the density of the matrix and improves the bond between the fibers and the matrix. Since silica fume has high water demand, a high-range water-reducing admixture is a necessity for mixes containing silica fume. Accelerating admixtures have been used for obtaining high early strengths.

As mentioned earlier, the reinforcement can be meshes, short filaments, or long continuous strands. If strands and meshes are used, special construction techniques are needed.

Matrices with high cement content and lower water-cement ratio provide better results, as in normal concrete. Water-cement ratios that were investigated ranged from 0.27 to more than 1.0. Water-reducing admixtures are needed for mixes with lower water-cement ratios. Fiber contents that were evaluated ranged from 0.25% to 5% by volume.

14.2.2 Workability of Plastic (Fresh) Mix

The addition of fibers typically reduces workability. Special mixers may be needed for stiff mixtures and higher fiber volume fractions. Figure 14-16 presents typical variations of workability measured using the Japan Standard Test JIS R5302 flow test [14.17]. The three lines in Figure 14-16 represent the behavior of mixes containing 0% and

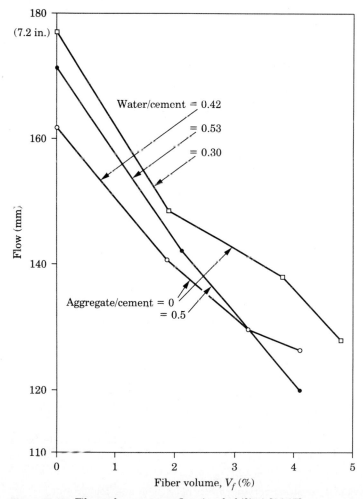

Figure 14-16 Fiber volume versus flow (workability) [14.17].

50% silica fume (by weight of cement) and methyl cellulose admixture. The loss in flow seems to be approximately linear for all three mixes, (Figure 14-16.)

14.2.3 Mechanical Properties

The unit weight of the carbon cement composite decreases with an increase in fiber content. The lower specific gravity of fibers compared to the cement matrix and the entrapped air created during mixing account for most of the decrease in density. The magnitude of reduction is usually not significant.

The addition of fibers increases both tensile strength and ductility in the tension mode. The tests conducted using 12 mm (0.5 in.) thick dog bone–shaped specimens indicate linear increases in tensile strength with respect to fiber content [14.17]. Tensile strengths could be doubled at a fiber content of about 4%. The strength increases were observed for all types of matrices with and without admixtures.

The fibers also provide ductility to the brittle cement matrix (Figure 14-17), [14.17]. Composites made with low-strength matrices have a long, stable postcrack stress-strain curve. In most cases there is an increase in strength with an increase in strain (Figure 14-17). High-strength matrices with lower water-cement ratios result in lower fracture strains than composites made with low-strength matrices (Figures 14-17, 14-18). A high-strength matrix provides a better bond, and hence more fibers fracture at failure, whereas, in composites with a low-strength matrix, most fibers pull out at failure. When fibers pull

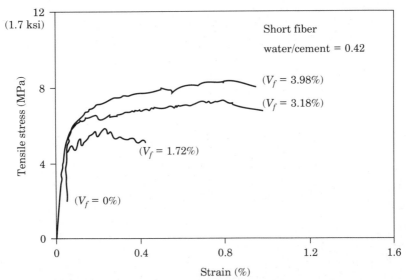

Figure 14-17 Tensile stress-strain curves: carbon fiber composite, higher water-cement ratio [14.17].

out they provide ductility but lower strength, whereas, when they break, their strength is utilized but the fracture strain decreases. The lengths of fibers also control the type of failure. Longer fibers tend to break more often than shorter fibers.

Adding fibers normally decreases the compressive strengths and the moduli of elasticity (Figures 14-19, 14-20). The decreases, attributed mainly to the entrapped air and less than optimal compaction, are not significant. The fibers contribute to the ductility by stabilizing the failure process. An increase in fiber content produces a consistent increase in strain capacity in compression.

The modulus of rupture (flexural strength) can also be enhanced using carbon fibers (Figures 14-21). It can be seen from Figure 14-21 that the modulus of rupture could be increased by as much as 150% [14.17]. The fibers provide even a more substantial contribution in the area of ductility, as shown in Figures 14-22 and 14-23 [14.17]. The toughness provided by fibers is relatively higher for lower-strength matrices, as in the case of tension specimens. The effect of fiber pull-out has a more substantial influence in bending ductility than in tension. Results shown in Figures 14-21 through 14-23 were obtained using 10 mm (0.4 in.) thick beams tested using a center point load over a span of 150 mm (6.0 in.).

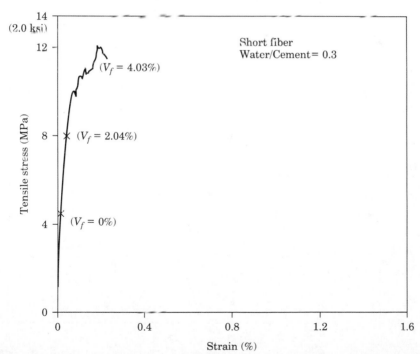

Figure 14-18 Tensile stress-strain curves: carbon fiber composite, low water-cement ratio [14.17].

The effect of specimen size on modulus of rupture was studied using beams with a thickness ranging from 6 to 60 mm (0.25 to 2.5 in.) [14.17]. The 460 mm (18.4 in.) long beams were loaded at middle third points over a simply supported span of 450 mm (18 in.). The results of strength variations are shown in Figure 14-24 [14.17]. The strengths decrease rather sharply in the thickness region of 6 to 20 mm (0.25 to 0.80 in.). The primary contributing factor to the reduction could be the strain capacity of the composite in tension. In deeper beams, the extreme tension zone could experience a strain magnitude larger than tensile fracture strains. This in effect reduces the thickness of the beam, contributing to the strengths. Hence the resulting flexural strength decreases. The strength reduction could be more pronounced if failure occurs by the fracturing of fibers rather than the pulling out of fibers. The above hypothesis is supported by the reduction in maximum surface strains at failure for thicker specimens (Figure 14-25) [14.17]. The formation of larger cracks reduces the strains experi-

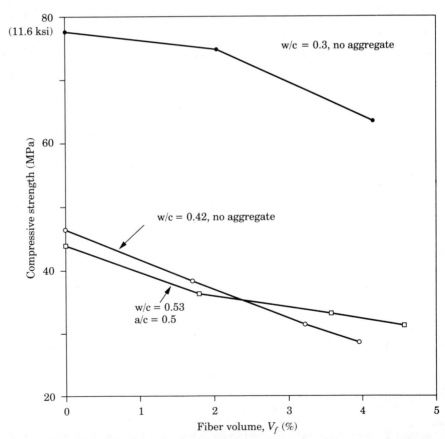

Figure 14-19 Fiber volume versus compressive strength: carbon fiber composite [14.17].

enced by the composite located between the cracks. Extensive studies of fracture mechanics have been conducted on the size effect (Chapters 2 and 3).

Studies were also conducted to measure the impact strength by dropping a steel ball on the specimen. Fiber volume fractions equal or higher than 3% provide large increases in impact strength [14.16]. The impact strength increase was found to be independent of fiber length.

14.2.4   Durability and Drying Shrinkage

The durability and effectiveness of carbon fibers in a cement matrix was evaluated by testing carbon fiber–reinforced prisms in bending after storing them in hot water, maintained at 167°F (75°C), for a period of 5 months [14.18]. The fibers were found to be durable and effective after the accelerated aging process. Both the flexural strength and the load-deflection behavior did not change after curing at high temperatures.

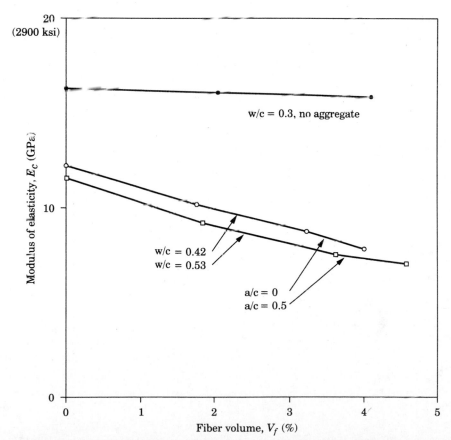

Figure 14-20 Fiber volume versus modulus of elasticity: carbon fiber composite [14.17].

The freeze-thaw durability of the composite was evaluated using ASTM C666, Procedure A [14.18]. The composite can be considered durable under freeze-thaw conditions since the durability factor after 300 freezing-and-thawing cycles was above 95%.

The addition of fibers decreases the drying shrinkage of the matrix [14.16, 14.18, and 14.20]. The drying shrinkage stabilizes at a very early age for autoclaved specimens [14.16]. In addition, the specimens cured in water undergo higher shrinkage strains compared to auto claved specimens [14.18]. The total shrinkage strains are normally limited to about 200 μin./in (200μm/m).

14.2.5 Water Absorption

Water absorption is an important property for cladding applications because it indirectly affects the dimensional stability and weight of the panel. An increase in fiber content results in a consistent decrease in water absorption [14.16].

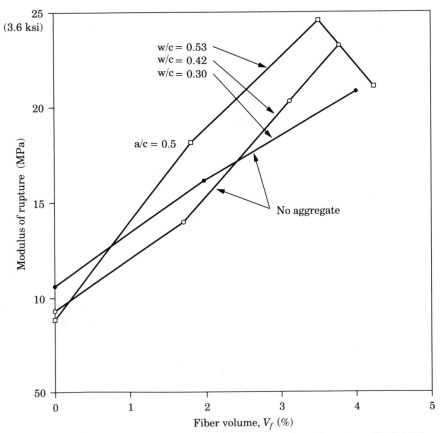

Figure 14-21 Fiber volume versus modulus of rupture: carbon fiber composite [14.17].

14.2.6   Effect of a Curing Regime on Engineering Properties

The effect of the curing procedure on the engineering properties is an important parameter because for certain applications the composite has to be cured in a short time using either autoclaving or steam curing. Results presented in Reference [14.21] provide a comprehensive comparison of normally used curing schemes.

The experimental composite was made using ASTM Type I cement, silica fume, water, high-range water-reducing admixture, and short carbon fibers. Shorter fibers that measured 1.5 mm (0.06 in.) were used at volume fractions of 3% and 5%. The longer, 3 mm (0.12 in.) fibers were used at a volume fraction of 3%. A water-binder (cement + silica fume) ratio of 0.3 was used for the specimens.

Flexural, compressive, and impact tests were carried out using $1.5 \times 1.5 \times 6$ in. ($38 \times 38 \times 150$ mm) prisms, 3×6 in. (75×150 mm) cylinders, and 6×2.5 in. (150×63 mm) cylindrical disks, respectively. Impact tests were conducted using ACI Committee 544 recommendations. After casting, the specimens were kept in their molds for 24 hours and covered with plastic sheets. They were removed from their molds after 24 hours, air dried for 12 hours, and subjected to one of the following curing schemes for 24 hours.

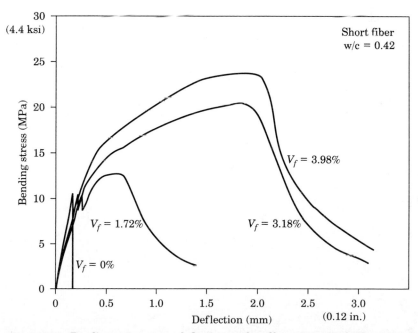

Figure 14-22   Bending stress versus deflection: carbon fiber composite, higher water-cement ratio [14.17].

- Air curing in a laboratory environment at a relative humidity of 50% and a temperature range of 70°–80°F (22°–26°C).
- Immersion in water maintained at a temperature of 70° –80°F (22°–26°C). This is designated as water curing.
- Storage in moist conditions with a relative humidity of 100% and a temperature range of 70°–80°F (22°–26°C). This is designated as moisture curing.
- Hot water curing, in which specimens were immersed in water maintained at 170°–180°F (76°–82°C).
- Steam curing in steam room maintained at 140°F (60°C).

After the 24-hour curing period, all the specimens were air cured for 14 days before testing.

Flexural strength results for the various mixtures and curing conditions are shown in Figure 14-26. The trends were similar for compressive and impact strengths. In the area of compressive strength, fibers provide better strength for water, air, and hot water curing. The plain matrix had a slightly higher strength for steam and moisture curing. Hot water curing provides the best overall results. Theo-

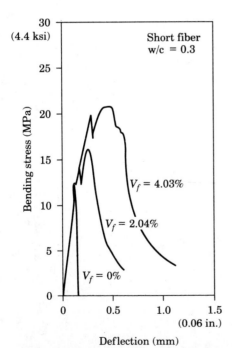

Figure 14-23 Bending stress versus deflection: carbon fiber composite, lower water-cement ratio [14.17].

retically, steam curing should provide the best results. However, the cylinder specimens stored in hot water might have retained more water in the 24 hour period for further curing, whereas in the steam curing most of the water could have evaporated.

Fiber-reinforced composite had a much higher flexural strength for all curing schemes. Here again hot water curing provided the best results, but the differences were much less substantial compared to behavior in compression.

Overall, steam or hot water curing seem to be the efficient forms of curing. It should be noted that autoclaving was also extensively used for curing [14.16, 14.18]. This is a special advantage when using carbon fibers since the building components (specimens) made with polymeric and cellulose fibers cannot be autoclaved.

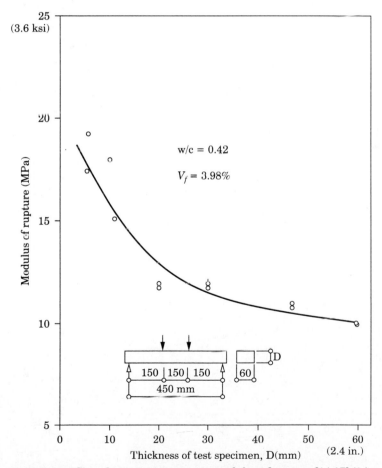

Figure 14-24 Size of test specimen versus modulus of rupture [14.17] (1 in. = 25.4 mm).

14.2.7   Recent Developments

Evaluations conducted using a lightweight matrix indicate that carbon fibers can be effectively used for lightweight claddings and similar building components [14.22]. More recently, high-modulus fibers have been developed and tested for use in a cement matrix. High-modulus fibers provide much better performance compared to low-modulus fibers in the area of strength increase [14.23].

14.3   Thin Sheets Reinforced with Cellulose Fibers

Cellulose is one of the natural fibers that is available in large quantities and that is less expensive than the synthetic fibers. As explained in Chapter 5, cellulose fibers are derived from wood. Fiber durability

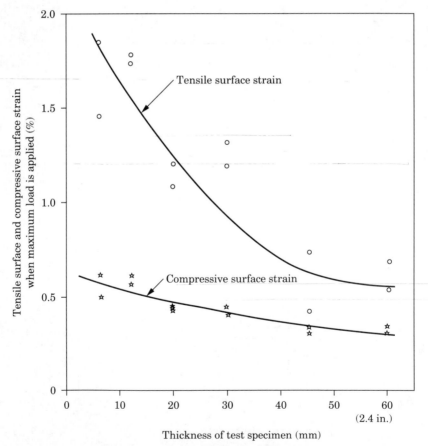

Figure 14-25   Size effect on strain distribution [14.17].

is the primary problem in using cellulose fibers with a cement matrix. However, researchers and industries looking for the replacement of asbestos fibers are making excellent progress in improving the longevity of cellulose fibers in cement matrices. In some instances cellulose fibers are also being used as a processing aid in a Hatschek kind of manufacturing of thin cement sheets.

14.3.1   Materials and Mixture Proportions

The primary components of cellulose cement sheets are cellulose fibers and portland cement. In some cases fillers such as silica fume are added to reduce the permeability and to improve the bond between the fibers and the matrix. Depending on the type of manufacturing process, the fiber content ranges from 1% to 12% by weight.

The mixture proportions depend on the type of manufacturing technique. In the premixing process, the fiber volume fraction is controlled by the mixer efficiency. The fiber contents are generally limited to about 2%. More fibers can be incorporated if a dewatering technique is used. It is advisable to keep the final water content minimal, irrespective of the manufacturing technique. A lower water content results in lower permeability, less water absorption, fewer dimensional changes, and better mechanical properties.

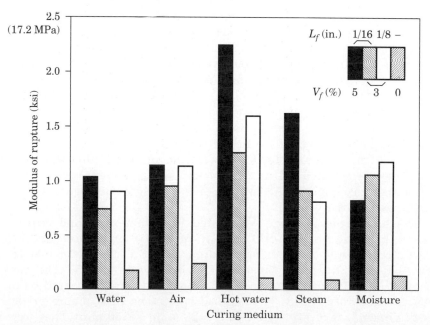

Figure 14-26   Effect of curing conditions on flexural strength [14.21] (1 in. = 25.4 mm).

TABLE 14.6 Mixture Proportions
for Mechanical and Kraft Pulp [14.24]

Fiber type	Fiber mass fraction	Water-cement ratio	Flow (%)
—	0	0.28	66
SSK*	1	0.35	65
SSK	2	0.40	62
NHK†	1	0.35	63
NHK	2	0.40	62
1000L‡	1	0.35	74
1000L	2	0.40	66

*SSK—southern softwood kraft
†NHK—northern hardwood kraft
†1000L—mechanical pulp

14.3.2 Properties of Sheets Made Using the Premixing Process

In the premixing process, the cellulose fiber composite is made like normal concrete (mortar), using molds. As mentioned earlier, the fiber weight fraction that can be incorporated is limited to about 2%. A water-cement ratio of about 0.4 is needed for the proper dispersion of fibers. High-range water-reducing admixtures could be used to improve the flowability of the wet matrix.

The following sections provide a brief discussion of the properties of cellulose-reinforced cement paste using softwood and hardwood kraft pulp and mechanically extracted fibers [14.24].

The average length of fibers was 8.0 mm (0.32 in.) for mechanical pulp. The fiber lengths were 3.0 and 0.9 mm (0.12 and 0.04 in.) for softwood and hardwood kraft pulp respectively. The matrix consisted of a pure cement paste made using ASTM Type I cement. The water-cement ratios for the various mixture proportions are shown in Table 14.6. The water-cement ratios were adjusted to obtain the same level of consistency of the freshly mixed composite. Table 14.6 also shows the flow, measured using the ASTM C-230 procedure [14.24].

The wet mixes were tested for flow at 1, 5, and 10 minutes after mixing; they were tested for air content and setting time by penetration resistance. Appropriate ASTM test procedures (C-230, C-185, C-403) were used for the tests.

The hardened composite was tested for compressive and flexural strengths using ASTM C-873 and ASTM C-1018 procedures, and impact strength was tested using the drop weight method recommended by ACI Committee 544: fiber-reinforced concrete [14.25]. The void content, specific gravity, and water absorption were measured using ASTM procedure C-642. The flexural specimens were square prisms with dimensions of 1.5 × 1.5 × 6 in. (38 × 38 × 150 mm). They were tested using third-point loading over a simply supported span of 4.5 in.

(114 mm). The cylinders used for compressive strength tests were 3 in. (75 mm) in diameter and 6 in. (150 mm) long. The impact test specimens were 2.5 in. (63 mm) thick and 6 in. (150 mm) in diameter.

The test specimens were cast using standard molds and external vibration. They were left in their molds, covered with plastic sheets for 24 hours, and moisture cured for 5 days. Then the specimens were left in the laboratory environment until the test age of 28 days.

The test on the workability of the mixture after 1, 5, and 10 minutes of mixing showed that the workability is about the same for all the fiber types and fiber content during the first 10 minutes [14.24]. Mechanical pulp had a higher initial and final set time compared to the plain mix and the mixes with kraft pulp. Chemicals present in the mechanical pulp might have some retarding effect, thus increasing the setting time. The air content increased with the increase in fiber content, as with other types of fibers.

The addition of cellulose fibers provided a considerable increase in flexural strengths (Figure 14-27) [14.24]. The compressive strength showed a slight reduction. The increase in air content and decrease in compaction efficiency could account for reduction in compressive strength. The flexural strength increase is contributed by the fibers. The fiber contribution is more substantial in the area of flexural toughness, as shown in Figure 14-27. The toughness values were calculated using the Japanese standards, in which toughness is measured as the area under the load-deflection curve up to a deflection of 0.03 in. (0.75 mm).

The impact strength results presented in Figure 14-28 show that the kraft pulps make an effective contribution to impact strength. The contribution of mechanical pulp is minimal. The impact strength is important in applications in which the thin sheets might have to resist small projectile loading caused by debris and particles carried by high winds.

The addition of cellulose fibers increases the air content and water absorption and reduces the density of the composite. This should be expected because the addition of fibers increases entrapped air, and cellulose fibers have an affinity to water.

In summary, cellulose fibers can be used to improve flexural strength and flexural toughness, and impact resistance in thin-sheet applications. A slight reduction in compressive strengths and an increase in water absorption should be accounted for in the design.

14.3.3 Properties of Sheets Made Using the Dewatering Process

In the slurry-dewatering process, up to 12% (by weight) of fibers can be easily incorporated in the cement matrix. In this process, fibers are

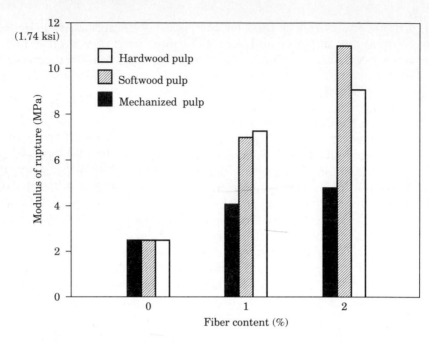

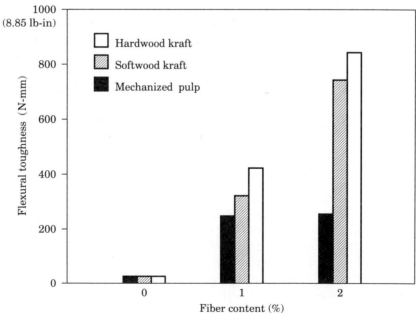

Figure 14-27 Flexural strength and toughness for wood pulp–reinforced composite [14.24].

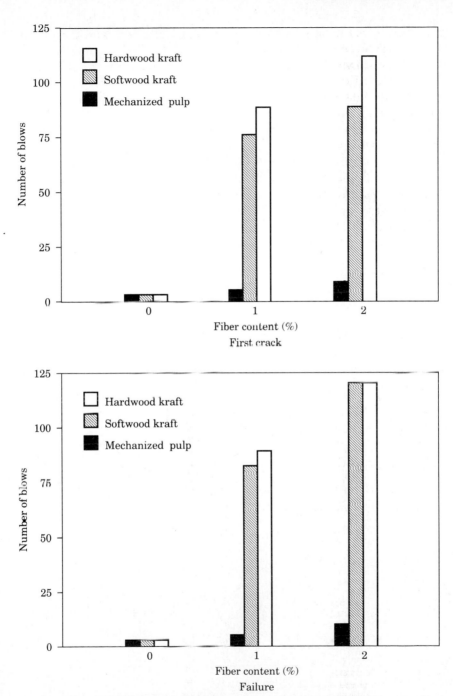

Figure 14-28 Impact strength for wood pulp–reinforced composite [14.24].

dispersed in the dilute cement slurry using agitation, and the excess water is removed using a vacuum process. The final product has low water-cement ratio and good compaction. Figure 14-29 shows flexural strength variations of cellulose fiber–reinforced composite containing from 2% to 12% fiber [14.26]. The sheets were manufactured using Pine Radiata kraft fibers (Canadian Standard Freeness of 700).

The specimens were either air cured after the initial 24 hours in the mold or autoclaved using steam at a pressure of 124 psi (0.8 MPa) for 8 hours. The mortar specimens contained equal amounts of cement and silica sand. From Figure 14-29, it can be seen that the increase in fiber content provides an increase in flexural strength up to a fiber content of 8%, and it is possible to obtain flexural strengths up to 3 ksi (21 MPa). The fiber addition resulted in a much higher increase in toughness compared to the increase in flexural strength (Figure 14-30) [14.26]. The toughness was measured using the area under the load-deflection curve.

A new generation of cellulose fibers was investigated in Reference [14.27]. This investigation dealt with three types of fibers: summer

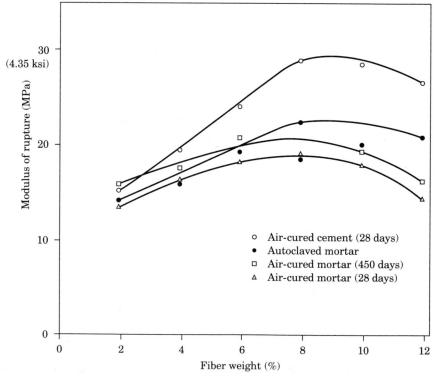

Figure 14-29 Fiber content versus flexural strength: unbeaten Pine Radiata kraft pulp [14.24, 14.26].

wood pulp, a combination of summer and spring wood pulp, and fibrillated fiber. The following abbreviations are used in the discussions.

> SSK— Summer wood pulp
>
> SSK-SUWD— Summer and spring wood pulp
>
> EF— Expanded fibrillated fibers
>
> -R— Refined in a laboratory beater to a Canadian Standard Freeness of 500

Summer wood fibers were found to be much denser and were expected to provide better mechanical properties. Fibrillation provides much finer fibers, improving filter characteristics in the Hatschek process. Fibrillation also provides more bonding area.

The test specimens were prepared using a simulated Hatschek process. Fibers were dispersed in water using high-speed mixing. The solid content was limited to about 5%. The ASTM Type I cement was added to the fiber mixture and blended well. The amount of cement needed was determined using the fiber-cement ratio required in the hardened composite. A flocculating agent was added, and the

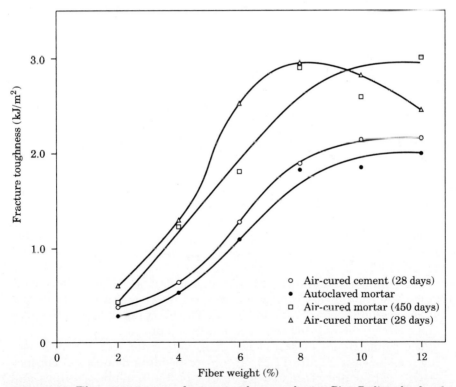

Figure 14-30 Fiber content versus fracture toughness: unbeaten Pine Radiata kraft pulp [14.24, 14.26].

TABLE 14.7   Mixture Proportions for Cellulose Fiber Composites [14.27]

Fiber type	Fiber content (% by weight)	Water-cement ratio	Wet density (lb/ft³)	Dry density (lb/ft³)
	4	0.33	126	118
SSK-R	8	0.34	120	108
	12	0.43	113	98
	4	0.34	124	119
SSK-SUWD	8	0.39	117	109
	12	0.57	109	96
	4	0.33	129	123
SSK-SUWD-R	8	0.36	122	112
	12	0.40	112	104
	4	0.30	132	123
EF	8	0.35	123	108
	12	0.42	117	101
SSK-SUWD/EF	8	0.36	117	109

1 lb/ft³ = 16.05 kg/m³
SSK-R—fiber refined in a laboratory beater (SSK southern softwood kraft contains 55% summer wood fibers.)
  SSK-SUWD—unrefinded (SSK-SUWD contains 86% summer wood fibers)
  SSK-SUWD-R—refined in a laboratory beater
  EF—expanded fiber
  SSK-SUWD/EF—unrefined (8%) + expanded fiber (1.5%)

mixture was poured through an assembly of permeable screens, the finest being ASTM No. 100 mesh. The material retained in the screens was dewatered using a vacuum dewatering pressure of 5 in. (125 mm) of mercury. The moist composite was pressed for 3 minutes under 1200 psi [8.5 MPa] to obtain the final product. A hydraulic press was needed for the composite containing the fine fibers.

The composite boards were left to harden for 4 hours and were cured at 23°C (73°F) and 100% relative humidity for 7 days. Flexural test specimens with dimensions of $2 \times 0.5 \times 12$ in. ($50 \times 12 \times 300$ mm) were sawed after the curing period.

Specimens were tested under dry (laboratory environment) and wet (soaked in water for 24 hours) conditions. The testing was done using a central load over a simply supported span of 10 in. (250 mm). Table 14.7 presents the details of mixture proportions for the various fibers tested [14.27].

The summer wood fibers, SSK-R, had a loss of 1% to 1.5% cement fines. The combined spring-summer wood fibers had a very high loss of fines ranging from 17.0% to 18.5% . The blending of fibrillated fiber and coarse fiber reduced the fines loss to about 3.8%.

The proportional elastic limit (PEL) values for wet and dry testing are shown in Figure 14-31. The corresponding modulus of rupture (MOR) values are presented in Figure 14-32. Toughness values and load-deflection behavior are shown in Figures 14-33 and 14-34, respectively.

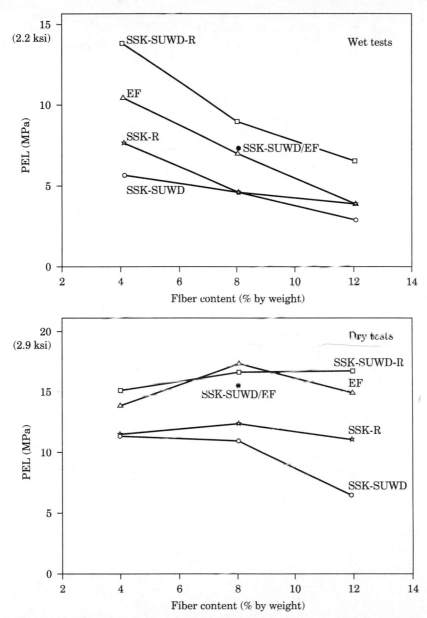

Figure 14-31   Proportional elastic limit versus fiber content: wet and dry tests [14.27].

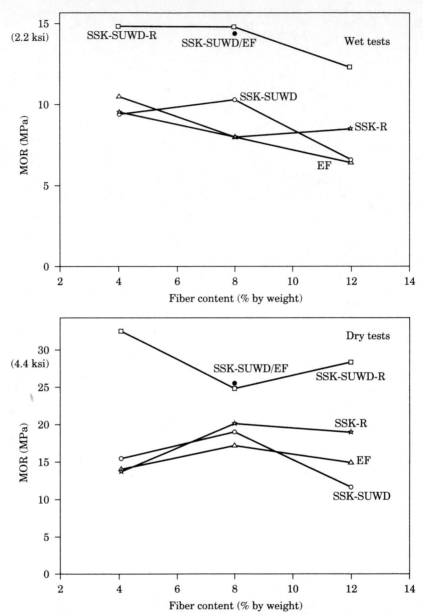

Figure 14-32 Modulus of rupture versus fiber content: wet and dry tests [14.27].

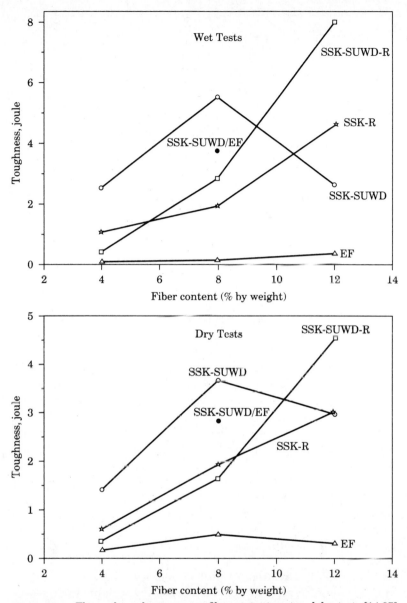

Figure 14-33 Flexural toughness versus fiber content: wet and dry tests [14.27].

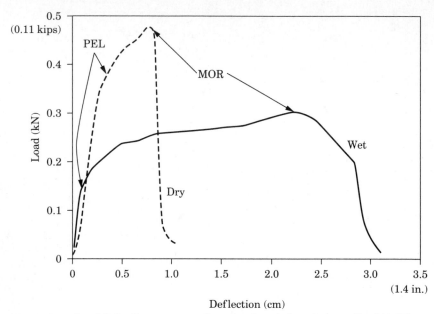

Figure 14-34   Load-deflection response of wood pulp–reinforced composites [14.27].

Based on the results shown in the figures and observations made during fabrication and testing the following observations can be made [14.27]. Summer wood fiber was found to provide an overall best performance. Even though the unrefined version provides good mechanical properties, the refining or adding of finer fibers is needed for processing. Expanded fiber has excellent potential for use as a processing aid. The addition of these fibers significantly reduces the loss of fines.

Performance among fiber types is not affected by wet or dry testing. But the absolute values reduce considerably under wet testing. On the other hand, wet samples have a higher energy-absorbing capacity, as shown in Figure 14-34. In wet samples, failure is dominated by fiber pull-out, resulting in lower strengths and higher ductilities compared to dry samples.

Higher fiber contents result in reduced density. There is no consistent trend in strength and stiffness. About 8% (by weight) fiber seems to provide optimal results overall.

14.3.4   Long-Term Durability of Cellulose Fibers

As mentioned earlier, the long-term durability of cellulose fibers in a cement matrix is still an unresolved problem. A number of researchers have investigated the problem [14.28–14.30].

TABLE 14.8 Properties of Fiber Cements [14.32]

Material		Elastic modulus (GPa)	Tensile strength (MPa)	Failure strain ($\times 10^{-6}$)	Relative density (saturated)	Water absorption (% dry weight)	Volume of pores (%)
A. Cellulose fibers in a calcium silicate matrix	Wet	11.7	7.6	730	1.85	21.3	32.6
	Dry	15.5	17.2	1170			
B. PVA and cellulose* fibers	Wet	15.0	6.4	20,000	1.94	20.4	32.7
	Dry	14.6	8.0	21,000			
C. Same as B but fully compressed	Wet	33.6	11.8	360	2.17	11.5	22.4
	Dry	37.0	16.6	490			
D. PVA and cellulose* fibers	Wet	15.3	7.1	3000	1.94	22.6	35.7
	Dry	15.3	5.6	<2500			
E. Continous polypropylene networks	Wet	28.0	17.0	20,000	2.11	9.5	18.3
F. Asbestos cement	Dry	21.0	23.0	2000	2.06	16.9	29.9

*Obtained from two sources; matrix composition could be different.
1 ksi = 6.89 MPa

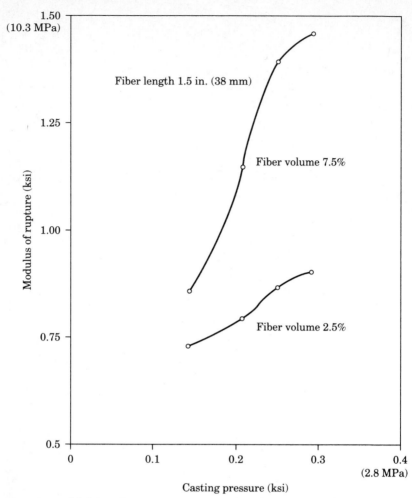

Figure 14-35   Modulus of rupture versus casting pressure: coir-reinforced composite [14.31].

The wetting-and-drying process has shown to decrease the ductility and hence the toughness of the composite considerably [14.28]. It is hypothesized that cellulose fiber becomes mineralized when lime compounds in the fiber lumen (the hollow interior portion of the fiber) precipitate. When the hollow spaces fill up and the bond between the fiber and the matrix strengthens by improved hydration, the fiber behavior becomes brittle. The brittleness of the fiber reduces the stress redistribution that could occur and hence reduces the postcrack strength and toughness.

Suggestions made to improve the durability include (1) reducing the alkalinity of the matrix using admixtures, (2) using silica fume as an additive, (3) coating the fibers with compounds, (4) better bleaching of

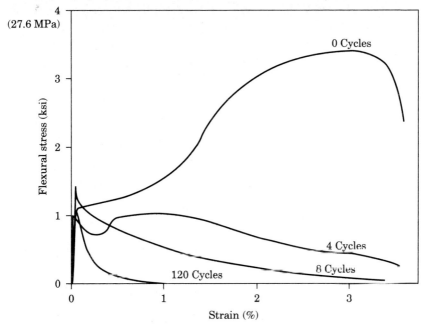

Figure 14-36 Flexural stress versus strain of sisal fiber–reinforced composites subjected to simulated weathering cycles [14.29].

fibers, and (5) a combination of techniques. Autoclaving, as opposed to moisture curing, was also found to improve the durability.

14.4 Thin Sheets Reinforced with Other Natural Fibers

A variety of natural fibers have been tried as reinforcement for cement matrices. Since the availability of some of these fibers is limited to certain regions, the development and use of these composites have a limited scope. Nevertheless, these composites have excellent potential to fill the needs for housing development so desperately needed in some of the developing countries.

The natural fibers used for reinforcement for thin sheets include sisal, coir, bamboo, jute, akwara, and elephant grass. Both short, discrete fibers and fabrics have been investigated. Almost all these fibers provide an increase in tensile and flexural strength and an improvement in ductility. The magnitude of increase depends on the fiber geometry, the mechanical properties of the fibers, the fiber volume fraction, and the manufacturing techniques. Figure 14-35, which shows the contribution of coir fibers [14.31] to the modulus of rupture, represents the typical behavior of natural fiber–reinforced composite.

As in the case of cellulose fibers, the primary problem with natural fibers is their durability in the alkaline environment of cement. A typical degradation of fibers can happen as shown in Figure 14-36, which graphs the loss of flexural ductility of samples subjected to accelerated weathering [14.29]. The specimens that were reinforced with sisal fibers were subjected to drying-and-wetting cycles. The solutions suggested to improve the durability are the same as those suggested for cellulose fibers. However, matrix modification is preferred for natural fibers because of the higher costs involved with the modification and surface treatment of fibers.

14.5   Comparative Behavior and Use of Thin-Sheet Products Made Using Various Fibers

Thin-sheet products reinforced with various types of fibers are becoming an integral part of construction. Their primary application is for nonstructural elements such as partitions and various types of exposed panels. However, the products are also being used as roofing elements that can support small live loads.

The industrial production of sheets reinforced with glass, polymeric, cellulose, and carbon fibers is steadily increasing. With the gradual phasing out of asbestos-reinforced products, the composites containing other types of fibers are sure to gain more use. The field use of carbon fibers has taken place in Japan, where they have been used extensively for cladding applications.

All the fibers discussed in this chapter, as well as glass fibers, can be used to obtain the desired properties. For certain applications, some types of fibers may be more preferable than others. In some instances, fabrics might be more favorable than discrete fibers. Table 14.8 shows typical strengths and other engineering properties of cement composites reinforced with various types of fibers [14.32]. This table can be used as a guide for choosing the type and configuration of fibers for a particular application.

14.6   REFERENCES

14.1   Naaman, A. E.; Shah, S. P.; and Throne, J. L. "Some Developments in Polypropylene Fibers for Concrete," *Fiber Reinforced Concrete-International Symposium,* SP-81, American Concrete Institute, Detroit, Michigan 1984, pp. 375–396.

14.2   Keer, J. G.; and Thorne, A.; "Performance of Polypropylene-Reinforced Cement Sheeting Elements," *Fiber Reinforced Concrete-International Symposium,* SP-81, American Concrete Institute, Detroit, Michigan 1984, pp. 213–231.

14.3   Swamy, R. N.; and Hussin, M. W.; "Woven Polypropylene Fabrics—An Alternative to Asbestos for Thin Sheet Applications," *Fiber Reinforced Cements and Concrete: Recent Development,* Elsevier, 1989, pp. 90–100.

14.4 Rongxi, S.; and Ruishan, Y.; "PVA Fiber Reinforced Cement Sheets: Production and Properties," *Fiber Reinforced Cements and Concretes: Recent Developments*, Elsevier, 1989, pp. 80–89.

14.5 Studinka, J. B. "Asbestos Substitution in the Fiber Cement Industry," *International Journal of Cement Composites and Lightweight Concrete*, Vol. 11, No. 2, May 1989.

14.6 Gale, D. N.; Shah, A; and Balaguru, P. "Oriented Polyethylene Fibrous Pulp Reinforced Cement Composites," *Thin-Section Fiber Reinforced Concrete and Ferrocement*, SP 124, American Concrete Institute, Detroit, Michigan, 1991, pp. 61–77.

14.7 Dave, N. J.; and Ellis, D. G.; "Polypropylene Fiber Reinforced Cement," *International Journal of Cement Composites*, Vol. 1, No. 1, May 1979, pp. 19–28.

14.8 Galloway, J. W.; Williams, R. I. T.; and Raithby, K. D. "Mechanical Properties of Polypropylene-Reinforced Cement Sheet for Grade Control in Reinforced Concrete," Transport and Road Research Laboratory Supplementary Report 658, 1981, 42 pp.

14.9 Hannant, D. J.; Zonsveld, J. J.; and Hughes, D. C. "Polypropylene Film in Cement-Based Materials," *Composites*, Vol. 9, No. 2, April 1978, pp. 83–88.

14.10 Vittone, A. "Industrial Development of the Reinforcement of Cement-Based Products with Fibrillated Polypropylene Networks on Replacement of Asbestos," RILEM Symposium, Developments in Fiber Reinforced Cement and Concrete, July 1986.

14.11 Baggot, R. "Polypropylene Fiber Reinforcement of Lightweight Cementitious Matrices,"*International Journal of Cement Composites and Lightweight Concrete*, Vol. 5, 1983, pp. 105–114.

14.12 Hannant, D. J.; Zonsveld, J. J. and Hughes, D.C. "Polypropylene Film in Cement Based Materials," *Composites*, Vol. 9, 1978, pp. 83–88.

14.13 Gardiner, T.; and Currie, B. "Flexural Behavior of Composite Cement Sheets Using Woven Polypropylene Mesh Fabrics," *International Journal of Cement Composites and Lightweight Concrete*, Vol. 5, 1983, pp. 193–197.

14.14 Ali, M.; Majumdar, A.; and Rayment, D. "Carbon Fiber Reinforcement and Cement," *Cement and Concrete Research*, Vol. 2, No. 2, 1972, pp. 201–212.

14.15 Sarker, S.; and Bailey, M. B. "Structural Properties of Carbon Fiber Reinforced Cement," RILEM Symposium 1975, Fiber Reinforced Cement and Concrete, Construction Press, pp. 361–371.

14.16 Ohama, Y.; Amano, M.; and Endo, M. "Properties of Carbon Fiber Reinforced Cement with Silica Fume,"*Concrete International*, American Concrete Institute, Detroit, Michigan, Vol. 7, No. 3, 1985, pp. 58–62.

14.17 Akihama, S.; Suenaga, T.; and Banno, T. "Mechanical Properties of Carbon Reinforced Cement Composites," *International Journal of Cement Composites and Lightweight Concrete*, Vol. 8, No. 1, 1986, pp. 21–33.

14.18 Akihama, S.; Suenaga, T.; and Nakagawa, H. "Carbon Fiber Reinforced Concrete," *Concrete International*, American Concrete Institute, Detroit, Michigan, Vol. 10, No. 1, 1988, pp. 40–47.

14.19 Soroushian, P.; and Bayasi, Z. "Development and Mechanical Characterization of Carbon Fiber Reinforced Cement Composites," Report No. MSU-ENGR-88-017, Michigan State University, East Lansing, Michigan, October 1988.

14.20 Soroushian, P.; and Nagi, M. "Carbon Fiber Reinforced Cement Cladding Panels: Material Development, Construction and Design," Report No. MSU ENGR-88-006, Michigan State University, East Lansing, Michigan, June 1988.

14.21 Soroushian, P.; Bayasi, Z.; and Nagi, M. "Effects of Curing Procedures on Mechanical Properties of Carbon Fiber Reinforced Cement," *Fiber Reinforced Cements and Concretes: Recent Developments*, Elsevier, 1989, pp. 167–178.

14.22 Ohama, Y.; Demura, K.; and Sato, Y. "Development of Lightweight Carbon Fiber Reinforced Fly Ash Cement Composites," Proceedings of the International Symposium on Fiber-Reinforced Concrete, Oxford and IBH Publishing Co,. New Delhi, India, 1987, pp. 3.23–3.32.

14.23 Ando, T.; Sakai, H.; Takahashi, K. Hoshijima, T.; Awata, M.; and Oka, S. "Fabrication and Properties for a New Carbon Fiber Reinforced Cement Product,"

Thin-Section Fiber Reinforced Concrete and Ferrocement, SP-124, American Concrete Institute, Detroit, Michigan, 1991, pp. 39–60.

14.24 Soroushian, P. and Marikante, S. "Reinforcement of Cement Based Materials with Cellulose Fibers," *Thin-Section Fiber Reinforced Concrete and Ferrocement*, SP-124, American Concrete Institute, Detroit, Michigan, 1991, pp. 99–124.

14.25 ACI Committee 544, "Measurement of Properties of Fiber Reinforced Concrete," American Concrete Institute, Detroit, Michigan, 1989, pp. 1–11.

14.26 Coutts, R. S. P. "Air-Cured Wood Pulp, Fiber-Cement Mortars,"*Composites,* Vol. 18, No. 4, 1987, pp. 325–328.

14.27 Vinson, K. D.; and Daniel, J. I. "Advances in the Development of Specialty Cellulose Fibers Specifically Designed for the Reinforcement of Cement Matrices," *Thin-Section Fiber Reinforced Concrete and Ferrocement*, SP-124, American Concrete Institute, Detroit, Michigan, 1991, pp. 1–18.

14.28 Sharman, W. R.; and Vautier, B. P. "Durability Studies on Wood Fiber Reinforced Cement Sheet," Third International Symposium on Developments in Fiber Reinforced Cement and Concrete, RILEM Symposium, FRC 86, Vol. 2, July 1986.

14.29 Gram, H. E. "Durability Studies of Natural Organic Fibers in Concrete, Mortar and Cement," Third International Symposium on Developments in Fiber Reinforced Cement and Concrete, RILEM Symposium, FRC 86, Vol. 2, July 1986.

14.30 Gram, H. E. "Durability of Natural Fibers in Concrete," Swedish Cement and Concrete Research Institute, Stockholm, Sweden, 1983.

14.31 Cook, D. J.; Pama, R. P.; and Weerasingle, H. L. S. D. "Coir Fiber Reinforced Cement as a Low Cost Roofing Material," *Building and Environment,* Vol. 13, 1983, pp. 193–198.

14.32 Keer, J. G. "Performance of Non-Asbestos Fiber Cement Sheeting," *Thin-Section Fiber Reinforced Concrete and Ferrocement*, SP-124, American Concrete Institute, Detroit, Michigan, 1991, pp. 19–38.

15

Slurry-Infiltrated Fiber Concrete

Slurry-infiltrated fiber concrete (SIFCON) is a special type of fiber-reinforced composite containing as much as 20% (by volume) of steel fibers [15.1]. In conventional FRC, the fiber volume fraction is generally limited to about 2%. Mixing and placing become difficult if the fiber volume exceeds 2%. Hence a different construction technique was devised to increase the fiber volume fraction, leading to the development of SIFCON. Because of its high fiber content, SIFCON has unique properties in the areas of both strength and ductility. Preparation techniques, mechanical properties, and some field applications of this composite are discussed in this chapter.

15.1 Preparation of SIFCON

SIFCON is cast using a preplacing technique in which fibers are placed in the mold or on a substratum and infiltrated with cement-based slurry. The fibers can be sprinkled by hand or by using fiber-dispensing units. The amount of fibers that can be incorporated depends on fiber aspect ratio, fiber geometry, and placement technique. More fibers can be incorporated if aspect ratios are low. The volume fraction can also be increased by using mild vibration. The fiber volumes typically range from 5% to 20%.

One of the important aspects in the fabrication of SIFCON is fiber orientation. In thin sections (less than 1 in., 25 mm), fibers tend to orient themselves in two dimensions. If the fibers are long, the two-dimensional effect can be witnessed even in thick sections. The fiber

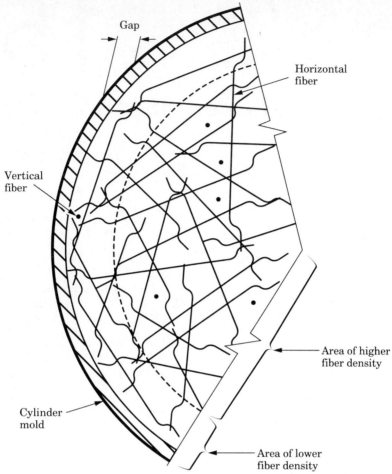

Figure 15-1 Fiber orientation "edge effects" in a molded SIFCON cylinder specimen. [15.1]

density at the edges of the mold can be much less, compared to other locations. This problem can be more significant when cylinders are prepared using preplaced-fiber techniques (Figure 15-1) [15.1]. One way to avoid the fiber orientation problem is to cast a slab and obtain the test specimens by a coring technique. Here again, attention should be paid to the orientation of fibers. If fibers are aligned along the diameter of the cylinder, a much higher strength can be expected compared to a cylinder in which fibers are aligned along the axis of the cylinder. Fibers tend to align perpendicular to the direction of placement (Figure 15-2) [15.1]. Fiber orientation has even more significance for specimens subjected to bending.

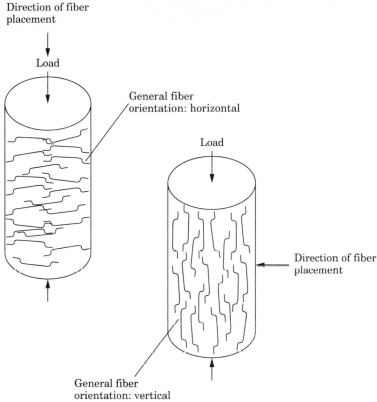

Figure 15-2 Orientation of fibers in cored SIFCON specimens as influenced by the coring direction with respect to the fiber placement direction [15.1].

15.2 Constituent Materials and Mix Proportions

The primary constituent materials of SIFCON are steel fibers and cement-based slurry. The slurry can contain only cement, cement and sand, or cement and other additives. In most cases, high-range water-reducing admixtures are used in order to improve the flowability of the slurry without increasing the water-cement ratio.

15.2.1 Fibers

A large variety of fibers have been investigated for use in SIF-CON. These include fibers with hooked ends, crimped fibers, surface-deformed fibers, and straight fibers [15.1–15.5]. In most applications in the United States, steel fibers with hooked ends have been used. The fiber lengths vary from 30 mm to 60 mm (1.2 to 2.4 in.). The length-diameter ratios range from 60 to 100. Crimped and straight fibers have also been used for some applications. Typically, a higher volume fraction of straight fibers can be incorporated compared to deformed fibers.

15.2.2 Matrix

The matrix of SIFCON does not contain coarse aggregates. The matrix compositions investigated include cement, cement–fly ash, cement–silica fume, cement-sand–fly ash, and cement-sand–silica fume. Matrices containing filler materials were found to have better shrinkage characteristics. Typically, silica fume addition increases the strength, whereas the addition of fly ash or sand results in no change or a slight decrease in strength.

15.2.3 Mix proportions

The primary variables are fiber content and matrix composition. A fiber volume fraction of about 10% seems to provide optimal strength values. However, the volume fraction is commonly controlled by the construction technique rather than by strength. Typically, fibers are sprinkled into the mold until it is completely filled. The volume fraction of fibers could differ depending on the fiber placement technique and the fiber geometry.

The recommended water-cement ratio for the slurry (matrix) is 0.3. High-range water-reducing admixtures can be used to improve the flowability of the slurry, which should be thin enough to flow through the fiber bed without leaving honeycombs. The cement-to-sand ratio should be limited to 1 : 2. Only fine sand should be used. If fly ash is used as an admixture, about 20% of the cement could be replaced with fly ash. If silica fume is used, the recommended dosage is 10% by weight of cement. The typical matrix strength varies from 9 to 13 ksi (60 to 90 MPa). Both Type I and Type III (ASTM) cements can be used.

15.3   Engineering Properties

15.3.1   Unit weight

The unit weight of SIFCON is typically higher than normal FRC because of the high fiber content. The matrix density is about 120 lb/ft^3 (1.9 g/cm^3). The addition of fibers results in an increase of density varying from 130 to 200 lb/ft^3 (2.1 to 3.2 g/cm^3). A unit weight increase is almost linearly proportional to the fiber content, as shown in Figure 15-3 [15.1].

15.3.2   Behavior in compression

The compressive strength of SIFCON can be as high as three times the matrix strength. Typical matrix strengths for plain cement slurry

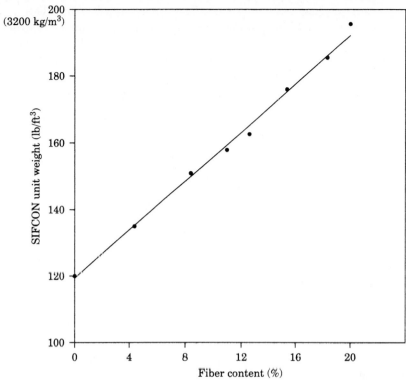

Figure 15-3 Effect of steel fiber content on the unit weight of SIFCON (cement-fly ash slurry) [15.1].

vary from 9 to 13 ksi (60 to 90 MPa). SIFCON prepared using plain cement slurries attained a compressive strength up to 27 ksi (190 MPa). The highest compressive strength reported for SIFCON is 30 ksi (210 MPa) [15.1]. The composite is also very ductile as compared to a plain matrix.

The compressive behavior of SIFCON was investigated using both cast and cored cylinders [15.1, 15.3, 15.4]. Other variables investigated include:

- Fiber orientation effect—parallel and perpendicular to the loading axis
- Fiber geometry—hooked ends, crimped and deformed surfaces
- Matrix composition—plain cement matrix, matrix containing silica fume, or fly ash, or sand, or their combinations.

The following is the summary of the results of the various investigations.

15.3.2.1 Compressive strength. Compressive strength of SIFCON depends on matrix strength, fiber orientation, fiber volume fraction, and fiber geometry. Since fibers themselves do not break, the tensile strength of fibers does not influence the compressive strength of SIFCON. Tables 15.1 and 15.2 present the range of compressive strengths obtained for various fiber types and matrix mix proportions [15.4]. Fiber addition provides a typical strength increase of about 100%. An increase in matrix strength resulted in an increase of SIFCON compressive strength. Fiber geometries showed less influence than matrix strength.

The composite strength is higher when fibers are oriented perpendicular to the loading axis. It should be noted that even though predominant alignment is perpendicular to the loading axis a few fibers are still aligned in other directions.

The effects of the water-cementitious and fly ash–cementitious ratios on the compressive strength of the composite are shown in Figures 15-4 and 15-5, respectively [15.3]. In all cases, an 11.6% volume of steel fibers with hooked ends was used. The fibers were 30 mm (1.2 in.) long and had a diameter of 0.5 mm (0.02 in.). From Figures 15-4 and 15-5, it can be seen that the matrix strength is the influencing factor for the composite compressive strength. A higher matrix strength could be obtained by using a lower water-cement ratio or lower fly ash content. The strength of the composite also increases with age as the matrix matures to attain higher strengths [15.3].

The typical failure mode of SIFCON in compression seems to be shear failure (Figure 15-6) [15.4]. This is true even for longer specimens with a length/diameter ratio of 4 (Figure 15-7) [15.4].

TABLE 15.1 Slurry Mix Designs and Strength Values [15.4]

Mix no.	Mix constituents	Relative weight of constituents	Water-Cementitious ratio	Strength range (ksi)
1	Type I cement	1.00		7.5
	Fly ash	.20	$\dfrac{\text{water}}{\text{cement} + \text{fly ash}}$	to
	Water	.36		17.0
	Superplasticizer	.03		
2	Type I cement	1.00		6.0
	Fly ash	.20	$\dfrac{\text{water}}{\text{cement} + \text{fly ash} + \text{silica}}$	
	Silica fume*	.20		to
	Water	.36		13.5
	Superplasticizer	.02		

* A slurry of approximately 50% water and 50% amorphous silica particles by weight (1 ksi = 6.89 MPa)

TABLE 15.2 Fiber Properties [15.4]

Fiber type	Length (mm)	Diameter (mm)	Volume fraction l/d	Volume fraction %
Crimped	25	0.9	28	20 to 23
Hooked end	30	0.5	60	10 to 12
Deformed	30	0.5	60	10 to 12

1 in. = 25.4 mm

Cored specimens sustain 15% to 30% more failure load than cast specimens [15.4]. The difference could be the result of better fiber packing in the cored specimens.

15.3.2.2 Stress-Strain Behavior in Compression Even though SIFCON has higher compressive strength than other cement composites, its uniqueness is much more important in the area of energy absorption and ductility. A great energy-absorbing capacity and a ductile mode

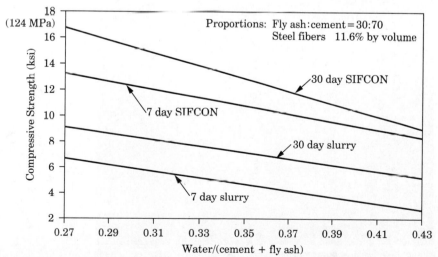

Figure 15-4 Compressive strength versus water/(cement + fly ash) ratios [15.3].

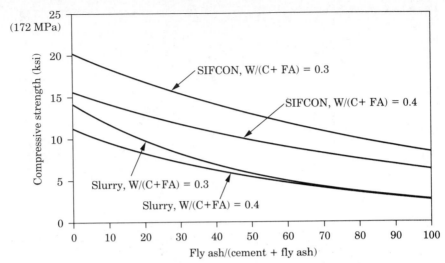

Figure 15-5 Influence of fly ash content on compressive strength [15.3].

of failure make SIFCON suitable for applications involving impact, blast, and earthquake loading.

Typical stress-strain (load-deformation) behaviors of SIFCON in compression are presented in Figures 15-8 through 15-11 [15.4]. The study of these figures leads to the following observations.

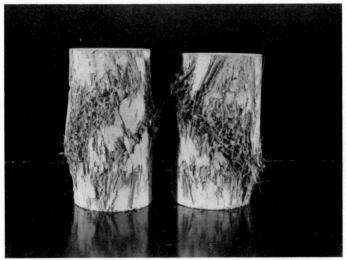

Figure 15-6 Tested compression specimens showed consistent shear failure (photo from Prog. A. E. Naaman).

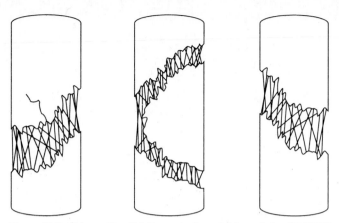

Figure 15-7 Schematic of failure modes of 12 × 3 in. SIFCON
compression cylinders.

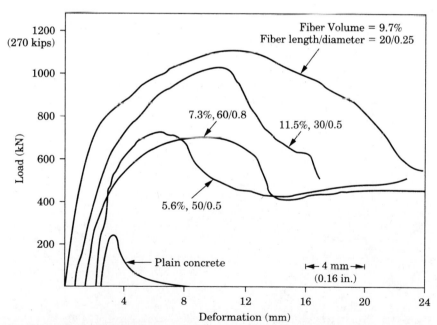

Figure 15-8 Typical load-deflection behavior in compression for 10.2 × 17.8 cm (4 × 7
in.) cylinder SIFCON specimens (28 day fog curing) containing 5.6% to 11.5% volume
hooked-end steel fibers [15.1].

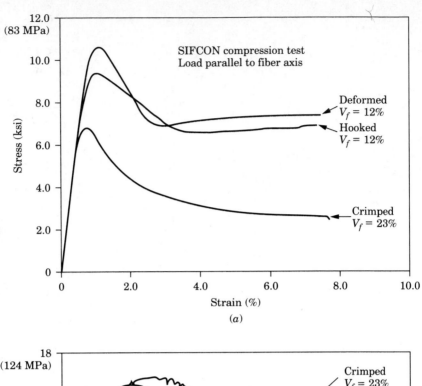

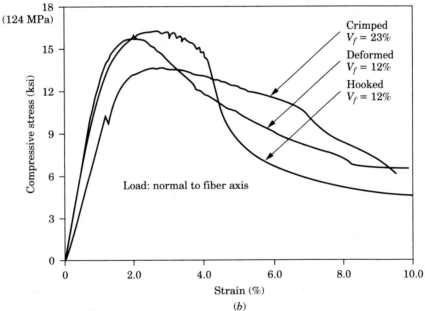

Figure 15-9 Typical effects of fiber type on the stress-strain curve of SIFCON in compression: (*a*) loading parallel to fiber axis, (*b*) loading nomal to fiber axis [15.4].

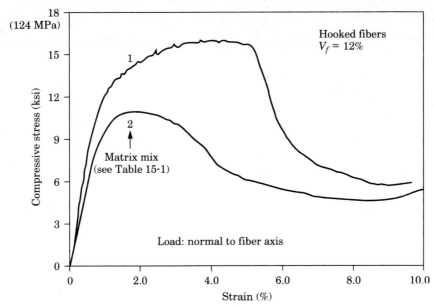

Figure 15-10   Typical stress-strain curves of SIFCON in compression for different matrix strengths [15.4].

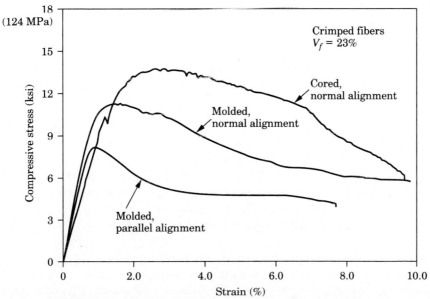

Figure 15-11   Typical stress-strain curves of molded and cored SIFCON specimens in compression [15.4].

- SIFCON has a large strain capacity.

- The energy absorption could be 1 to 2 orders of magnitude higher than that of a plain matrix.

- Fiber orientation that is perpendicular (normal) to the loading axis results in not only higher strength but also higher ductility. This should be expected because, as the concrete cracks and expands along the diameter, the fibers provide pseudo confining effect, improving both the load capacity and the ductility. Fibers with better bonding capacity provide better results because they can transfer more loads across the cracks.

- The modulus of elasticity of SIFCON is slightly higher than the matrix modulus. The postlinear elastic part of the stress-strain curve is much longer for SIFCON. This indicates that SIFCON specimens sustain much more microcracking before reaching the peak load than plain matrix specimens.

15.3.3   Behavior in tension

The tensile strength of SIFCON is about 2 ksi (14 MPa), compared to the matrix strength of 1 ksi (7 MPa). Tension tests were conducted primarily using thin, long dog bone–shaped specimens (Figure 15-12c) [15.2]. The variables investigated include [15.2, 15.5, 15.6]:

- Fiber type and geometry consisting of straight, hooked-end, and surface-deformed fibers

- Fiber volume fraction ranging from 5% to 13.8%

- Matrix formulation using admixtures and water-cement ratios ranging from 0.26 to 0.45

The test specimens were either cut from SIFCON plates [15.2, 15.5] or cast using Plexiglas molds [15.6]. Fibers were oriented randomly in cut specimens and were placed parallel (as much as possible) to the loading direction in cast specimens. The fibers were placed by hand and subjected to vibration to obtain compaction of fibers.

The following sections provide a brief review of the results.

15.3.3.1   Tensile strength.   As mentioned earlier, the tensile strength of SIFCON is about twice the strength of the matrix. Table 15.3 presents the tensile strengths obtained for various combinations of matrix composition, fiber types, and fiber volume fractions. The results indicate the following trends.

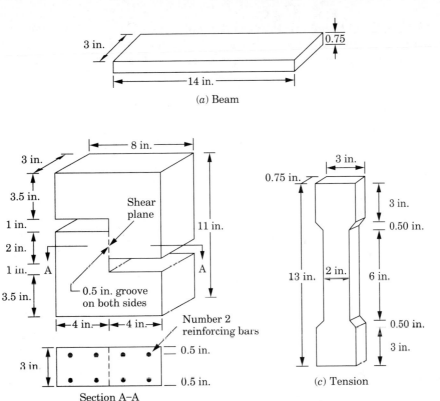

Figure 15-12   Test specimens [15.2] (1 in. = 25.4 mm).

A matrix containing either fly ash or silica fume provides higher strength than a plain matrix. A matrix containing silica fume provides the highest strength. Since failure occurs by the debonding of fibers and the spalling of matrix, a denser matrix can be expected to provide a better strength.

Lower water-cement ratios result in better matrix bonding and hence provide higher tensile strengths. The differences are higher for fibers with relatively weak bond strength. For example, in the case of hooked fibers that have good mechanical bond, lower water-cement ratios do not improve the strength significantly.

Tensile strengths are greater for fiber volume fractions higher than 10% compared to volume fractions lower than 10%. The variations are not significant in the ranges of 6% to 9% and 11% to 13.8%.

Overall, tensile strengths higher than 2 ksi (14 MPa) can be expected for matrices containing silica fume and a fiber content of about 8% and matrices containing fly ash and a fiber content of about 12%.

15.3.3.2 Stress-strain behavior in tension. As in the case of compression, SIFCON exhibits high ductility in the tension mode. Typical stress-strain curves obtained using various fiber types and volume fractions are presented in Figures 15-13 through 15-17 [15.2, 15.5, 15.6]. Figure 15-13 shows the stress-strain curves for specimens containing hooked-end fibers and a matrix made with silica fume [15.2]. The strains were computed over a gauge length of 4 in. (100 mm). Figures 15-14 and 15-15 present results for specimens made using a plain cement matrix and three different types of fibers. Figures 15-16 and 15-17 present results for specimens with fibers oriented primarily along the loading axis and a matrix containing fly ash.

In all cases, the stress-strain (load-deformation) curves have three distinct regions. The first part, which is primarily elastic, is very steep. After the initiation of microcracks, the curve becomes nonlinear,

TABLE 15-3   Tensile Strengths of SIFCON [15.2, 15.5, 15.6]

Sample no.	Matrix constitutents	Water-cement ratio*	Fiber type, l/d	Volume fraction %	Tensil strength (MPa)	Remarks
1	C+10% SF	0.3	HE, 50/.5	5	13.6	19mm thick specimen
2	C+10% SF	0.3	HE, 50/.5	8	15.7	
3	C only	0.45	DF, 30/.5	8.5	7.0	
4	C only	0.45	ST, 25/.4	8.5	4.0	
5	C only	0.45	HE, 30/.5	8.5	9.2	
6	C only	0.45	HE, 30/.5	13.5	14.1	35mm thick specimen
7	C only	0.35	HE, 60/.8	7.4	6.7	
8	C only	0.35	EE, 25/.5	9.9	7.8	
9	C only	0.35	HE, 50/.8 + EE	6.5 + 4.0	6.9	
10	C only	0.45	HE, 60/.8 + EE	6.1 + 4.2	10.7	
11	C+20% FA	0.35	HE, 30/.5	11.7	15.6	37mm thick specimen
12	C+20% FA	0.35	DF, 30/.5	12.6	10.9	
13	C+25% FA	0.26	HE, 30/.5	12.1	15.7	
14	C+25% FA	0.26	DF, 30/.5	13.8	16.1	

* Weight of fly ash was included with the weight of cement for water-cement ratio.

C = cement l = length of fiber (mm)

SF = silica fume d = diameter of fiber (mm)

FA = fly ash   1 in. = 25.4 mm

HE = hooked ends   1 ksi = 6.89 MPa

DF = deformed   EE = deformed

ST = straight

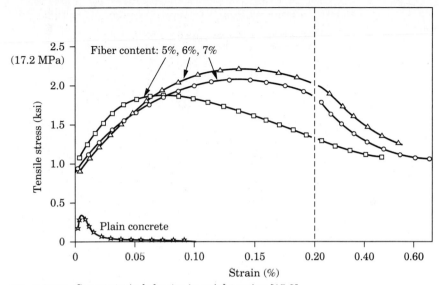

Figure 15-13 Stress-strain behavior in axial tension [15.2].

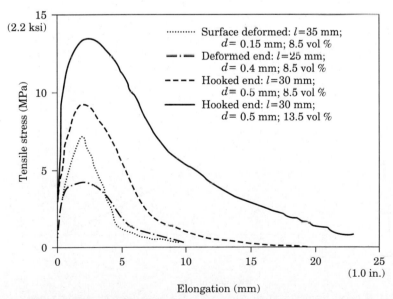

Figure 15-14 Stress-displacement curves of various mixes with w/c ratio of 0.45 [15.2] (1 in = 25.4 mm).

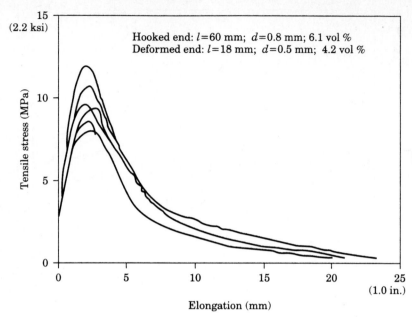

Figure 15-15 Stress-displacement curves for a mix with a fiber blend (nonvibrated) and w/c = 0.45 [15.5] (1 in. = 25.4 mm).

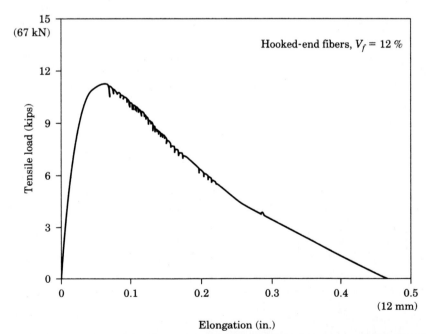

Figure 15-16 Typical load-elongation curve of SIFCON in tension: whole range [15.6].

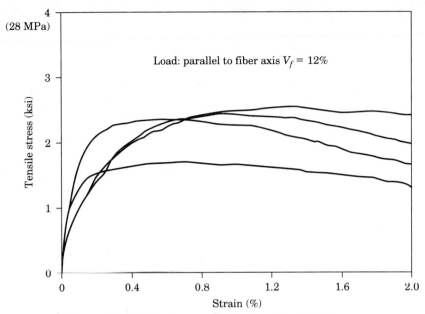

Figure 15-17 Stress- Strain behavior; magnified at peak load [15.6]

representing postcrack behavior. A well-defined descending branch exists for all the fiber and matrix combinations investigated.

Specimens with fibers oriented parallel to the loading direction and specimens containing a cement–silica fume matrix exhibit a longer plateau near the peak load. The descending parts of the stress-strain curves are also less steep for these specimens. The specimens could sustain peak load at strains as great as 2%.

The ductility measured as area under the stress-strain curve up to 2% strain could be as much as 1000 times the area under the curve for a plain matrix.

15.3.4 Behavior under Flexural Loading

In most field applications, SIFCON is subjected to bending stress, at least partially. Hence the behavior under flexural loading plays an important role in field applications. Flexural tests have been conducted using SIFCON beams both under static and cyclic loading [15.2, 15.7, 15.8].

Extensive experimental results reported in References [15.2 and 15.7] deal with 0.75 in. (19 mm) thick SIFCON beam specimens cut from a slab. The specimens were 3 in. (75 mm) wide and 14 in. (350 mm) long (Figure 15-12a) and were tested over a simply supported span of 12 in. (300 mm) using a center point load. The variables investigated are shown in Table 15.4. The investigation was designed to evaluate the various fiber lengths, fiber contents, and matrix compositions.

The beams were tested using a Universal Testing Machine equipped with servocontrol. The load-deflection responses were obtained using x-y plotters. For each variable, three beams were tested. The first beam was subjected to monotonic loading. The load was applied in deflection control up to a maximum deflection of 0.75 in. (19 mm), which was the thickness of the beam. The typical load-deflection response is shown in Figure 15-18a. The second beam was subjected to a cyclic loading ranging from 10% to 100% of ultimate load. Since the beam is very ductile, for every cycle, the load was increased until the peak load was reached, (Figure 15-18b). The test was terminated when the maximum deflection reached 0.75 in. (19 mm). The third specimen was subjected to reverse cyclic loading, in which the load varied from the peak load in one direction to the peak load in the other direction (Figure 15-18c). The test was terminated when the deflection in one direction reached 0.35 in. (9 mm). The strength and ductility characteristics observed in these tests are discussed in the following two sections.

15.3.4.1 Flexural strengths.   Table 15.5 presents the flexural strengths for four fiber lengths and various fiber contents together with the compressive strengths of slurry obtained using companion cube specimens. The flexural strength was computed using the average maximum

TABLE 15.4   Fiber Content and Slurry Composition Used for Flexural Test Specimens (Strength Values Are Reported in Table 15-5) [15.7]

Type of specimen	Fiber length (mm)	Fiber content (%)	Type of slurry
	30	6,8,10,12	
	40	4,6,8,10	Cement + 10% Silica Fume
	50	4,5,6,8	
	60	5,6,8,10	
Beam			
	30,40 50,60	8	Cement
	30	8	Cement + Sand (1 : 1, 1 : 1.5, 1 : 2)
Freeze-thaw beam specimens	30,40 50,60	7	Cement + 10% Silica Fume

For all mixes: water-cement ratio = 0.3, amount of high-range water-reducing admixture = 4.8% by weight of cement.

1in. = 25.4 mm

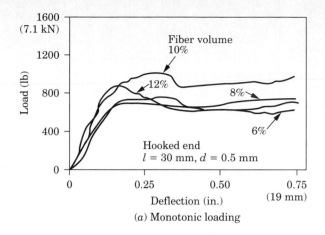

(a) Monotonic loading

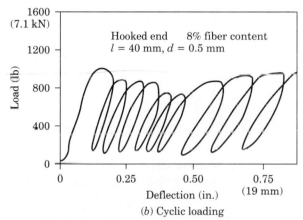

(b) Cyclic loading

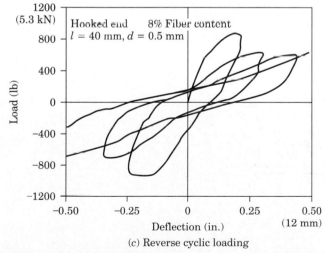

(c) Reverse cyclic loading

Figure 15-18 Load-deflection curves: 0.75 in. (19mm) thick specimens [15.7].

load of all three specimens. In the case of specimens subjected to cyclic loading, the maximum load was taken as the maximum reached in the first cycle. The flexural strength, f_{tmax} was computed using the classical bending theory equation for a homogeneous elastic section:

$$f_{tmax} = \frac{M}{I} \times \frac{d}{2}$$

where M = maximum moment
I = moment of inertia
d = total thickness of the beam

The beam behavior at ultimate load is neither elastic nor homogeneous; hence the apparent stresses should be used only as a relative indicator.

Figure 15-19 presents the variation of flexural strength for the various fiber contents and fiber lengths. A study of Table 15.5 and Figure 15-19 leads to the following observations:

- The flexural strength of SIFCON is an order of magnitude greater than the flexural strength of normal fiber-reinforced concrete.
- For a constant fiber length, the flexural strength increases with the volume fraction of fiber only up to a certain limit. After a certain

TABLE 15.5   Flexural Strength of SIFCON [15.7]

Fiber (length/dia.) (mm)	Fiber volume (%)	Apparent maximum stress (ksi)	Compressive strength of slurry (ksi)
	6	8.0	11.9
30/.50	8	9.0	12.7
	10	13.3	11.0
	12	9.1	11.7
	4	6.8	9.6
40/.50	6	9.8	11.6
	8	10.9	11.1
	10	11.1	9.7
	4	5.3	11.6
50/.50	5	8.5	11.0
	6	11.4	9.2
	8	10.7	10.9
	5	7.2	11.9
60/.80	6	7.8	12.1
	8	10.5	11.5
	10	9.2	9.2

1 ksi = 6.89 MPa; 1 in. = 25.4 mm

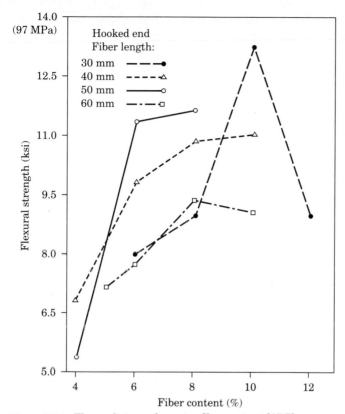

Figure 15-19   Flexural strength versus fiber content [15.7].

fiber loading, the bond strength decreases because of the lack of matrix in between the fibers, thus reducing the flexural strength. The optimum fiber content seems to be in the range of 8% to 10%.

- The optimum fiber volume seems to decrease with an increase in fiber length. For the same fiber volume, longer fibers provide a slight increase in flexural strength. The maximum apparent flexural strength seems to level off around 11 ksi [76 MPa].

Table 15.6 presents the comparison of flexural strengths for specimens made with and without silica fume and specimens made using mortar. All the specimens had an 8% fiber content. As can be seen from Table 15.6, the use of silica fume increases the flexural strength substantially. The percentage increase in flexural strength is about the same as the percentage increase in the compressive strength. The increase in flexural strength can be explained by the fact that silica fume results in a much denser matrix. The increase in the denseness of the matrix possibly provides as much improvement in bond between the matrix and the fiber as in the compressive strength. The predominant failure pattern is by the pulling out of the fibers.

The addition of sand beyond a cement-to-sand ratio of 1 : 1 decreases the compressive strength, whereas the flexural-strength reduction starts when the cement-to-sand ratio exceeds 1 : 1.5. The decrease in compressive strength is more rapid than flexural strength. The limited results indicate that sand can be added to cement up to a ratio of 1 : 1.5 (cement : sand) without adversely affecting the flexural strengths. These results are consistent with the results reported for compressive strength.

15.3.4.2 Load-Deflection Behavior in Flexure. The load-deflection behavior of SIFCON (Figure 15-18) is quite different from the load-deflection behavior of typical FRC beams. The curves have a short linear elastic response and a considerable plateau at the peak. The beams can also sustain a high percentage of peak loads (more than 80% of peak load) even at large deflections. The load-carrying capacity does not diminish even under reverse cyclic loading (Figure 15-18c).

While fiber length and fiber volume fraction influence strength (Figure 15-19), the ductility is not affected by either of the variables, (Figure 15-20a). [15.2, 15.7]).

Plain cement slurry, cement plus silica fume slurry, and cement plus sand slurry resulted in the same type of load-deflection response. The beams made with all three matrix types were highly ductile.

Specimens that were 4 in. (100 mm) and 6 in. (150 mm) thick show a lesser rotation capacity [15.8, 15.9], compared to 0.75 in. (19 mm) specimens. This results in a faster drop in load-carrying capacity after peak load (Figure 15-20) [15.2]. The explanation is that higher strain capacities are needed for thicker beams to produce the same amount of rotation.

15.3.5 Behavior in Shear

The shear behavior of SIFCON was evaluated using double-L (push-off) specimens (Figure 15-12b) [15.2]. These specimens were cast using a special mold (Figure 15-21a). Mild steel reinforcements were placed both on the legs and heads of L-shapes to improve their strengths. After curing, grooves were cut along the shear plane to induce shear failure.

The tests were conducted using a standard compression-testing machine. The movement along the shear plane was measured using a dial gauge (Figure 15-21b). The dial gauge was mounted on steel angles attached to the specimen using epoxy.

TABLE 15.6 Comparison of Flexural Strengths for Various Matrix Compositions [15.7]

| Fiber length (mm) | Flexure strength (ksi) | | | | | $\frac{SF+C}{C}$ | | Strength ratio $\frac{C+S}{C}$ | | | | | |
| | Silica fume + cement | Cement | Mortar, cement : sand | | | | | 1:1 | | 1:1.5 | | 1:2 | |
			1:1	1:1.5	1:2	FS	CS	FS	CS	FS	CS	FS	CS
30	9.0	6.5	6.7	7.8	6.1	1.38	1.59	1.03	1.08	1.20	0.79	0.93	0.62
40	10.9	7.4				1.48	1.48						
50	10.7	7.8				1.37	1.60						
60	10.5	6.6				1.58	1.47						

$\dfrac{SF+C}{C} = \dfrac{\text{silica fume} + \text{cement}}{\text{cement}}$

$\dfrac{C+S}{C} = \dfrac{\text{cement} + \text{sand}}{\text{cement}}$

FS = flexural-strength ratio

CS = compressive-strength ratio

1 ksi = 6.895 MPa

1 in = 25.4 mm

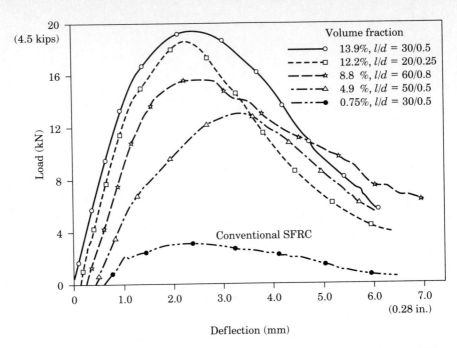

Figure 15-20a Typical load-deflection behavior in flexure for SIFCON containing 4.9% to 13.9% volume hooked-end steel wire fiber (deformed ends) and for conventional SFRC: 10 × 10 × 35 cm (4 × 4 × 14 in.) beam specimens, third-point loading, 30 cm (12 in.) span [15.1] (1 in. = 25.4 mm).

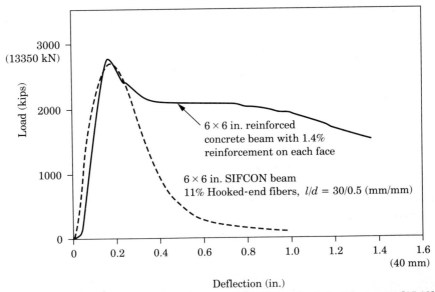

Figure 15-20b Flexural strength comparison of SIFCON and reinforced concrete [15.19] (1 in. = 25.4 mm).

The primary test variable was fiber length, which varied from 30 mm to 60 mm (1.2 to 2.4 in.). The fiber volume fraction was 6% for all specimens. The matrix consisted of cement plus silica fume slurry. The cement-to-silica fume ratio was 1 : 0.1 and the water-cement ratio was 0.3. A high-range water-reducing admixture was used to improve the flowability of the slurry.

Figure 15-22 shows the variation of load with respect to slip at the shear plane. The shear strength, f_S, was computed by dividing the maximum load P by the area of the shear plane A, which was 4 in.^{2} (2581 mm^2) for all specimens. The shear strengths for 30, 40, 50, and 60 mm (1.2, 1.6, 2.0, 2.4 in.) long fibers were 4.43, 4.08, 4.84, and 4.61 ksi (30.5, 28.1, 33.3, 31.8 MPa) respectively. The fiber length does not seem to affect shear strength. The average shear strength of SIFCON was 4.5 ksi (30.9 MPa), compared to about 0.8 ksi (5.5 MPa) for plain concrete.

In Figure 15-22, each line represents the average of two specimens. The diagram is split into two parts in order to present the behavior at large deformation (slip). Initial cracking occurred at the shear plane at about 0.5 ksi (3.4 MPa). The shear capacity continued to increase after initial cracking. The peak load was reached at about 0.04 in. (1.00 mm) of slip over a contact length of 2 in. (50 mm). The specimens sustained large deformations before failure. Figure 15-23 shows a photo of a typical specimen after testing. The specimen remained as one piece even though the slurry at the interface completely disintegrated.

In summary, the shear strength of SIFCON is about four times the shear strength of plain concrete. It has excellent ductility. At a fiber content of 6% for hooked fibers, fiber length has very little influence in both strength and ductility.

15.3.6   Drying shrinkage strain

Figure 15-24 shows the shrinkage behavior of SIFCON up to about 140 days [15.1]. The behavior of plain slurry is also shown in Figure 15-24 for comparative evaluation. The strains were measured using $3 \times 3 \times 11.25$ in. ($75 \times 75 \times 286$ mm) prisms cured in a fog room for 28 days and then placed in 50% relative humidity. The temperature during exposure was 74°F (23°C).

From Figure 15-24 it can be seen that the shrinkage of SIFCON stops after about 28 days, whereas the shrinkage of plain slurry continues up to 160 days. The magnitude of SIFCON shrinkage strain, which ranges from 0.0002 to 0.0005 in./in. (0.005 to 0.0125 mm/mm), is slightly lower than that reported for typical normal-weight concrete.

The addition of sand reduces the shrinkage of the matrix considerably. Hence SIFCON made with cement-sand slurry can be expected

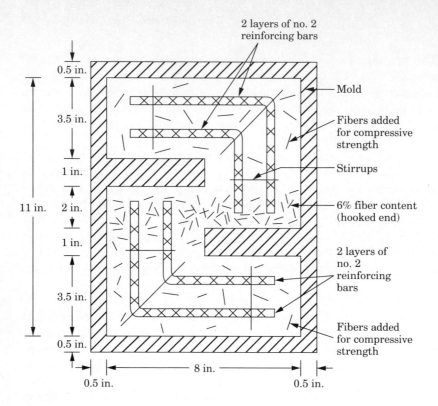

Figure 15.21a Mold for shear specimens [15.2] (1 in. = 25.4 mm).

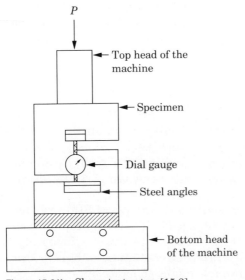

Figure 15-21b Shear test setup [15.2].

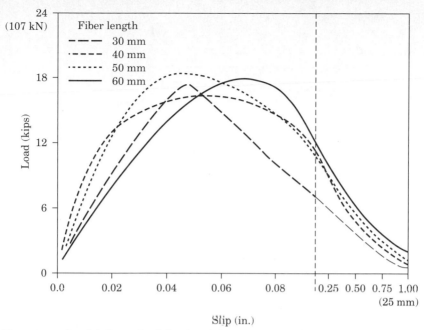

Figure 15-22 Load-deformation behavior of shear specimens; fiber content = 6% [15.2] (1 in. = 25.4 mm).

Figure 15-23 Typical shear specimen after failure [15.2].

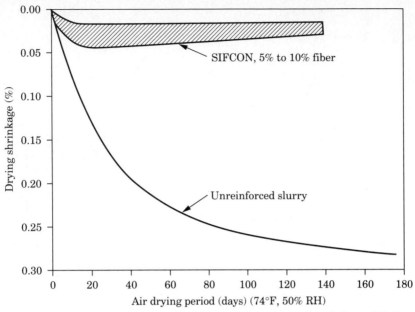

Figure 15-24   Drying shrinkage of SIFCON and plain, unreinforced slurry [15.1].

to have much less shrinkage strain than that shown in Figure 15-24, in which the slurry was made using only cement.

15.3.7   Resistance to freezing and thawing

The durability of SIFCON under freezing and thawing was evaluated using the rapid freeze-thaw procedure similar to the one recommended in ASTM C666, Procedure A. Essentially, four sets of specimens containing four different fiber lengths were subjected to 300 cycles of freezing and thawing. The temperature range was $0° \pm 1°F$ $(-17.8° \pm 0.5°C)$ to $40° \pm 1°F$ $(4.4° \pm 0.5°C)$, and each cycle had a period of 4.5 hours. Two inch (50 mm) slurry cubes were also placed in the freeze-thaw machine.

Half of the slurry cubes disintegrated in the freeze-thaw chamber. None of the prism specimens disintegrated; however, there was considerable scaling on the surfaces. After the freeze-thaw cycles, the specimens were tested in flexure. Because of the scaling, the flexural strength reduced considerably. The reduction in strength ranged from 26% to 43% compared to virgin specimens. The percentage strength reduction could have been high because the specimens were thin. There was no difference in the ductility of the specimens [15.7].

Based on the results of freeze-thaw tests, it is recommended that thin SIFCON sections exposed to freezing and thawing should be protected by some kind of overlay or coating.

15.3.8   SIFCON under fatigue loading

Tests conducted using $4 \times 4 \times 14$ in. ($100 \times 100 \times 350$ mm) beams under third-point loads, indicate that SIFCON beams can sustain more than 90% of static strength up to 2×10^6 load cycles. The specimens tested contained a 16% volume fraction of crimped fibers.

15.3.9   Resistance to impact and blast loading

Investigations conducted by a number of researchers indicate that SIFCON has a great energy-absorbing capacity under impact and blast loading [15.9 through 15.18]. Its performance is also better than reinforced concrete for missile loads.

15.4   SIFCON With Steel Bar Reinforcement

A few investigators evaluated the flexural behavior of SIFCON beams and beam-column connections that were reinforced with mild steel reinforcement [15.19, 15.20]. Tests conducted using beams 6×6 in. (150×150 mm) in cross section, over a span of 24 in. (600 mm), indicate that SIFCON beams have a lower ductility compared to reinforced concrete beams containing tension and compression reinforcement, Figure 15-25 [15.19]. However, SIFCON beams reinforced with mild steel bar reinforcement have much higher flexural strength and flexural toughness. This kind of special composite can be used in applications for which very high strength and ductility are needed.

A reinforced SIFCON beam-column joint was tested under monotonic and cyclic loading to evaluate its applicability for earthquake-resistant structures. The SIFCON sections were located slightly away from the beam-column joints and were purposely made weak to induce failures at the SIFCON locations. The fiber volume fraction was 5%. Figure 15-26 shows the comparison of the normal reinforced concrete response and the SIFCON response under cyclic loading [15.20]. It can be seen that SIFCON sections absorb much more energy and sustain much higher loads compared to normal reinforced concrete.

15.5   Applications

SIFCON is relatively a new product, yet the composite has been used in a number of areas. The following are some of the successful applications:

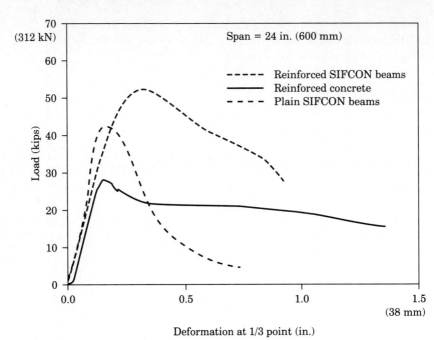

Figure 15-25 Flexural strength comparison of SIFCON, reinforced concrete, and reinforced SIFCON [15.19].

- Security vaults that need protection against blasting, drilling, and fire (torching)
- Explosive-resistant containers for storing materials that could accidently explode
- Repair of structural components such as damaged prestressed concrete beams
- Bridge deck rehabilitation
- Pavement rehabilitation
- Abrasive-resistant surfaces in places such as scrap yards
- Precast products for rapid repairs
- Refractory applications

15.5.1 Security vaults

In this application, the product must have excellent resistance against blasting, torching, drilling, and chipping. Both reinforced concrete and steel have certain weaknesses. For instance, steel walls can be torched,

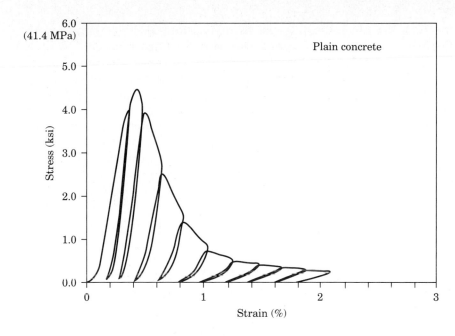

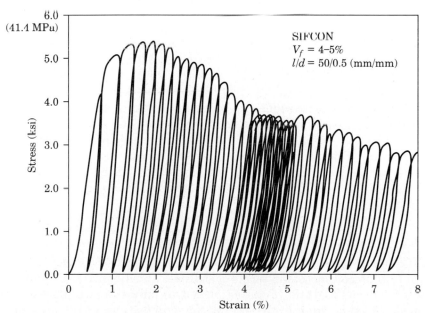

Figure 15-26 Typical stress-strain responses of plain concrete and SIFCON developed for the ductile connection [15.20].

whereas the concrete walls can be drilled and blasted to gain entry. But SIFCON has an advantage over both steel and concrete. SIFCON walls cannot be torched because concrete will resist deterioration by heat and will also slow down heat conduction. The composite can resist blast loading because of its high ductility. Chipping and drilling is very cumbersome because of the fiber intrusions. Hence, SIFCON is being used for various types of safes and vault doors. The mix compositions used by the various manufacturers are not divulged both for business and security reasons. In most cases, the preplaced fibers are mixed with clean coarse aggregates before infiltration with cement-based slurry.

15.5.2 Explosive-resistant containers

SIFCON has been used for making containers to store various kinds of ammunition. The primary concern in this application is to limit the spread of explosions from container to container. SIFCON provides good resistance in terms of containing the exploded materials in one chamber. The primary weak points were found to be connections between the walls. Successful connections have already been developed using the test results of prototype boxes.

SIFCON was also evaluated for missile silo structures [15.21]. A $\frac{1}{8}$ scale model of a hardened silo structure was constructed (Figure 15-27) at the New Mexico Research Institute in Albuquerque, New Mexico, USA. The silo was 20 ft (6.1 m) long. The wall consisted of 6 in. (15 cm) thick SIFCON sandwiched between two 0.25 in. (6 mm) thick steel plates. First, the inner and outer liners were placed in position and the fibers were placed between the liners. The slurry was infiltrated with the aid of vibration. The liner was cast in lifts of 5 ft (1.5 m) in order to obtain properly compacted composites.

The completed model silo was instrumented, buried underground, and tested for resistance to explosive loading. The performance was found to be excellent.

15.5.3 A repair material for structural components

SIFCON serves as an excellent repair material because it is compatible with reinforced concrete in terms of stiffness and dimensional changes caused by temperature. It can be placed in hard-to-reach places and provides good bonding to the parent concrete because of the presence of fibers. The matrix can be modified to suit the particular repair. For example, rapid strength gain can be obtained using accelerators.

SIFCON was used to repair prestressed concrete beams spanning an interstate highway in the state of New Mexico. The beams had

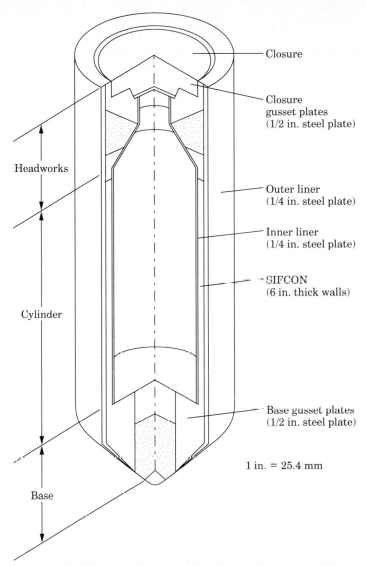

Figure 15-27 Schematic diagram of ⅛ scale model (6.1 m [20 ft] long) of a silo structure containing SIFCON [15.1].

been damaged by a vehicle passing under the bridge. Some of the pretensioned tendons had been exposed by the damage. The beams were restored, using SIFCON, without removing them. Restoration in-place not only resulted in a large cost saving but also reduced the time of repair by a few months. The repaired beam, which is more than eight years old, is functioning well.

15.5.4   Bridge rehabilitation

SIFCON has been used for a number of bridge rehabilitation projects. Typically, the area to be repaired is chipped off and cleaned thoroughly. The fibers are placed in position and infiltrated with slurry. In most cases, the infiltration is achieved by gravity alone.

In certain applications, a thin layer of coarse aggregates were placed on the top and troweled into place to form a wearing surface. The repaired sections, some of which are more than six years old, are functioning satisfactorily.

15.5.5   Pavement rehabilitation

Pavement rehabilitation is similar to bridge deck rehabilitation except that the repair surfaces are normally large and the loading pattern is primary compressive. In certain cases, a thin, bonded overlay is placed on the existing surface to act as a wearing surface.

The construction sequence is the same as in bridge deck repairs. If the area to be repaired is deep, SIFCON can be used for only the top 1 in. (25 mm). The bottom part can be repaired using normal concrete. SIFCON has been successfully used for overlays ranging in thickness from 0.75 to 2 in. (19 to 50 mm).

15.5.6   Precast products

Precast products made out of SIFCON include precast slabs, small vaults, and special units such as basketball backboard. Precast slabs that are 1 to 2 in. (25 to 50 mm) thick were used primarily as the wear-resistant surface. The slabs, which provide good impact resistance, have also been used in airport taxiways and gate areas.

15.5.7   Refractory applications

The concept of SIFCON has also been successfully used for refractory applications [15.22, 15.23]. In this application, stainless steel fibers were used with a calcium aluminate cement matrix. SIFCON was found to perform well under the high-temperature shock and mechanical loading. The service life of components such as a soaking pit cover made using SIFCON was found to be much longer than the components made with other materials.

15.5.8   Potential new applications

SIFCON has potential for a large number of applications for which high strength or high ductility or both are needed. These include a large variety of defense installations, explosive and penetration-resistant structures, and earthquake-resistant structures.

15.6   References

15.1 Lankard, D. R. "Preparation, Properties and Applications of Cement-Based Composites Containing 5% to 20% Steel Fiber," *Steel Fiber Concrete*, Elsevier, 1986, pp. 199–217. (Also published by Swedish Cement and Concrete Research Institute Stockholm, 1985.)

15.2 Balaguru, P.; and Kendzulak, J. "Mechanical Properties of Slurry Infiltrated Fiber Concrete (SIFCON)," *Fiber Reinforced Concrete Properties and Applications*. SP-105, American Concrete Institute, Detroit, Michigan, 1987, pp. 247–268.

15.3 Mandragon, R. "SIFCON in Compression," *Fiber Reinforced Concrete Properties and Applications*. SP-105, American Concrete Institute, Detroit, Michigan, 1987, pp. 269–282.

15.4 Homrich, J.; and Naaman, A. E. "Stress-Strain Properties of SIFCON in Compression," *Fiber Reinforced Concrete Properties and Applications*. SP-105, American Concrete Institute, Detroit, Michigan, 1987, pp. 283–304.

15.5 Reinhardt, H. N.; and Fritz, C., "Optimization of SIFCON Mix," *Fiber Reinforced Cements and Concretes—Recent Developments*, Elsevier, 1989, pp. 11–20.

15.6 Naaman, A. E.; and Homrich, J. R. "Tensile Stress-Strain Properties of SIFCON," *ACI Materials Journal*, Vol. 86, No. 3, 1989, pp. 244–251.

15.7 Balaguru, P.; and Kendzulak, J. "Flexural Behavior of Slurry Infiltrated Fiber Concrete (SIFCON) Made Using Condensed Silica Fume," *Fly Ash, Silica Fume, Slag, and Natural Pozzalans in Concrete*, SP-60, American Concrete Institute, Detroit, Michigan, 1986, pp. 1215–1230.

15.8 Lankard, D. R. "Slurry Infiltrated Fiber Concrete (SIFCON)," *Concrete International*, Vol. 6, No. 12, 1984, pp. 44–47.

15.9 Mondragon, R. "SIFCON Testing of the Generic Hard Silo 1/8th Scale Model," AFWL-TN-86-22, New Mexico Engineering Research Institute, Albuquerque, New Mexico, May 1985.

15.10 Carson, J. "The Ballistic Performance of SIFCON," AFWL-TR86-103, Air Force Weapons Laboratory, Kirtland AFB, New Mexico, July 1986.

15.11 Carson, J.; and Morrison, D. "The Response of SIFCON Revetments to a Mark 83 General Purpose Bomb," AFL-TR-86-42, Air Force Weapons Laboratory, Kirtland AFB, New Mexico, December 1986.

15.12 Cheney, S.; Carson, J.; and Swift, H. "SIFCON Impact Performance," Proceedings, International Symposium on the Interaction of Non-nuclear Munitions with Protective Structures, Mannheim, FRG, March 1987, pp. 557–581.

15.13 Schneider, B. "Defensive Fighting Position (DFP)—Component Test Report," AFWL-TR-88-42, Air Force Weapons Laboratory, Kirtland AFB, New Mexico, December 1987.

15.14 Schneider, B. "Design Criteria for Fragment Resistant SIFCON Panels," NMERI-31, New Mexico Engineering Research Institute, Albuquerque, New Mexico, Task Report for the Naval Civil Engineering Laboratory, Port Hueneme, California, February 1988.

15.15 Schneider, B. "The Use of SIFCON in Building Security, NMERI-38, New Mexico Engineering Research Institute, Proceedings, ASCE Specialty Conference: Structures for Enhanced Safety and Physical Security, Arlington, Virginia, March 1989.

15.16 Marchand, K.; and Nash, P. "Test and Analysis of Localized Response of CRC and SIFCON," Proceedings, 58th Shock and Vibration Symposium, October 1987.

15.17 Marchand, K., et al. "Development of Alternate Barrier Concepts for Munition Storage Applications," Southwest Research Institute Project 06-8986, Task for the U.S. Army Construction Engineering Research Laboratory, Champaign, Illinois, October 1988.

15.18 Marchand, K.; Nash, P.; and Cox, P. "Tests and Analysis of the Localized Response of Slurry Infiltrated Fiber concrete (SIFCON) and Conventionally Reinforced Concrete (CRC) Subjected to Blast and Fragment Loading," *Journal de Physique*, September 1988.

15.19 Marchand, K. "Tests and Analysis of the Localized Response of SIFCON and CRC Subjected to Blast and Fragment Loading," Presented at SIFCON 1989, held in NMERI, Albuquerque, New Mexico, October 1989.

15.20 Naman, T.; Wight, J.; and Abdou, H. "SIFCON Connections for Seismic Resistant Frames," *Concrete International*, Vol. 9, 1987, pp. 34–39.

15.21 Schneider, B.; Mondragon, R.; and Kirst, J."Task Report NMERI, JA 8-69 (8.361.01)," New Mexico Engineering Research Institute, 1984, pp. 1–83.

15.22 Lankard, D. R.; and Lease, D. H. "Highly Reinforced Precast Monolithic Refractories," *Bulletin of American Ceramic Society*, Vol. 61, No. 7, 1982, pp. 728–732.

15.23 Lankard, D. R. "Factors Affecting the Selection and Performance of Steel Fiber Reinforced Monolithic Refractories," *Bulletin of American Ceramic Society*, Vol. 63, 1984, pp. 919–925.

16

The Use of FRC for Structural Components

The use of FRC for load-bearing structural components has been investigated by a number of researchers [16.1–16.38]. The structural components investigated include columns, beams, corbels, and beam-column connections. The primary contribution of fibers is in the area of ductility. Hence most of the studies were focused on evaluating the cyclic behavior and the energy-absorption characteristics. Fiber addition was also found to improve the fatigue performance of reinforced-concrete beams [16.6].

It should be noted that the structural components also contain regular bar reinforcement in addition to discrete fibers. In some instances, it was found that reinforcement congestion can be reduced by using less mild-steel reinforcement in combination with discrete fibers. This is particularly true in the area of connections and shear (stirrup) reinforcement. The amount of fiber reinforcement is normally limited to a 2% volume. So far, only steel fibers have been investigated for use in structural components.

This chapter provides a review of the behavior of fiber-reinforced beams, columns, corbels, and beam-column connections. The design concepts are discussed briefly. The continuous bar reinforcement can be just mild-steel or prestressed reinforcement or a combination of both. The basic properties of the composite needed for design, including stress-strain behavior are also presented. The bond-slip properties of steel bars embedded in fiber concrete are discussed because of their importance to the cyclic behavior of beam-column connections.

16.1   Behavior under Flexure

The addition of steel fibers in reinforced concrete provides a substantial improvement in ductility. The load-deflection curve of fiber-reinforced beams has longer plateau at the peak load, and the descending part of the curve is less steep compared to normal reinforced-concrete beams. This increase in energy-absorption capacity is even more notable under cyclic and impact loads. There is notable improvement in cracking characteristics. The cracks are more uniformly distributed, with substantial reduction in maximum crack width. First-crack strengths are also higher, especially for prestressed concrete beams. The fibers also provide slight increase in moment capacity and flexural stiffness. The increase in stiffness reduces both deflection and total crack width along the tension face of the beam at service loads.

A comprehensive study conducted using normal-weight and light-weight aggregate concrete and steel fibers with hooked ends provides insight to both qualitative and quantitative comparisons of plain and fibrous concretes [16.17]. The variables studied in this investigation include the compressive strength of concrete, the yield strength of mild steel, fiber length and volume fraction, and the presence of compression steel reinforcement. The compressive strength of concrete varied from 2 to 6 ksi (14 to 42 MPa). Reinforcing bars with three yield strengths, namely, 40, 60, and 75 ksi (276, 414, and 518 MPa) and three fiber lengths of 1.2, 1.6, 2.0 in. (30, 40, and 50 mm), were evaluated. The fiber volume fraction for most beams was 1.75%. All the beams were tested over a simply supported span of 36 in. (90 cm) with loads applied at middle-third points.

The effect of fibers on strength and ductility is shown in Figure 16-1 [16.17]. The beam with plain concrete and no compression steel has the lowest strength and ductility. The addition of compression steel (area of compression steel = 0.5 times the area of tension steel) increases the strength and ductility by improving the compressive strain capacity of the concrete. The beams with fibers had better ductility than beams with only continuous compression reinforcement. The fiber-reinforced beam sustained a large residual load even at a deflection of 3 in. (75 mm) and spalling at the compression zone. The presence of compression reinforcement in fiber concrete further improves the ductility. It was observed that reduced spalling and improved integrity of fiber concrete in the failure zone prevented the buckling of compression bars, thus providing improved ductility [16.17]. It was also reported that cracks were more uniformly distributed and that first-crack moment was higher for the fibrous concrete beams.

Figure 16-2 shows the influence of reinforcement yield strength and fiber length. Moment capacity increased with the increase in the yield strength of the reinforcing bars, which should be expected because of

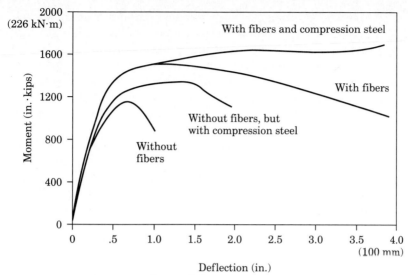

Figure 16-1 Comparison of moment-deflection curves of plain and FRC beams [16.17].

the higher tensile-force capacity. Shorter fibers resulted in a slight decrease in moment capacity. In all cases, the beams exhibited excellent ductility characteristics. The compressive strength of concrete was about 5 ksi (34 MPa).

Tests conducted using high-strength concrete with a compressive strength of about 10 ksi (69 MPa) indicate that the ductility contri-

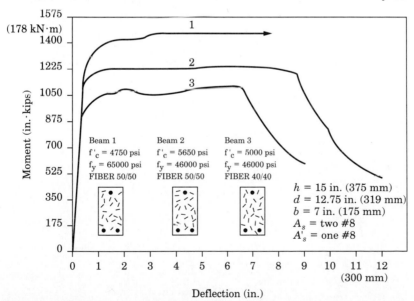

Figure 16-2 Moment versus deflection for fiber-reinforced concrete beams [16.17] (1ksi = 6.89MPa).

bution of fibers is greater for high-strength concrete than normal-strength concrete. This should be expected because high-strength concrete typically has a steep descending stress-strain curve in compression. The rapid decrease in compressive strength in the postpeak load region results in sudden failure of the beam. The presence of fibers reduces this sudden drop and results in improved ductility.

Figures 16-3, 16-4, and 16-5 present the effect of fiber volume fraction, compressive strength of the concrete, and yield strength of the steel on the behavior of beams. From these figures it can be seen that the addition of fibers increases the ductility. The magnitude of improvement is greater for the higher-strength concrete and the higher yield-strength reinforcement. For volume fractions greater than 1%, increases in the moment capacity can be expected. A slight improvement in stiffness can also be expected for fiber volume fractions of 1% or greater. The fibrous beams can sustain more than 90% of peak moment (load) even at large deformations.

Fibers were also found to be very effective in lightweight concrete beams. Improvement obtained in ductility is similar to the improvement attained in normal-weight concrete.

A study conducted using a slab reinforced with 2 in. (50 mm) long hooked-end steel fibers in conjunction with wire meshes indicates that the fibers improve both strength and ductility [16.16]. First-crack strength is also increased, but to a lesser extent than ultimate

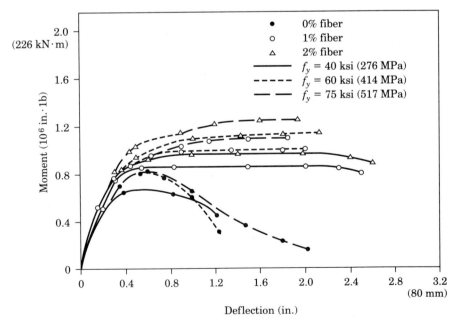

Figure 16-3 Comparison of moment-deflection curves for beams constructed with 2000 psi (14 MPa) concrete [16.17].

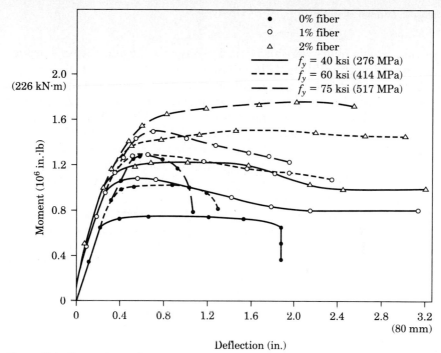

Figure 16-4 Comparison of moment-deflection curves for beam sections constructed with 4000 psi (28 MPa) concrete [16.17].

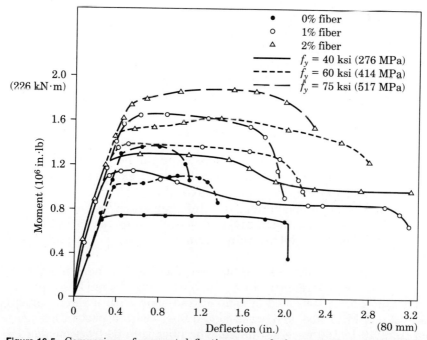

Figure 16-5 Comparison of moment-deflection curves for beam sections constructed with 6000 psi (42 MPa) concrete [16.17].

strength. The fibers were also found to improve the ductility of T-beams reinforced with mild steel [16.19].

Tests conducted using straight steel and brass-coated fibers and cement mortar show that a substantial increase in moment capacity can be expected at a fiber volume fraction of 1% [16.8]. The increase in the aspect (length/diameter) ratio in the range of 10 to 75 provided a consistent increase in the moment capacity. A reduction of moment capacity is reported at an aspect ratio of 100, compared to 75. The decrease could be due to mixing and compaction problems rather than the behavior of the composite.

Results obtained using 2 in. (50 mm) long and 0.02 in. (0.5 mm) diameter crimped fibers and a cement–fly ash–sand mortar also show that fibers are effective in a number of areas [16.17]. The beams were also reinforced with steel bars. The fibers increase the first-crack strength, reduce the maximum crack width, reduce the number of cracks, and improve the ductility. The moment increase was limited to about 10% for fiber volume fraction of 1%. Concrete strains at failure were as high as 0.0066 in./in. (mm/mm) as compared to about 0.003 in./in. (mm/mm) for plain concrete.

16.2   Behavior under Shear

One of the very promising applications of fiber-reinforced concrete is in the area of shear reinforcement. The random distribution of fibers provides a close spacing of reinforcement in three dimensions that cannot be duplicated with bar reinforcement. The fibers intercept and bridge cracks in all directions. This process not only increases shear capacity but also provides substantial postpeak resistance and ductility. Laboratory tests have established that both strength and ductility can be enhanced using fibers. The test variables investigated include the shear span-to-depth ratio (a/d ratio), fiber type, fiber volume fraction, and the compressive strength of concrete. The shear span a is defined as the distance between the load and the nearest support, and d is the depth of the beam measured from the extreme-compression fibers to the center of gravity of the flexural tension steel. Fibers have been also used to augment stirrup reinforcement. When used in conjunction with stirrups, fibers make it possible to increase the spacing of stirrups thereby reducing reinforcement congestion in high-shear areas.

Results obtained using straight steel fibers and cement mortar indicate that shear strength can be doubled using fibers [16.8]. In this study the matrix consisted of cement and sand with a water-cement ratio of 0.6. Mild steel and brass-coated fibers with aspect ratios ranging from 10 to 100 were evaluated. The fiber volume fraction was 1% The 4×6 in. (100×150 mm) beams had two 0.5 in. (12 mm) diameter

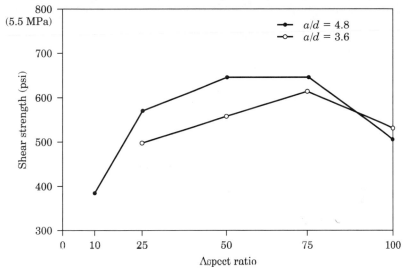

Figure 16-6 Effect of aspect ratio of fiber on ultimate shear stress [16.8].

steel bars for flexural reinforcement. They were tested on a simply supported span of 60 in. (152 cm). The loads were positioned to obtain shear span-to-depth ratios (a/d) ranging from 2.0 to 4.8. The plain concrete (mortar) beams failed in shear for a/d ratios of 2.0, 2.4, and 3.6. The maximum shear stress was 438 psi (2.8 MPa). The results are presented in Figure 16-6. The mild steel fibers provided a maximum increase of 44%, and the brass-coated fibers recorded a 105% increase. The increase in the aspect ratio of the fiber resulted in an increase of shear strength up to a value of 75. Fibers with an aspect ratio of 100 had lower strengths than fibers with an aspect ratio 75. Higher aspect ratios provide higher bond strength, thus generating more resistance. When fibers are too long, they tend to ball and to reduce compaction, resulting in poor distribution. These could be the reasons for obtaining lower strengths for fibers with an aspect ratio of 100.

The primary variables for results reported in Reference [16.2] are fiber size, shape, volume fraction, and shear span-to-depth ratio (a/d). The beams had an overall dimension of $4 \times 6 \times 78$ in. ($100 \times 150 \times 2000$ mm). Conventional flexural reinforcement was provided to induce shear failure. Plain matrix failed in shear for beams with a/d ratio of 4.8. When a sufficient volume fraction of fibers was added, the failure mode changed from shear failure to moment failure. For each volume fraction of fibers, the shear span-to-depth (a/d) ratios were decreased until a shear failure was induced. The variations of shear strength with respect to a/d are shown in Figure 16-7 [16.2]. A volume fraction of 1.76% provided considerable increase in shear

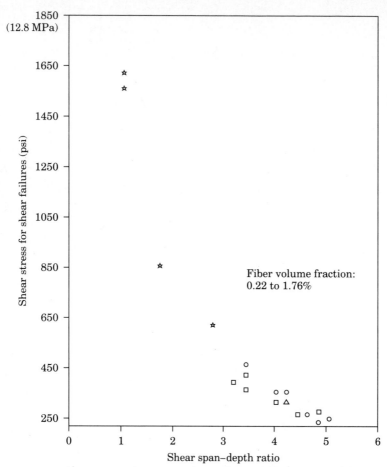

Figure 16-7 Shear strength versus *a/d* ratio for steel fiber-reinforced beams [16.2, 16.20].

strength. The results, however, show considerable scatter when *a/d* ratios are less than 3.0.

Results reported in Reference [16.23] provide a comparison between straight and hooked-end fibers. Volume fractions of 0.75% and 1.00% hooked-end fibers recorded failure loads of 14.8 and 16.4 kips (65.6 and 72.7 kN), whereas 1.5% and 2.0% of straight fibers recorded failure loads of 14.0 and 15.0 kips (62.1 and 66.7 kN). The shear capacity for beams with stirrups was 16.2 kips (72.1 kN), and the plain concrete had a capacity of 8.0 kips (35.6 kN). The hooked-end fibers provide a greater contribution both in strength and ductility because of the end anchors of the hooked ends. These fibers also provide better mechanical properties, as explained in other chapters.

The shear strength increase provided by hooked-end fibers is shown in Figure 16-8 [16.20, 16.23]. The results were obtained using $6 \times 12 \times$

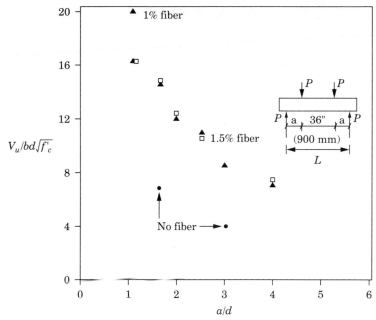

Figure 16-8 Shear strength versus a/d ratio for reinforced beams with hooked-end steel fibers [16.20].

72 in. (150 × 300 × 1830 mm) beams tested over a simply supported span. For a shear span-to-depth ratio of 1.5, the shear strength increase was 130% and 140% for fiber volume fractions of 1.0% and 1.5% compared to plain concrete beams.

Tests conducted using prototype beams, 12 × 21.5 × 276 in. (300 × 540 × 7000 mm), showed that steel fibers with hooked ends can be used as effective shear reinforcement [16.24]. The beams with 1.1% volumes of hooked-end steel fibers provided an increase of 67% shear capacity. The beams had no stirrup reinforcement. Still they failed in flexure.

The contribution of fibers was also evaluated for T-beams [16.19, 16.25]. The variables were fiber content, longitudinal (flexural) reinforcement ratio, and stirrup spacing. An a/d ratio of 4.5 was kept constant. The ultimate shear strength was found to increase almost linearly for fiber volume fractions ranging from 0.8% to 1.2%. The addition of hooked-end fibers was found to augment the nominal amount of stirrups and to enhance the shear strength contribution from dowel action.

Fibers were also found to be effective in I cross sections [16.26]. Fiber addition prevented shear failure for a/d ratios between 3.24 to 4.68.

Overall, fibers are effective in providing improvement in shear capacity. Deformed fibers are more efficient than straight fibers. A minimum of 1% volume of fibers can replace stirrups in most instances.

The fibers can also be used in conjunction with nominal stirrup reinforcement to reduce reinforcement congestion.

16.3 Behavior under Torsion

The behavior of fiber-reinforced concrete beams under pure torsion has been investigated by a few researchers [16.9, 16.27–16.29]. The primary test variables were fiber type and volume fraction, aspect (length/diameter) ratio of the fibers, and the amount of hoop reinforcement.

When fibers are added to plain concrete, both torsional strength and energy-absorbing capacity increase [16.29]. When straight steel fibers are used without incorporating longitudinal or hoop reinforcement, the torsional strength increases from 10% to 26% for volume fractions ranging from 1% to 3% (Table 16.1) [16.29]. The increase in torsional strength with respect to aspect ratio is not substantial. (Table 16.1 and Figure 16-9 [16.29]. However, the areas under the torque-twist curve

TABLE 16.1   Torsional Strength of Fiber-Reinforced Concrete [16.29]

Volume fraction, V (%)	*Length of fiber, l_f (mm)	Compressive strength (MPa)	Modulus of rupture (MPa)	Ultimate torsional strength (N·m)	Increase in torsional strength (%)
Plain concrete†		26.4	2.67	843.7	—
		26.2	2.80	814.2	—
		30.6	2.60	824.0	—
1	15	36.3	3.32	904.8	9
		32.3	3.32	902.5	9
		36.6	3.55	892.7	8
1	22	35.0	3.70	922.8	12
		34.1	3.78	922.0	12
		35.0	4.04	915.2	11
1	30	34.1	3.92	922.1	12
		35.2	3.83	963.3	16
		33.6	3.95	902.5	9
1	44	32.3	3.99	995.7	20
		33.9	3.76	921.9	13
		32.7	3.80	959.3	16
2	30	29.7	4.38	981.0	19
		31.5	4.23	981.0	19
		31.1	4.50	985.9	19
3	30	33.7	4.92	1054.6	27
		33.9	4.98	1044.8	26
		35.0	5.31	1049.7	27

*Diameter of fiber = 0.57 mm
†Split cylinder strength = 2.2 MPa
1 in. = 25.4mm
1 ksi = 6.89 MPa
1 lb = 4.4 N

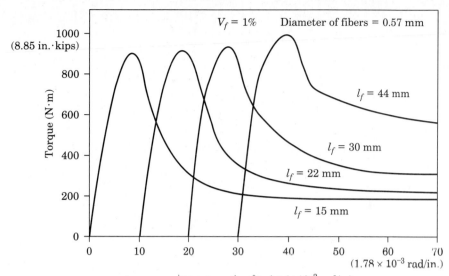

Figure 16-9 Torque-twist curves for specimens reinforced with fibers having different aspect ratios [16.29] (1 in. = 25.4 mm).

increase substantially with an increase in aspect ratio, resulting in a more ductile behavior. The improvement in ductility with the increase in fiber volume fraction is also more substantial than the strength increase (Figure 16-10) [16.29]. An increase in fiber volume fraction provided a consistent increase in both strength and ductility.

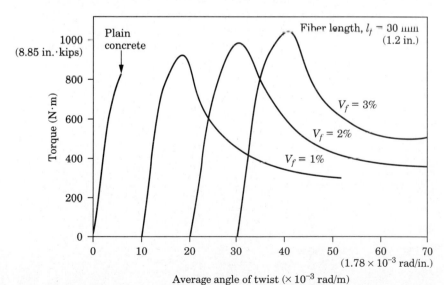

Figure 16-10 Torque-twist curves for specimens with different volume fractions [16.29].

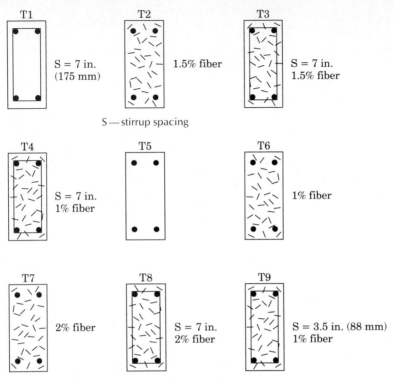

Figure 16-11 Test beam cross sections [16.9].

Tests conducted using fibers in beams reinforced with longitudinal and hoop reinforcement indicate that both strength and ducility can be improved considerably using hooked-end steel fibers [16.9]. The beams were 6×12 in. (150×300 mm) in cross section and were reinforced with four longitudinal bars (Figures 16-11 and 16-12). Some of the beams had hoop reinforcement spaced at 3.5 and 7.0 in. (87 and 175 mm). The experimental program was designed to evaluate the contribution of fibers in the presence and absence of hoop reinforcement.

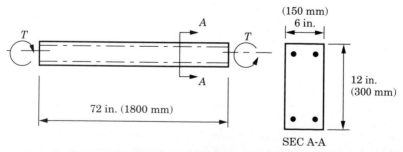

Figure 16-12 Dimension of test beams [16.9].

The influence of fibers in the absence of hoop reinforcement is shown in Figure 16-13. The strength increase is about 50% for a fiber volume fraction of 2%. The torque-twist curve has a considerable plateau at the peak load, indicating a high-energy absorption capacity. The improvement provided by fibers is much more substantial if hoop reinforcement is present (Figures 16-14 and 16-15) [16.9]. The fibers seem to provide a synergistic effect in the presence of hoop reinforcement. Figure 16-15 shows the influence of hoop reinforcement spacing. The fiber contribution to ductility seems to reduce at smaller stirrup spacing. In general, fibers were found to induce a greater number of finer cracks in the postpeak region [16.9]. Mathematical expressions used for estimating fiber contribution are discussed in the design concepts section, 16.9.

16.4 Behavior under Combined Bending, Shear, and Torsion

Very rarely are structural elements subjected to either pure bending, or shear, or torsion. In most cases bending and shear force occur together. In some instances the structural element will also be subjected to torsion in addition to bending and shear. In the case of torsion, the mem-

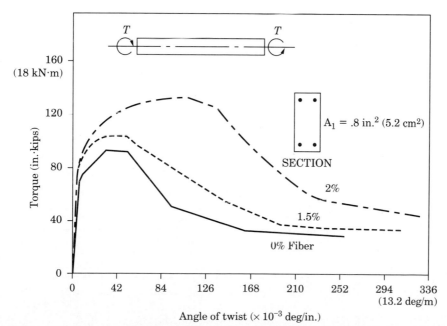

Figure 16-13 Influence of percentage of fibers on the behavior of reinforced fibrous concrete beams in torsion: no stirrups [16.9].

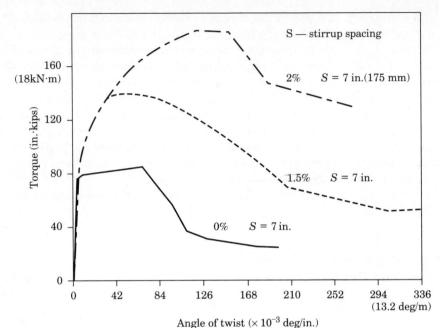

Figure 16-14 Influence of percentage of fibers on the behavior of reinforced fibrous concrete beams with hoop reinforcement in torsion [16.9].

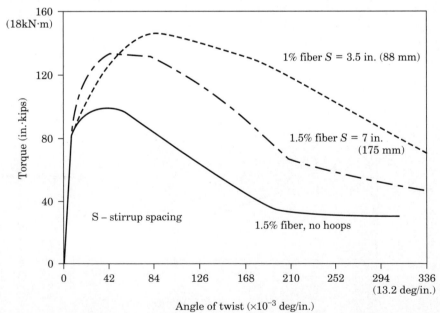

Figure 16-15 Influence of hoop spacing on the behavior of reinforced fibrous concrete in torsion [16.9].

bers could be subjected to either equilibrium or compatibility torsion. Compatibility torsion arises when structural elements are twisted at the support or at the load points. For example a spandral beam will be twisted by the beams at supports. The beams connected to the spandral beam rotate owing to bending and transfer this bending moment at the supports as torsion to the spandral beam. If the spandral beam rotates, the torque will be partially relieved. The final torque will depend on the torsional stiffness of the spandral beam. Since there is a considerable release owing to rotation and since the torque develops only to satisfy the compatibility of rotations, this torsion is considered less crucial in design than the equilibrium torsion. However, members subjected to compatibility torsion should be designed to maintain their integrity in sustaining other forces such as bending moment and shear force.

When fibers are added to members subjected to compatibility torsion, they were found to reduce stiffness degradation. The cracks were finer and more uniformly distributed, thus reducing sudden drops in stiffness at wider cracks. The members with fibers could also sustain more rotation without loosing their load carrying capacity [16.20].

The contribution of fibers to structural elements subjected to equilibrium torsion, moment, and shear were investigated using both straight and hooked-end steel fibers [16.21, 16.22]. The performance of straight fibers was evaluated using 0.5 mm (0.02 in.) diameter, 38 mm (1.5 in.) long (aspect ratio, l/d = 75) fibers at volume fractions of 0.5, 1.0, and 1.5%. Control specimens with no fibers were also tested. Rectangular beams with overall dimensions of $125 \times 300 \times 2500$ mm ($5 \times 12 \times 100$ in.) were reinforced with both longitudinal and hoop reinforcement. The testing was conducted under pure torsion, torsion and bending, and torsion, bending, and shear.

The fiber addition provided an increase of 33% and 44% respectively, for cracking and ultimate torque when only torsion was applied. Under combined bending and torque, the increases were 53% and 21% for cracking and ultimate torque. The increases were 82% and 65% when the beams were subjected to combined bending, shear, and torsion. The higher increase under combined loading shows the effectiveness of randomly distributed fibers under complex loading patterns that could induce tensile stresses in any of the three mutually perpendicular directions. Continuous reinforcement cannot be effectively placed to counteract the principal tensile stresses developed by the combined loading.

The torque-rotation curves for three loading conditions—torsion, torsion and bending, and torsion, bending, and shear—are shown in Figure 16-16 [16.22]. The addition of fibers increases the rotational capacities of the elements. The fibers are more effective under combined loading than in pure torsion (Figure 16-16) [16.22]. The torsional capacity and maximum rotation increases substantially for beams with 1.5% fiber under torsion, bending, and shear, as compared to the other two loading patterns involving just torsion or torsion and bending.

Fibers with hooked ends increased the ductility much more than straight fibers. Tests conducted using 1.2 in. (30 mm) long, 0.02 in. (0.5 mm) diameter fibers with hooked ends were found to be very effective even at a volume fraction of 0.5% for rectangular beams that were 6 × 12 in. (150 × 300 mm) subjected to either torsion, moment and torsion, or torsion, moment, and shear. The loading ratios for the various tests are shown in Table 16.2 [16.21]. The torque-rotation characteristics are shown in Figure 16-17. As with straight fibers, hooked fibers seem to provide the best results under combined loads. The rotation capacities of beams even with 0.5% fiber are about five times the rotation capacity of plain concrete. Since all beams had longitudinal and hoop reinforcement, the increase in strength and ductility provided by fibers was significant.

16.5   Deep Beams and Corbels

Deep beams and corbels can be considered as special beams in which the shear force is much more predominant than bending moment. Fibers can be effectively used in these structural members, especially to improve the ductility.

The behavior of deep beams is very much influenced by the shear span-to-depth (a/d) ratio. Failure patterns change considerably around a/d ratios of 1.5. Fibers have been evaluated for a/d ratios ranging from 0.3 to 2.4 [16.23–16.33]. Details of test specimens used in conjunction with surface-deformed brass-coated, 0.4 mm (0.016 in.) diameter by 38 mm (1.5 in.) long fibers are shown in Table 16.3 [16.31]. The overall dimensions of the beam cross section were 2 × 8 in. (50 × 200 mm). The volume fraction is approximately 1.2% and 1.8%. The failure and crack loads shown in Table 16.4 [16.31] show that fibers make a definite contribution to the strength development both before and after cracking. The improvement in first-crack load is higher than the failure load for all three a/d ratios. The magnitude of increase does not vary with a/d ratio.

The addition of hooked-end fibers was also found to improve the performance both in strength and ducility for deep beams with and without openings [16.32, 16.33].

Corbels can be considered as special types of cantilever deep beams and are often used to transfer heavy beam reactions to columns. Corbels are typically reinforced with flexural reinforcement at the top and stirrups to resist shear forces. Fibers have been tried as a replacement

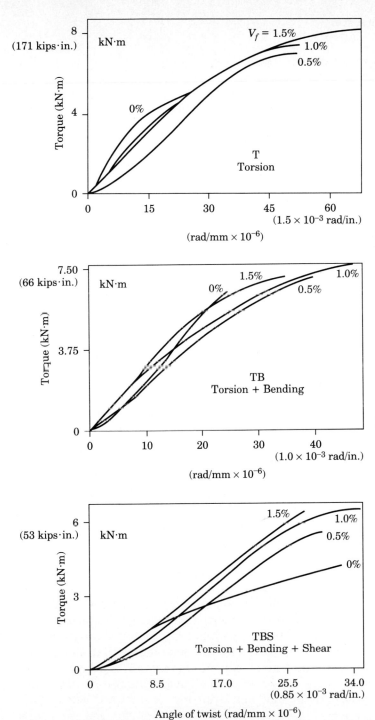

Figure 16-16 Effect of fiber content and type of loading on torque-rotation behavior [16.22].

TABLE 16.2 Loading Ratios for Beams Shown in Figure 16-17 [16.21]

Beam	Type of loading	Fiber content, V_f (%)	Loading ratio T/M	T/V (mm)	M/V (mm)
T0	Pure torsion	0.0	∞	∞	—
T1		0.5	∞	∞	—
T2		1.0	∞	∞	—
MT1	Combined moment and torsion	0.0	1.00	∞	∞
MT2		1.0	1.00	∞	∞
MT3		1.0	0.35	∞	∞
MT4		1.0	0.25	∞	∞
MT5		0.5	0.25	∞	∞
MTS1	Combined moment, torsion, and shear	0.0	1.00	400	400
MTS2		0.5	1.00	400	400
MTS3		1.0	1.00	400	400
MTS4		0.0	0.35	175	500
MTS5		0.5	0.35	175	500
MTS6		1.0	0.35	175	500
MTS7		1.0	0.25	175	500

T—torsion
M—Moment
V—Shear
1 in. = 25.4 mm

for the stirrup reinforcement. The behavior of corbels containing shear reinforcement in the form of stirrups was compared to the behavior of corbels containing only fibers [16.34, 16.35]. The fibers used were surface-deformed steel fibers. The nominal size was 60 mm by 0.65 mm (2.4×0.026 in). The general configuration of the corbel is shown in Figure 16-18 [16.35]. The primary test variables were the a/h ratio (Figure 16-18), the shear reinforcement in the form of stirrups, and the shear reinforcement in the form of fibers.

The test results showed that the failure is catastrophic when no shear reinforcement is provided. The addition of fibers was found to change the mode of failure and improve the ductility. A fiber volume fraction of 1.4% was found to improve both strength and ductility. Some of the corbels with fibers exhibited good elasto-plastic behavior at fiber volume fractions that are greater than twice the bar reinforcement (Figure 16-19). It was also found that the contribution of stirrup reinforcement can be duplicated by fibers. Fibers were found to be more effective for inducing a ductile flexural failure instead of a brittle diagonal splitting or inclined shear failure.

Investigations conducted using surface-deformed, crimped and hooked steel, and fibrillated polypropylene fibers show that fiber type and geometry play an important role in the corbel behavior [16.36].

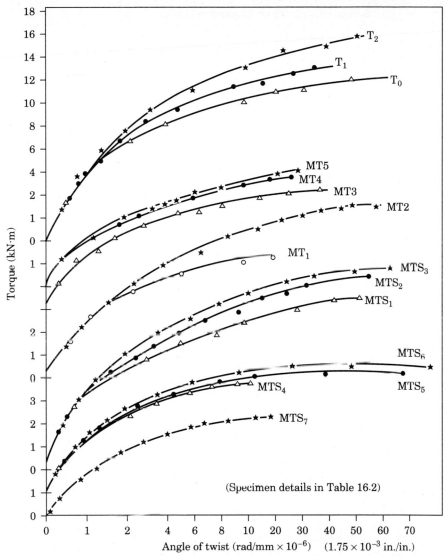

Figure 16-17 Effect of fiber content and type of loading on torque-rotation behavior [16.21] (1 kN · in. = 0.11 kN · m).

16.6 Beam-Column Connections

One of the most promising areas of structural application for fiber-reinforced concrete is structural connections. The fibers provide the necessary ductility to induce flexural failure rather than sudden shear failure. This aspect is extremely important for structures located in earthquake-prone zones. Investigations have been carried out using both straight and hooked-end steel fibers [16.4, 16.12–16.15]. All the

TABLE 16.3   Details of Deep Beam Specimens [16.31]

Beam	Width, b (mm)	Effective depth, d (mm)	Length (mm)	Shear span, a (mm)	a/d	Fiber content by weight (%)
PCB1	50	170	1000	410	2.4	0
F3.0B1	50	170	1000	410	2.4	3.0
F4.5B1	50	170	1000	410	2.4	4.5
PCB2	50	170	725	273	1.6	0
F3.0B2	50	170	725	273	1.6	3.0
F4.5B2	50	170	725	273	1.6	4.5
PCB3	50	170	455	137	0.8	0
F3.0B3	50	170	455	137	0.8	3.0
F4.5B3	50	170	455	137	0.8	4.5

1 in. = 25.4 mm

typical connections in a framed structure—cross, T, and knee types have been studied. Since cyclic loading is of primary importance, all the investigations were carried out using low-cycle high-amplitude loading.

Typical reinforcement arrangements for cross, T, and knee types of connections are shown in Figure 16-20 [16.14]. The contribution of fibers can be demonstrated using moment-rotation curves or energy absorption measured using hysteresis loops generated under cyclic loading. Typical moment-rotation curves for a knee joint reinforced with straight steel and brass-coated fibers are shown in Figure 16-21 [16.15]. Fibers provide a large increase in ductility for different moment–axial force ratios. As expected, higher volume fractions and larger aspect ratios resulted in higher moment and rotation capacity.

TABLE 16.4   Test Results of Deep Beams (Beam Details in Table 16.3) [16.31]

Beam	a/d	Failure load point (kN)	Cracking load (kN)	Max. average shear stress, v (MPa)	Max. moment at failure, M_f (k·Nm)	Theoretical max. moment M_u (k·Nm)	M_f/M_u
PCB1	2.4	47.0	20.6	2.76	9.64	12.48	0.77
F3.0B1	2.4	65.6	43.2	3.86	13.45	13.44	1.00
F4.5B1	2.4	72.4	38.0	4.26	14.84	14.98	0.99
PCB2	1.6	90.6	36.0	5.33	12.34	12.48	0.99
F3.0B2	1.6	101.8	57.0	5.99	13.87	13.44	1.03
F4.5B2	1.6	108.4	74.8	6.37	14.74	14.98	0.98
PCB3	0.8	149.0	92.0	8.76	10.25	12.48	0.82
F3.0B3	0.8	162.4	112.8	9.55	11.17	13.44	0.83
F4.5B3	0.8	215.2	193.2	12.66	14.8	14.98	0.99

1 lb = 4.4 N; 1 ksi = 6.89 MPa

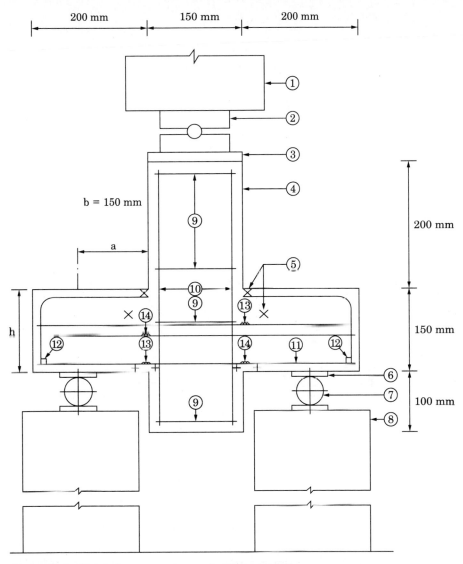

1. Hydraulic loading jack
2. Spherical ball seat
3. Loading plate
4. Corbel
5. Demec discs
6. Bearing plate
7. Roller support
8. Floor supports
9. 6mm dia. bars
10. 10 mm dia. bars
11. 10 or 12 mm dia. main bars
12. 10 mm or 12 mm welded cross bars
13. ERS gauges (near face)
14. ERS gauges (far face)

Figure 16-18 Reinforcement details and test setup for corbel [16.35] (1 in = 25.4 mm).

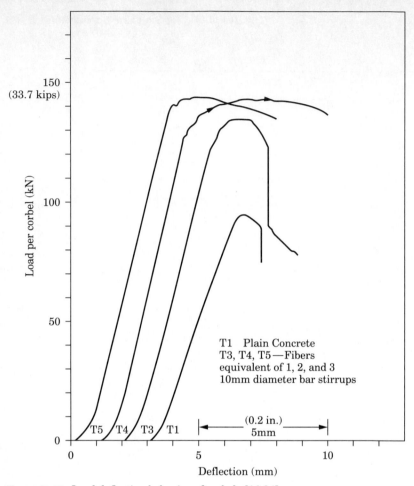

Figure 16-19   Load-deflection behavior of corbels [16.34].

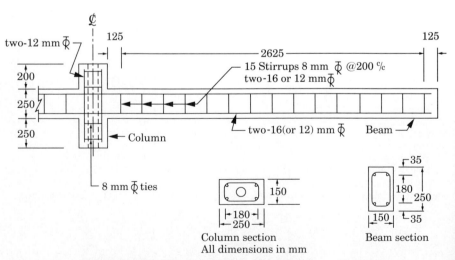

Figure 16-20   (a) reinforcement of cross connections [16.14] (1 in. = 25.4 mm).

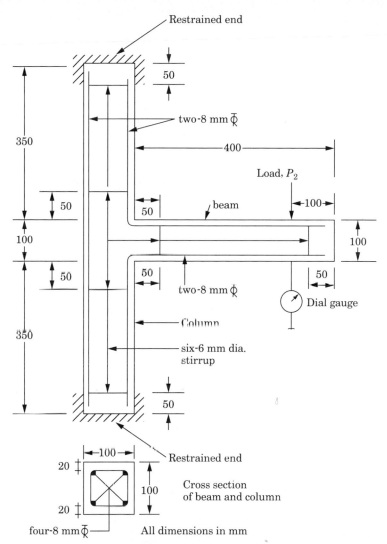

Figure 16-20 (*b*) reinforcement of T connections [16.14] (1 in = 25.4 mm).

Typical load-rotation curves under cyclic loading for plain and fibrous concretes are shown in Figure 16-22 [16.13]. From this figure it can be seen that fiber-reinforced specimens can sustain a higher percentage of initial load after the cyclic load applications. Fiber-reinforced specimens also sustained higher deformations. The fibers used were 1.2 in. (30 mm) long, 0.02 in. (0.5 mm) diameter hooked-end steel fibers. The fiber volume fraction was 1.5%.

The fiber-reinforced specimens were also found to show less damage at junctions, greater initial stiffness, less cracking, and larger mo-

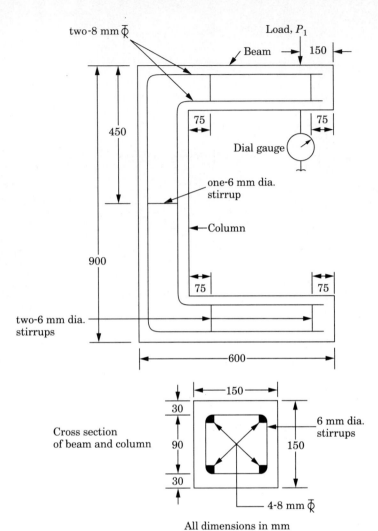

All dimensions in mm

Figure 16-20 (*c*) reinforcement of knee connections [16.14] (1 in. = 25.4 mm).

ment and shear capacities. All the investigators concluded that fibers make a definite contribution to strength increase and ductility. Design methods and specifications are needed to promote the use of fibers in construction.

16.7 Columns

The addition of fibers to columns that are subjected to only compressive loads is not beneficial [16.11]. The increase in compressive strength capacity obtained by the addition of fibers is negligible. On the other

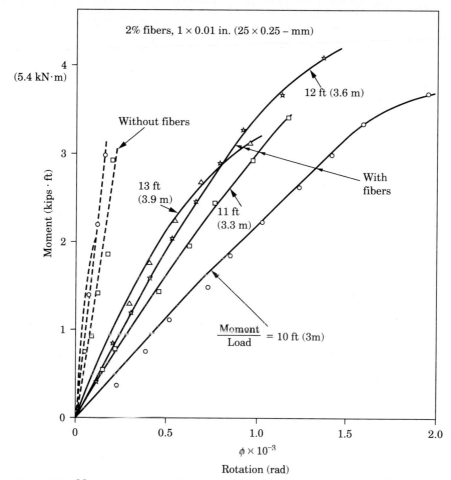

Figure 16-21 Moment versus rotation at junction of beam and column [16.15].

hand if the columns are subjected to lateral shear force, the fiber provides a substantial contribution to ductility [16.11, 16.37].

The effectiveness of the fibers in resisting later shear forces was
demonstrated using a short column subjected to axial force and lateral
shear [16.37]. The test specimen, load setup, and load configurations
are shown in Figures 16-23, 16-24, and 16-25 respectively [16.37]. The
primary test variables were axial load, amount of tie reinforcement,
and volume fraction of fibers. Typical results obtained for 0, 1, and
2% fiber volume fractions are shown in Figure 16-26 [16.37]. From
this figure it can be seen that fiber-reinforced specimens have a much
greater load-retaining capacity under cyclic (shear) loading. It was
also observed that cracks were distributed over a larger portion of
column, resulting in greater ductility. Improvement from a 0% to 1%

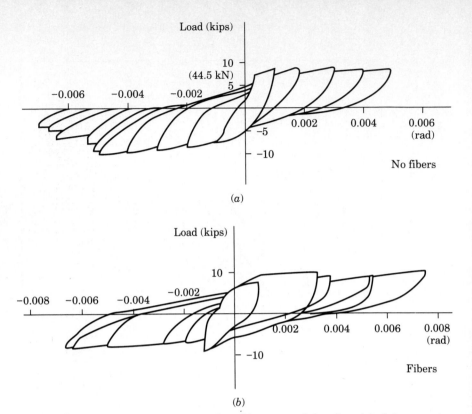

(a)

(b)

Figure 16-22 Beam-tip load versus rotation of beam at column face: (a) plain concrete, (b) FRC [16.13]

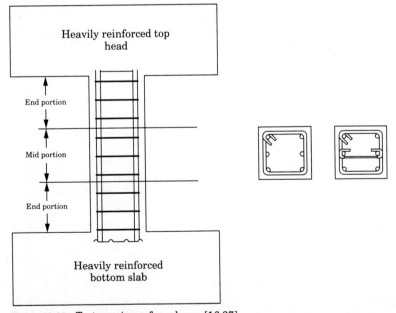

Figure 16-23 Test specimen for column [16.37].

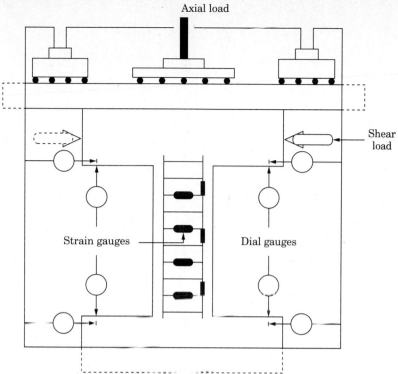

Figure 10-24 Loading setup and placement of dial gauges and strain gauges for column [16,37].

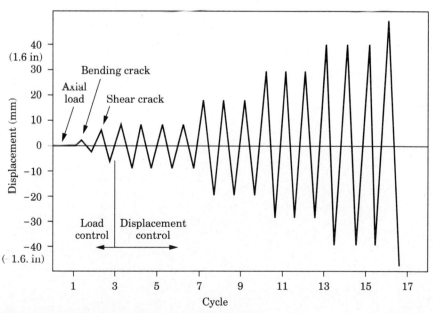

Figure 16-25 Loading sequence for the column [16.37].

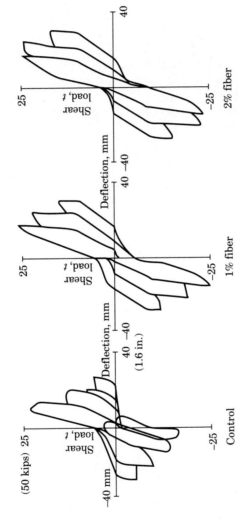

Figure 16-26 Typical shear loading and deformation relationships for the column [16.37].

fiber volume was much greater than the improvement provided by the increase from 1% to 2%.

16.8   Prestressed Concrete Beams

The limited results available show that fibers can be used to improve the performance of prestressed concrete beams [16.18]. Tests conducted using T-beam cross sections and high-strength concrete show that first-crack moment and ductility were increased by adding fibers. Figure 16-27 shows the maximum deflection at failure with respect to fiber content. The 2 in. (50 mm) long, 0.2 in. (0.5 mm) diameter hooked-end steel fibers provided a consistent improvement both in ductility and first-crack moment with an increase in fiber volume fraction.

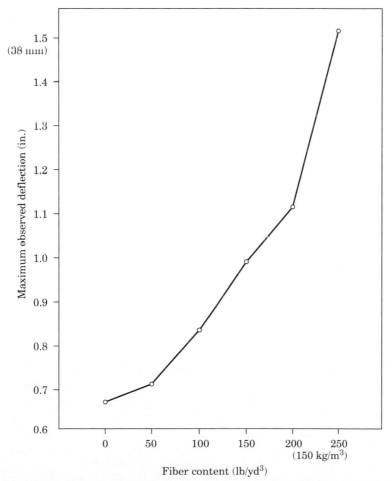

Figure 16-27   Maximum measured deflection prior to failure for a prestressed concrete beam [16.18].

For a given moment, the deflection, reinforcement stresses, and crack widths were lower for beams containing fibers. Spacing and crack widths decreased with an increase in fiber content (Figure 16-28) [16.18]. Narrower and more uniform crack spacing resulted in a consistent decrease in the maximum crack widths with increases in fiber content ranging from 0 to 250 lb/yd^3 (0 to 150 kg/m^3, approximately 0% to 1.7%).

Fibers also have good potential to reduce the distress caused by the handling of beams at early ages of maturity. For example when long girders are stored in the casting yard, they typically develop tension cracks at the top flanges. Fibers could be effectively used to minimize both cracking and crack growth. In partially prestressed beams, which are allowed to crack under service loads, fibers can be used to improve the cracking characteristics. The cracking and decrease of stiffness become critical when the beams are subjected to fatigue loading because the decrease in stiffness can lead to higher stress ranges in the prestressed tendons, creating a premature failure.

16.9   Design Concepts

Design concepts used for normal reinforced concrete can also be used for fibrous reinforced concrete with appropriate modifications. A number of researchers have proposed modified design or analysis procedures for structural elements subjected to axial compression, bending, shear, torsion, and combined effects. American Concrete Institute Committee 544: Fiber Reinforced Concrete is currently developing

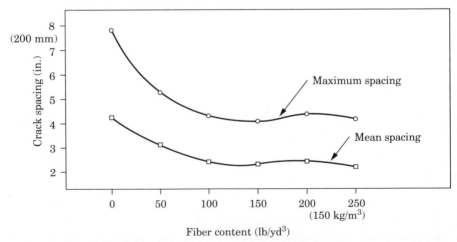

Figure 16-28   Crack spacing at failure for a prestressed beam [16.18].

a comprehensive design guide. This section highlights the special contributions made by fibers that can be incorporated in the analysis or design so that the behavior of the composite can be predicted with more accuracy. The reader is assumed to have a basic background in reinforced concrete design. A number of excellent text books are available for the design and analysis of reinforced concrete for the reader who needs help in this area.

The analysis and design of reinforced concrete involves the judicious addition of concrete and reinforcement to develop a specified resistance, against forces such as moment or torsion. The concrete contributes to resistance in compression. The addition of fibers does not appreciably increase the compressive strength but alters the deformation characteristics especially in the postpeak region. The strength design methods, which are based on ultimate (moment, shear, or torque) capacity depend not only on strength but also on the deformation characteristics, and hence fiber addition becomes significant in design. The magnitude of deformation at peak and near postpeak loads is extremely important because without sufficient deformation, the reinforcements will not be able to develop the required resistance. For example, in beams subjected to bending, the concrete in the compression zone has to sustain a certain amount of strain (say, 0.003 in./in. [mm/mm]) in order to develop sufficient strain in the tension steel. In most cases, the primary contribution of fiber is to enable the concrete to sustain larger deformations at or near peak loads, but in certain cases, such as shear, fibers also contribute to an increase in strength because of their ability to transfer loads across cracks.

The change in design concepts needed is explained by analyzing the basic assumption used for the analysis of members subjected to flexure, shear, and torsion.

16.9.1 Flexure

The basic assumptions used for flexural analysis of reinforced concrete beams are as follows:

1. The strain distribution across the thickness of beams is linear.

2. The maximum usable strain at the extreme concrete compression fiber is limited to a certain value. The American Concrete Institute Code recommends a value of 0.003 in./in. (mm/mm).

3. The tensile strength contribution of the plain concrete is negligible and need not be considered in design.

These assumptions are used to set up two equations involving force and moment equilibrium. Then they can be solved to obtain the moment capacity for a beam of given dimension, the area of steel, and the properties of concrete and steel. In most instances the concrete contribution is calculated using a rectangular stress block assumption. That is, the contribution of concrete is assumed to be $0.85f_c'\beta_1 bc$, where f_c' is the compressive strength, b is the width of the beam, c is the depth of neutral axis, and β_1 is a factor that depends on strength of concrete.

The amount of tension reinforcement is limited by most codes in order to ensure a ductile failure by yielding of the reinforcing steel rather than by crushing of concrete.

In the case of fiber-reinforced concrete, the major differences are

- The concrete could sustain strains much higher than 0.003 in./in. (mm/mm) without failing catastrophically

- The tensile strength contribution of fiber concrete could be substantial.

A higher strain capacity could be used to design beams with larger reinforcement areas. For example, the area of steel could be adjusted to balance the compression force created by the concrete. This would permit simultaneous failing of concrete and steel, which is not allowed by the codes for fear of brittle failure. With the addition of fibers, brittle failure could be avoided, and hence a balanced failure condition could be used for design. However, in most practical design cases, reinforcement congestion would occur if beams are designed for balanced conditions. Hence the ductility provided by the fiber should be used for generating rotation capacity for applications involving earthquake-type loading rather than increasing the tension reinforcement. Unfortunately, the computation of ductility factors for reinforced fibrous concrete beams has not yet been addressed by the researchers.

The tensile force contribution of fiber depends on the type, geometry, and volume fraction. For fiber volume fractions less than 2%, the tensile strength of fiber-reinforced concrete could be assumed to be about the same as that of the matrix. However, once the tensile crack develops, the contribution of a plain concrete matrix reduces rapidly. Since strength calculations are done near failure when the crack widths could be very large, zero contributions from plain matrix are quite valid. The fiber-reinforced matrix on the other hand can sustain tensile forces even at large deformations, especially with high volume fractions. Typically, uniform tensile stress is assumed to exist for the entire tension zone. A uniform tensile strength of 300 psi (2 MPa) results in about a 15% increase in moment capacity. This increase is confirmed by experimental results obtained using 1% and 2% volumes steel fibers.

If the fiber volume fraction is less than 1%, it is advisable to use the design method recommended for reinforced concrete containing a plain concrete matrix. In special circumstances, if a higher volume fraction of fiber is used, then the contributions of both the increase in compressive strain and tensile strength should be considered in the design. Detailed design procedures for both singly reinforced beams and beams reinforced with compression and tension steel can be found in Reference [16.17].

16.9.2   Shear

In normal reinforced-concrete design the shear strength of the reinforced-concrete section is assumed to be the summation of the concrete contribution V_c and shear reinforcement (stirrup) contribution V_s. If a fiber-reinforced matrix is used, the reinforcement contribution can be assumed to be the same because this resistance is governed by the yielding of stirrup reinforcement.

The concrete contribution V_c can be quite different if fibers are added to the matrix. The behavior of plain concrete in shear is very complex. For design purposes, the shear stress at which failure occurs is assumed to be proportional to the square root of the the 28 day compressive strength. In some instances the strength developed by the dowel action of tension bars is also included. Since the contribution of fiber is more predominant in the tension mode and since shear failures are initiated by principal tensile stresses, various investigators have suggested that modulus of rupture should be used for the computation of shear capacity. In general, if more than 1% volume of deformed fibers is used, the fibers provide a substantial contribution to shear capacity and should be considered in the design. However, data is not currently available to formulate a general procedure because of the variability among fiber types. The design has to be based on the results obtained for particular fiber types. A fiber volume fraction of less than 1% is not recommended because lower volume fractions could result in possible weak zones where only a few fibers are present.

16.9.3   Combined bending, shear, and torsion

This is a complicated loading scheme even for regular reinforced concrete. The analysis and design methods are still being refined. The addition of fibers was found to make a definite contribution. For the quantification of the fiber contribution, a number of equations have been proposed. Here again the equations depend on fiber type and fiber geometry. The only common conclusion is that the sand-heap analogy (plastic theory) works well for fiber-reinforced concrete.

The design of beams subjected to combined bending, shear, and torsion is very complex. Not only the longitudinal and shear reinforcement requirements have to be calculated for the combined effects, but also the reinforcement must be placed in proper locations across the cross section and along the span. More experimental investigations are needed before specifications can be written for an increase in the spacing of stirrups or a reduction in the longitudinal reinforcement needed for torsion.

16.9.4   Columns

The primary purpose of adding fibers to concrete used in columns is to improve the ductility. This aspect is very important in earthquake-resistant design. Current practice involves the careful design of spiral reinforcement to confine the concrete and to ensure ductile failure of columns. Fiber can provide some of the needed confinement effect. Reinforcement congestion that is often encountered in earthquake-resistant design can be reduced by using fibers. Once again, sufficient documentation is not available for specifying a certain volume fraction of fiber in exchange for increasing the spacing of spirals or stirrups.

16.10   Nonlinear Analysis and Computation of Ductility

As mentioned repeatedly in earlier sections, the primary contribution of fibers for structural applications is in the area of ductility. The enhancement of ductility can be effectively utilized in all three modes of loading: bending, shear, and torsion. The fact that fibers increase ductility is well established, but how to estimate this improved ductility and incorporate it into design is not established to the point where a designer can specify a certain fiber type and volume fraction to achieve the level of ductility needed.

A mathematical model is needed to compute moment-rotation capacities and load-displacement variations for given material properties. In order to develop these mathematical models, basic stress-strain characteristics are needed for both compression and tension. Such relationships are available in the literature, but the mathematical models to compute structural-element behavior are scarce [16.38]. Computational procedures are needed for both monotonic and cyclic loading. Once it is established that fibers can provide predictable and repeatable ductility, design procedures can be formulated.

Stress-strain relationships under compression and tension are discussed in this section that can be used to develop the aforementioned models. The accuracy of the models can be evaluated using the experimental results presented in earlier sections. A brief discussion on the bond characteristics of reinforcement bars embedded in fiber reinforced concrete is provided because of its contribution to the behavior of structural elements.

16.10.1 Stress-strain behavior
in compression

Typical stress-strain curves for normal- and high-strength fiber reinforced concrete are shown in Figures 16-29 and 16-30 [16.39]. The
fibers used were 30 mm (1.2 in) long by 0.5 mm (0.02 in) diameter hooked-end steel fibers. The high-strength concrete contained condensed silica fume.

The basic parameters of the stress-strain curves are initial modulus
of elasticity, peak stress and the corresponding strain, and the rate
of drop of the postpeak curve. The addition of fibers increases the
compressive strength, stiffness, strain at peak load, and the shape
of the descending part of the curve. The magnitude of these changes
depends on the type, geometry, and volume fraction of fiber and on
the matrix composition. For example, more fibers may be needed to
produce the same postpeak performance for high-strength concrete
compared to normal-strength concrete.

A number of empirical relationships are available for plain concrete,
but only a few equations are available for fiber-reinforced concrete
[16.39, 16.40]. The equation with the least empirical constants can be
written as

$$\frac{f_c}{f_c'} = \frac{\beta(\epsilon/\epsilon_o)}{(\beta - 1) + (\epsilon/\epsilon_o)^\beta} \tag{16.1}$$

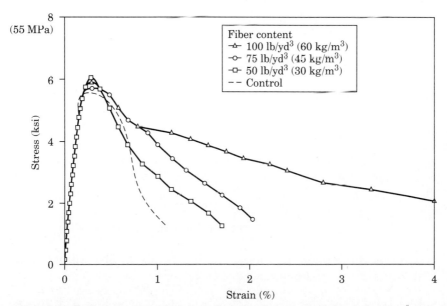

Figure 16-29 Typical stress-strain curves for normal strength fiber-reinforced concrete
[16.39].

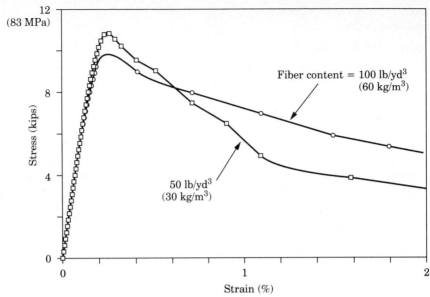

Figure 16-30   Stress-strain curves for high-strength fiber-reinforced concrete [16.39].

where
f_c = stress corresponding to a strain of ϵ
f_c' = compressive strength (or peak stress)
β = material parameter
ϵ = strain corresponding to stress f_c
ϵ_o = strain corresponding to peak stress

The equation predicts the stress-strain response for a given compressive strength f_c', strain at peak load ϵ_o, and the material parameter, β. Normally the compressive strength f_c' is specified. The peak strain for plain concrete varies from 0.002 to 0.0025 in./in. (mm/mm), whereas for fiber concrete this strain could vary from 0.002 to 0.003 in./in. (mm/mm). In the absence of reliable experimental data the following equation can be used for estimating ϵ_o. This parameter was found to be much less influential than β in Equation (16.1).

$$\epsilon_o = 0.002 + 0.5 \times 10^{-6} R \tag{16.2}$$
$$\leq 0.003 \text{ in./in. (mm/mm)}$$

where R is the reinforcing index, which depends on fiber volume fraction V_f, length l, and diameter ϕ, and is defined as

$$R = \frac{V_f l}{\phi} \tag{16.3}$$

Equation (16.2) is valid only for hooked-end fibers with volume fractions less than 1.5%. For straight fibers, the value would be smaller.

The material parameter β can be estimated using the equation

$$\beta = 1.09 + 0.71(R)^{-0.93} \tag{16.4}$$

for hooked fibers and

$$\beta = 1.09 + 7.5(R)^{-1.39} \tag{16.5}$$

for straight fibers.

Empirical equations for β were obtained using the experimental results reported in References [16.39, 16.40]. Equation (16.1) was found to predict the stress-strain behavior with reasonable accuracy for a wide range of strengths and for fiber contents less than 1.5% volume [16.39].

16.10.2 Stress-strain behavior in axial tension

A detailed discussion of axial tension behavior was presented in Chapter 3. A numerical equation that is useful for structural analysis is needed for design purposes. A typical stress-strain curve in tension is shown in Figure 16-31. The ascending portion is linear up to about 80% of tensile strength. The tensile strength for concrete varies from

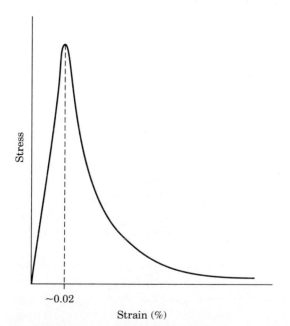

Strain (%)

Figure 16-31 Stress-strain curve in tension.

600 to 1500 psi (4 to 10 MPa) depending on the compressive strength. The strain corresponding to peak load is about 0.0002 in./in. (mm/mm) for plain concrete. The corresponding strain for fiber-reinforced concretes is normally greater. The descending part of the curve is very steep for plain concrete. With the addition of fiber, the drop of the descending part becomes less steep. The stress retention at greater strains also increases with the increase in fiber content.

Even though a number of analytical models are available for the prediction of the axial tension behavior of cement composites reinforced with greater volumes (2%) of fibers (Chapter 3), numerical equations are not available for concretes containing coarse aggregates and fiber volumes less than 2%. The ascending part could be represented as a straight line without loss of accuracy. The peak stress or axial tensile strength could be assumed to be about 6.9 f_c' where $\sqrt{f_c'}$, is the compressive strength. Even though fibers increase the tensile strength, the percentage increase is negligible for lower volume fractions. The descending part could be best represented using exponential curves, such as

$$f_t = Ae^{-\alpha\epsilon} \tag{16.6}$$

where f_t and ϵ are stress and strain, A is the tensile strength, and α is a coefficient. The coefficient α can be determined using experimental results and regression analysis.

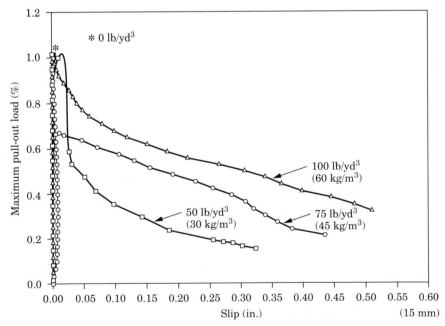

Figure 16-32   Pull-out load-slip relationship: Monotonic loading [16.43].

16.10.3 Bond characteristics of reinforcement bars embedded in fiber-reinforced concrete

The characteristics of bond between reinforcing bars and fibrous concrete have been studied for both monotonic and cyclic loading conditions [16.41–16.43]. The addition of fibers was found to improve the bond-slip behavior considerably in the postpeak region (Figure 16-32) [16.42]. The bond-slip characteristics were obtained using bars embedded in prismatic concrete members. The concrete surrounding the bars was in tension, simulating the condition of the reinforcement in the tension zone.

The fiber also makes a substantial contribution under the cyclic loading mode (Figure 16-33) [16.43]. The envelope curve of load-slip relationship under positive cyclic loading (in which loads varied from zero to a predetermined tensile load) was found to be the same as the load-slip curve under monotonic loading. Considerable deterioration of bond strength was observed for specimens subjected to reverse cyclic loading. The load-retention capacity for successive cycles of reversed loads reduced considerably. Nevertheless, the retention was much greater than that recorded for a plain matrix.

The most important observation is that ductile failure conditions can be created even if the failure is generated by the pull-out of the bars. As mentioned earlier, ductile failure is a design condition for structures located in earthquake-prone locations.

16.11 The Behavior of FRC under Biaxial Loading

In most structural applications, biaxial and triaxial stress states exist in some part of the structure. The understanding of the material behavior under multiaxial loading is important to ensure the safety of the structure under all loading conditions. Limited results are available on the biaxial behavior of FRC [16.44], indicating that fibers provide an increase in strength compared to plain concrete.

Typical strength variations for biaxial loading conditions in compression-compression, compression-tension, and tension-tension are shown in Figure 16-34 [16.44]. The fibers used were straight steel fibers with an aspect ratio of 100. Measurements were made using $75 \times 75 \times 300$ mm ($3 \times 3 \times 12$ in.) concrete specimens. From Figure 16-34 it can be seen that of 0.5% fibers provides a noticeable improvement. An increase of fiber content beyond 0.5 percent does not seem to provide corresponding increases in strength. The stress-strain measurements and observations of the failure mode indicate that the fiber contribution of ductility is substantial under biaxial conditions, as it is with uniaxial conditions.

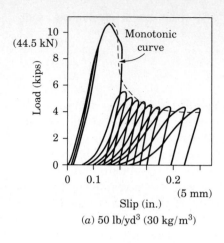

(a) 50 lb/yd³ (30 kg/m³)

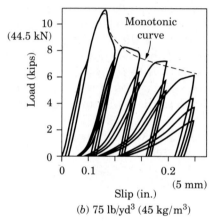

(b) 75 lb/yd³ (45 kg/m³)

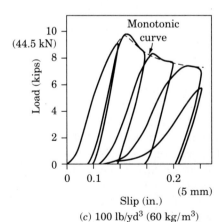

(c) 100 lb/yd³ (60 kg/m³)

Figure 16-33 Pull-out load-slip relationship: Cyclic loading [16.43].

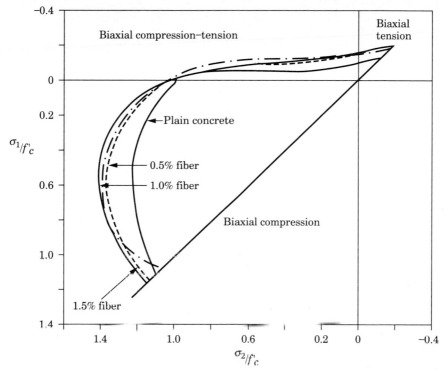

Figure 16-34 Biaxial strength of plain concrete and fiber concrete: Fiber aspect ratio = 100 [16.44].

16.12   References

16.1   Shah, S.; and Rangan, V. D. "Effects of Reinforcements on Ductility of Concrete," *Journal of Structural Engineering, ASCE*, Vol. 96, No. ST6, June 1970, pp. 1167–1184.

16.2   Batson, G.; Jenkins, E.; and Spatney, R. "Steel Fibers as Shear Reinforcement in Beams," *ACI Journal*, Vol. 69, No. 10, 1972, pp. 640–644.

16.3   Yamada, M. "Shear Strength, Deformation and Explosion of Reinforced Concrete Short Columns," *Fiber Reinforced Concrete*, SP-44, American Concrete Institute, Detroit, Michigan, 1974, pp. 617–638.

16.4   Henager, C. H. "Steel Fibrous, Ductile Concrete Joint for Seismic-Resistant Structures," *Fiber Reinforced Concrete, International Symposium* SP-81, American Concrete Institute, Detroit, Michigan, 1974, pp. 371–386.

16.5   Henager, C. H.; and Doherty, T. J. "Analysis of Reinforced Fibrous Concrete Beams," *Journal of Structural Division, ASCE*, Vol. 102, No. ST1, 1976, pp. 177–188.

16.6   Kormeling, H. A.; Reinhardt, H. W.; and Shah, S. P. "Static and Fatigue Properties of Concrete Beams Reinforced with Bars and Fibers," *ACI Journal*, Vol. 77, No. 1, 1980, pp. 36–43.

16.7   Swamy, R. N.; and Al-Ta'an, Sa'ad A. "Deformation and Ultimate Strength in Flexure of Reinforced Concrete Beams Made with Steel Fiber Concrete," *ACI Journal*, Vol. 78, No. 5, 1981, pp. 395–405.

16.8   Jindal, R. L. "Shear and Moment Capacities of Steel Fiber Reinforced Concrete Beams," *Fiber Reinforced Concrete, International Symposium* SP-81, American Concrete Institute, Detroit, Michigan, 1984, pp. 1–16.

16.9   Craig, J. R.; Dunya, S.; Riaz, J.; and Shirazi, H. "Torsional Behavior of Reinforced Fibrous Concrete Beams," *Fiber Reinforced Concrete, International Symposium* SP-81, American Concrete Institute, Detroit, Michigan, 1984, pp. 17–49.

16.10 Batson, G.; Terry, T.; and Chang, M. "Fiber Reinforced Concrete Beams Subjected to Combined Bending and Torsion," *Fiber Reinforced Concrete, International Symposium* SP-81, American Concrete Institute, Detroit, Michigan, 1984, pp. 51–68.

16.11 Craig, J. R.; McConnell, J.; Germann, H.; Dib, N.; and Kashani, F. "Behavior of Reinforced Fibrous Concrete Columns," *Fiber Reinforced Concrete, International Symposium* SP-81, American Concrete Institute, Detroit, Michigan, 1984, pp. 69–105.

16.12 Jindal, R. L.; and Hassan, K. A. "Behavior of Steel Fiber Reinforced Concrete Beam-Column Connections," *Fiber Reinforced Concrete, International Symposium* SP-81, American Concrete Institute, Detroit, Michigan, 1984, pp. 107-123.

16.13 Craig, J. R.; Mahader, S.; Patel, C. C.; Viteri, M.; and Kertesz, C. "Behavior of Joints Using Reinforced Fibrous Concrete," *Fiber Reinforced Concrete, International Symposium* SP-81, American Concrete Institute, Detroit, Michigan, 1984, pp. 125–167.

16.14 Sood, V.; and Gupta, S. "Behavior of Steel Fibrous Concrete Beam-Column Connections," *Fiber Reinforced Concrete Properties and Applications*, SP-105 American Concrete Institute, Detroit, Michigan, 1987, pp. 437–474.

16.15 Jindal, R.; and Sharma, V. "Behavior of Steel Fiber Reinforced Concrete Knee-Type Beam-Column Connections," *Fiber Reinforced Concrete Properties and Applications*, SP-105 American Concrete Institute, Detroit, Michigan, 1987, pp. 475–492.

16.16 Clarke, R. P.; and Sharma, A. "Flexural Behavior of Fibro-Ferrocrete One-Way Slabs," *Fiber Reinforced Concrete Properties and Applications*, SP-105 American Concrete Institute, Detroit, Michigan, 1987, pp. 493–516.

16.17 Craig, R. "Flexural Behavior and Design of Reinforced Fiber Concrete Members," *Fiber Reinforced Concrete Properties and Applications*, SP-105 American Concrete Institute, Detroit, Michigan, 1987, pp. 517–564.

16.18 Balaguru, P.; and Ezeldin, A. "Behavior of Partially Prestressed Beams Made with High Strength Fiber Reinforced Concrete," *Fiber Reinforced Concrete Properties and Applications*, SP-105 American Concrete Institute, Detroit, Michigan, 1987, pp. 419–434.

16.19 Swamy, R. N.; and Bahia, H. M. "The Effectiveness of Steel Fibers as Shear Reinforcement," *Concrete International,* vol. 7, No. 3 1985, pp. 35–40.

16.20 Batson, G. "Use of Steel Fibers for Shear Reinforcement and Ductility," *Steel Fiber Concrete,* Elsevier, 1985, pp. 377–419. (also published by Swedish Cement and Concrete Institute, Stokholm, 1985)

16.21 Al-Ausi, M. A.; Abdul-Wahab, H. M. S.; and Khidair, R. M. "Effect of Fibers on the Strength of Reinforced Concrete Beams under Combined Loading," *Fiber Reinforced Cements and Concretes: Recent Developments,* Elsevier, 1989, pp. 664–675.

16.22 Kaushik, S. K.; and Sasturkar, P. J. "Simply Supported Steel Fiber Reinforced Concrete Beams under Combined Torsion, Bending and Shear," *Fiber Reinforced Cements and Concretes: Recent Developments,* Elsevier, 1989, pp. 687–698.

16.23 Bollana, R. D. "Steel Fibers as Shear Reinforcement in Two Span Continuous Reinforced Concrete Beams," M.S. Thesis, Clarkson University, Potsdam, New York, May 1980.

16.24 Williamson, G. R. "Steel Fibers as Web Reinforcement in Reinforced Concrete," Proceedings, U.S. Army Science Conference, Vol. 3, 1978, West Point, New York.

16.25 Batson, G.; and Alquire, C. "Steel Fibers as Shear Reinforcement in Reinforced Concrete T Beams, "Proceedings, International Symposium on FRC, Madras, India, December 1987, pp. 113–124.

16.26 Muhidin, N. A.; and Regan, P. E. "Chopped Steel Fibers as Shear Reinforcement in Concrete Beams," Proceedings, Conference on Fiber-Reinforced Materials: Design and Engineering Applications, Institute of Civil Engineers, London, 1977, Paper 17, pp. 149–163.

16.27 Mindess, S. "Torsional Tests of Steel Fiber Reinforced Concrete," *International Journal of Cement Composites,* Vol. 2, No. 2, 1980, pp. 85–89.

16.28 Narayanan, R.; and Green, K. R. "Fiber Reinforced Concrete Beams in Pure Torsion," Proceedings, Institution of Civil Engineers, London, Part 2, Vol. 69, 1980, pp. 1043–1044.

16.29 Mansur, M. A.; and Paramasivam, P.; "Steel Fiber Reinforced Concrete Beams in Pure Torsion," *International Journal of Cement Composites,* Vol. 4, No. 1, 1982, pp. 39–45.

16.30 Mansur, M. A.; and Paramasivam, P. "Fiber Reinforced Concrete Beams in Torsion, Bending and Shear," *ACI Journal,* Vol. 82, No. 1, 1985, pp. 33–39.

16.31 Roberts, T. M.; and Ho, N. L.; "Shear Failure of Deep Fiber Reinforced Concrete Beams," *International Journal of Cement Composites,* Vol. 4, No. 3, 1982, pp. 145–52.

16.32 Shanmugam, N. E.; and Swaddiwudhipong, S. "The Ultimate Load Behavior of Fiber Reinforced Concrete Deep Beams," *The Indian Concrete Journal,* Vol. 58, No. 8, August 1984, pp. 207–211.

16.33 Shanmugan, N. E.; and Swaddiwudhipong, S. "Behavior of Fiber Reinforced Concrete Deep Beams Containing Openings," *Fiber Reinforced Cement and Concretes: Recent Developments,* Elsevier, 1989, pp. 479–488.

16.34 Fatthui, N. I.; and Hughes, B. P. "Reinforced Steel Fiber Concrete Corbels with Various Shear Span-to-Depth Ratios," *ACI Structural Journal,* Vol. 86, No. 6, 1989, pp. 590–596.

16.35 Fattuhi, N. I.; and Hughes, B. P. "Ductility of Reinforced Concrete Corbels Containing Either Steel Fibers or Stirrups," *ACI Materials Journal,* Vol. 86, No. 6, 1989, pp. 644–651.

16.36 Hughes, B. P.; and Fattuhi, N. I. "Reinforced Steel and Polypropylene Fiber Concrete Tests," *The Structural Engineer* (London), Vol. 67, No. 4, 1989, pp. 68–72.

16.37 Yashior, H.; Tanaka, Y.; Ro, Y.; and Hirose, K. "Study of Shear Failure of Steel Fiber Reinforced Concrete Short Columns in Consideration of Arrangement of Ties," *Fiber Reinforced Cements and Concretes: Recent Developments,* Elsevier, 1989, pp. 489–498.

16.38 Soroushian, P.; and Reklaoui, A. "Flexural Design of Reinforced Concrete Beams Incorporating Steel Fibers," *Fiber Reinforced Cements and Concretes: Recent Developments,* Elsevier, 1989, pp. 454–466.

16.39 Ezeldin, A. S.; and Balaguru, P. "Stress-Strain Behavior of Normal and High Strength Fiber Reinforced Concrete under Compression," ASCE *Materials Engineering Journal,* 1992, in press.

16.40 Fanella, D. A.; and Naaman, A. E. "Stress-Strain Properties of Fiber Reinforced Mortar in Compression," *ACI Journal,* Vol. 82, No. 4, 1985, pp. 475–583.

16.41 Spencer, R. A.; Panda, A. K.; and Mindess, S. "Bond of Deformed Bars in Plain and Fiber Reinforced Concrete under Reversed Cyclic Loading," *International Journal of Cement Composites,* Vol. 4, No. 1, 1982, pp. 3–18.

16.42 Ezeldin, A.; and Balaguru, P. "Bond Behavior of Normal and High Strength Fiber Reinforced Concrete," *ACI Materials Journal,* Vol. 86, 1989, pp. 515–524.

16.43 Ezeldin, A.; and Balaguru, P. "Bond Performance of Reinforcing Bars Embedded in Fiber Reinforced Concrete and Subjected to Monotonic and Cyclic Loads," ASCE Materials Engineering Congress, Denver, Colorado, August 1990, pp. 145–154.

16.44 Abdull-Ahad, R. B.; and Abbas, J. M. "Behavior of Steel Fiber Reinforced Concrete under Biaxial Stresses," *Fiber Reinforced Cements and Concretes: Recent Developments,* Elsevier, 1989, pp. 126–135.

17

Field Performance
and Case Studies

Fiber-reinforced concrete has been in use for almost twenty-five years. This chapter deals with the field performance of some of the FRC installations. The objective is to provide information on both success and failure so that future projects can be designed to avoid some of the pitfalls encountered in existing projects. The case studies presented are chosen to represent the wide spectrum of applications. Engineers can obtain detailed information for a particular project if they are contemplating a similar project.

The material presented in this chapter covers concrete reinforced with low volume fractions (usually less than 2 percent) of discrete fibers. Information on other cementitious products is presented in Chapters 12 through 15. Because the use of polymeric fibers started only in the 1980s, most of the information available pertains to steel fiber–reinforced concrete. The various applications of FRC can be grouped into the following categories:

- Pavements and overlays
- Bridge deck overlays
- Industrial floors
- Hydraulic structures
- Marine structures
- Repairs and rehabilitation
- Miscellaneous, such as safety vaults and manhole covers

17.1   Field Performance
and Lessons Learned

The following is the gist of reports available on the performance of FRC that was placed as early as 1971.

Fibrous concrete used for slab and pavement applications generally performed better than comparable plain concrete having identical thickness, foundation (subgrade) condition, and concrete flexural strength.

In the early applications, the thickness of the slabs was reduced, taking into account the improved flexural strength of FRC. The reduction in thickness resulted in problems for some applications. The primary problem was found to be the curling and breaking of corners. Other than the external loads, internal stresses resulting from drying shrinkage, stresses developed because of temperature gradient, and restraints provided at the joints were assumed to contribute to this problem. Slab movement owing to temperature change was also found to cause more problems in thinner sections. The solutions suggested to avoid this problem include the use of double polyethylene sheeting, other types of induced debonding with the existing layer, and the use of sliding dowels to join adjacent slabs.

Another general problem is the development of full-depth transverse cracks. The primary suspected cause for this is the use of higher cement and sand contents, thereby increasing the shrinkage. Successful techniques used to eliminate or minimize this problem are (a) reducing the cement content and using the highest possible coarse aggregate content, (b) replacing part of the cement by fly ash, and (c) using admixtures to reduce shrinkage. Reduction in the water-to-cement ratio is also very effective. The workability of concrete with low water-cement ratio could be easily improved using high-range water-reducing admixtures.

Other problems encountered in the use of FRC are

- Loose fibers on the surface
- Protruding fibers
- Load transfer to adjacent slabs
- Extension of existing cracks (or joints) in the case of overlays
- Rusting of exposed fibers

Loose and protruding fibers became a major problem when FRC was used in residential streets and at naval air stations. At naval air stations it was reported that fibers were sometimes sucked in by the aircraft, causing major problems. Besides that, the fibers were blown out onto people by aircraft exhaust. On residential streets, exposed and protruding fibers were causing problems for children and pedestrians for obvious reasons. Techniques have been developed to avoid this problem; for example, a "roller bug" was found to be very effective in pressing down the fibers at the surface. A thin protective coat applied right after the final grading was also found to be very effective.

In some instances, FRC slabs were cast to span existing joints and cracks, and it was assumed that the FRC would transfer the stresses across. In most cases, it was found that thin overlays cannot provide this function.

Proper load-transfer mechanisms should be provided using keyways and dowels. The omission of such provisions leads to warping, curling, and breaking at corners. In some cases long cracks have also developed.

The slabs that measure about 10 in. (250 mm) thick and that were designed conservatively in terms of load transfer and the spacing of construction joints are performing very well. Some of these slabs were subjected to heavy aircraft loads. In spite of the heavy loads, the pavements needed little or no maintenance.

There is still a lack of information for designing pavement systems. Knowledge is lacking in the areas of joint spacing, joint design (including methods for load transfer), thickness design, fiber volume fraction needed for a particular application, and techniques for surface finishing.

17.2 Case Studies: Pavements

Some of the fiber-reinforced concrete pavements in service are shown in Table 17.1 [17.1]. The table provides a representative list of projects around the world.

17.2.1 Airport pavements

Fiber-reinforced concrete has been used to produce both overlays and full-depth slabs for airport pavements [17.2–17.14]. The overlay application is more popular than the full-depth slab application. The following short descriptions of certain projects provide an overview of the use of FRC in various situations. More details regarding any particular project can be obtained from the cited references.

Vicksburg, Mississippi, USA.   Two experimental sections of fibrous-concrete pavement were constructed by the Waterways Experiment Station June 1970–March 1971 [17.2–17.4]. The slabs were intended to carry heavy, wheeled items such as a Boeing 747.

TABLE 17.1 Fiber-Reinforced Concrete Pavements [17.1]

Location	Project	Agent	Year built	Length or Size	Thickness	Steel	Special features
					Project description		
Ashland, OH	I-71 truck weigh station	Ohio DOT	1971	500'	4"	Wirand 265#/yd^3	on 5" bitum. agg. base
Cedar Rapids, IA	Danbury St.	City of Cedar Rapids	1972	175' × 28'	3"	175#/yd^3 Wirand 1" × .016 dia.	direct overlay of old Portland cement concrete (pcc)
	Fifth Ave.	City of Cedar Rapids	1972	200' × 24'	2^1/$_2$"	200#/yd^3 1" × .016 dia.	over old asphalt
	Airport taxiway	City of Cedar Rapids	1972	95' × 75'	3"	150#/yd^3 2^1/$_2$" × .025 dia.	
Vicksburg, MS	Street	WES	1972	1000' × 20'	5"	200#/yd^3 3/4" × .010 × .014"	slab on grade; no joints
Detroit, MI	8-Mile Road	Michigan Dept. of Highways	1972	1100'	3" or less	Wirand and Fibercon 120 and 200#/yd^3 1" × .016 × .010	overlay (direct) of old concrete; much of pavement less than designed thickness
Tampa, FL	Taxiway, Tampa Int'l Airport	C. of E. (WES)	1972	175' × 75' 50' × 50'	6" 4"	Fibercon 200#/yd^3 1" × .01 × .02	overlay of old concrete (direct)

(continued)

TABLE 17.1—cont'd

Location	Project	Agent	Year built	Length or Size	Thickness	Steel	Special features
				Project description			
Fayetteville AK	City street	City of Fayetteville	1972	200′ × 29′	2½″	100#/yd³ 3/4″ × .012 dia.	40′ slabs
Greene County, IA	County Road E53	Greene Co. & Iowa DOT	1973	3.03 mi.	3″ 2″	60#, 100#, 160#/yd³ 1″ & 2½″ Wirand & Fibercon	33 test sections with variable thicknesses, steel contents, cement contents, and degrees of bond
Beaumont, TX	SH124-S	Texas Highway Commission	1973	24′ × 80′	3″	192#/yd³	overlay on old PCC
Roseville, MN	Trunk Hwy. 51 Snelling Ave	Minn. DOT	1974	2526 yd²	2″ 3″	160#/yd³ Fibercon or 55#/yd³ of glass fibers 1″ × .016 × .010	bonded overlay on old PCC
Fort Hood, TX	Tank parking area	U.S. Army	1974	27,000 yd²	4″	204#/yd³ ½″ wires	over existing asphalt pavement
JFK Intl. N.Y.C., NY	Runway end (N end runway 4L-22R)	Port Authority of NY & NJ	1974 1975	2350 yd² 650 yd²	5″ 8½″ 5″	175#/yd³	over asphalt overlay on PCC

(continued)

TABLE 17.1—cont'd

Location	Project	Agent	Year built	Project description			
				Length or Size	Thickness	Steel	Special features
Cannon Int'l Airport, Reno, NV	Apron terminal	Airport Authority of Washoe County	1975	35,000 yd²	4"	200#/yd³ Fibercon 1" × .010 × .022	bonded to PCC
LaGuardia Airport, NY	Finger aprons	Port Authority of NY & NJ	1975 1977	2 @ 40' × 45' 2 @ 40' × 45'	6" 6"	175#/yd³ 175#/yd³	overlay on asphalt overlay on asphalt
Newark Airport, NJ	Runway end	Port Authority of NY & NJ	1976	225'×120'	5"	175#/yd³ deformed (Bekaert)	overlay on asphalt on lime cement fly ash base
McCarran Int'l Airport, Las Vegas, NV	Transient aircraft apron	Clark County	1976	65,000 yd²	6"	160#/yd³ Fibercon 1" × .01 × .022	overlay of existing asphalt
Ft. Myers, FL, U.S. 41	Econocrete test road	Florida DOT	1977		2" & 3"	158#/yd³ Fibercon, 100#/yd³ Bekaert	bonded to econocrete base and unbonded
Norfolk NAS, VA	Apron overlays	U.S. Navy	1977	36,000 yd²	5"	160#/yd³ Fibercon	resurfacing (separated) of old 6" PCC
Calgary, Alberta, Canada	City street	City of Calgary	1978	180'	4" 5" 6"	75#, 125#/yd³, 3/4'-wire fibers made by Stelco	test section by Univ. of Calgary

(continued)

TABLE 17.1—cont'd

Location	Project	Agent	Year built	Project description			
				Length or Size	Thickness	Steel	Special features
McCarran Int'l Airport, Las Vegas, NV	Terminal apron	Clark County	1979	87,000 yd²	7″	85#/yd³ Bekaert 2″ × .02″ diam.	subbase of flexible pavement 2″ A/C on 12″ gran.
Norfolk NAS, VA	Apron overlay	U.S. Navy	1979	36,000 yd²	5″	160#/yd³ Fibercon	resurfacing (separated) of old 6″ PCC
Cannon Int'l Airport, Reno, NV	Taxiway	Washoe County	1980	2900′×75′	8″	85#/yd³ Bekaert	on new gran. subbase
Fallon NAS, Las Vegas, NV	Apron	U.S. Navy	1980	600′×600′	5″	85#/yd³ Bekaert	on 1″ asphalt over WWII PCC
Salt Lake City Airport, UT	Terminal apron	Salt Lake City	1980	4370 yd²	8″	85#/yd³ Bekaert	overlay-old PCC overlay-old A/C surface on LCB
Norfolk NAS, VA	Apron overlay	U.S. Navy	1980	55,000 yd²	5″	80#/yd³ Bekaert	resurfacing (separated) of old 6″ PCC
Newark Airport, NJ	Runway end	Port Authority of NY & NJ	1980	3556 yd²	3″	85#/yd³ Bekaert	bonded to existing FRC
Stapleton Int'l Airport, Denver, CO	Apron	City of Denver	1981	84,000 yd² 38,900 yd²	7″ 8″	66#/yd³ Bekaert	overlay over PCC 15″ CTSB

(continued)

499

TABLE 17.1—cont'd

Location	Project	Agent	Year built	Length or Size	Project description			Special features
					Thickness	Steel		
Fallon NAS, NV	Apron	U.S. Navy	1981	600' × 640' 600' × 600'	5"	85#/yd³ Bekaert		on 1" asphalt over WWII PCC
JFK Int'l Airport, N.Y.C., NY	Taxiway & holding area	Port Authority of NY & NJ	1981	1000' × 50'	9"	85#/yd³ Bekaert		on flexible after removing 4" A/C surface
Salt Lake City Int'l Airport, UT	Taxiway	Salt Lake City	1981	30,000 yd²	7 & 8"	85#/yd³ Bekaert		taxiway and apron, overlay and reconstruction

The full-depth FRC slab was 6 in. (150 mm) thick in the middle and was thickened to 9 in. (225 mm) at the transverse edges. Steel fibers that were 25 mm long and 0.41 mm in diameter were used at 148 kg/m^3 (250 lb/yd^3). A control slab made of 10 in. (250 mm) thick plain concrete was also cast for comparative evaluation. The FRC was on a much weaker subgrade compared to the plain concrete, which was cast on a subgrade with a higher modulus. In spite of this, the FRC developed its first crack after 350 simulated traffic loadings, whereas the plain concrete developed its first crack after 40 traffic loadings. Plain concrete was shattered after 700 loadings and failed completely after 950 loadings. The FRC pavement was found to be in excellent condition even after 8735 loadings.

A 4 in. (100 mm) thick overlay that was cast over the failed plain concrete also provided excellent service. The first crack did not develop until 900 traffic loadings and the pavement was reported to be in good condition after 6900 loadings.

Two more test sections that were 7 in. (175 mm) and 4 in. (100 mm) thick were constructed on strong subgrades and tested using 90.7 and 108.9 tonne dual tandem assembly. The 7 in. (175 mm) thick slab failed completely after the 3000 and 1010 coverages of lighter and heavier loads, respectively. The 4 in. (100 mm) thick slab was considered to have failed at 1770 and 740 coverages with lighter and heavier loads, respectively. Considerable flexural ductility and load-retention capacity was observed during the loading.

Lockbourne Air Force Base, Ohio, USA, 1970 [17.4].   Two FRC slabs were cast on a concrete base of 9 in. (225 mm) thickness. A 6 in. (150 mm) thick, 35 × 46 ft (10.7 × 14 m) slab was placed as a paving apron, and a 6 in. (150 mm) thick, 5 × 22 ft (1.5 × 6.7 m) slab was cast for use as a taxiway. Steel fibers that were 0.010 × 0.022 × 1 in. (0.25 × 0.5 × 25 mm) were used at 180 lb/yd^3 (106 kg/m^3). The slabs were cast over a 0.2 mm (8 mil) thick polyethylene sheet placed on the base, thus producing an unbonded overlay.

The fiber-reinforced slabs were found to perform much better than the adjacent plain concrete slabs, which developed a longitudinal crack and a number of short traverse cracks.

Detroit, Michigan, USA, 1971 [17.11].   An 8 in. (200 mm) thick, 20 × 30 ft (6.1 × 9.1 m) steel fiber–reinforced concrete slab was cast near the gate area used by Boeing 747 aircraft. The FRC slab was cast adjacent to a 12 in. (300 mm) thick plain concrete slab and tied to it by deformed rebars. The slabs were reported to be performing well, after a period of nine months.

Tampa, Florida, USA, 1971 [17.13].   In this project two FRC overlays were constructed on a taxiway parallel to one of the primary runways. The overlays that were 4 and 6 in. thick (100 and 150 mm) were cast on a base pavement with a minimum of surface preparation. No attempt was made to either bond or debond the pavement. The concrete mixture had a fly ash content of 225 lb/yd^3 (133 kg/m^3) supplementing a cement content of 517 lb/yd^3 (306 kg/m^3). The fiber content was 200 lb/yd^3 (120 kg/m^3). The steel fibers had a cross section of 0.01 × 0.22 in. (0.25 × 5.5 mm) and were 0.75 in. (19 mm) long.

In this application, reflecting cracking was observed over the existing cracks. Cracks also occurred over the joints of the base pavement. The cracks over the base pavement cracks did not widen significantly, whereas the cracks over the joints opened considerably, causing fiber fracture or pull-out. The overlays were reported to be in service until 1984.

Cedar Rapids, Iowa, USA, 1972 [17.9].   The overlay used in this application had a thickness varying from 1 to 4 in. (25 to 100 mm). The FRC was cast over previously existing cracks and joints. Two fiber lengths, 1.0 and 2.5 in. (25 and 62 mm), were used at fiber contents of 200 and 150 lb/yd^3 (118 and 89 kg/m^3) respectively. Since the overlay was relatively thin, reflective cracks developed over the existing cracks and joints.

New York, New York, USA, 1974 [17.10].   The overlay constructed in this project was 5.5 in. (138 mm) thick. The overlay was completely debonded from the existing pavements using two 0.15 mm (6 mil) thick polyethylene sheets. The construction joints were either keyed or doweled. The fiber content used was 175 lb/yd^3 (100 kg/m^3). The fibers were 0.025 in. (0.62 mm) in diameter and 2.5 in. (62 mm) long. The overlays provided good service even though some cracking and shattering occurred at the intersections.

Las Vegas, Nevada, USA, 1976 and 1979.   Two aprons, 6 and 7 in. (150 and 175 mm) thick, were built in 1976 and 1979 using steel fibers. Lanes were 25 ft (7.5 m) and had transverse joints at 50 ft (15 m) intervals. No dowels, keyways, or tie bars were placed across the joints. About 10 percent of the panels were reported to have developed corner breaks in the 7 in. (175 mm) thick apron. Some of the corner breaks in the 6 in. (150 mm) slab developed spalling. Some of the transverse joints also opened up considerably. For the 1976 project, straight steel fibers with dimensions of 0.010 × 0.022 × 1 in. (0.25 × 0.55 × 25 mm) were used at 160 lb/yd^3 (95 kg/m^3). For the 1979 project-hooked-end steel fibers were used at 85 lb/yd^3 (50 kg/m^3). These fibers were 0.02 in. (0.5 mm) in diameter and 2 in. (50 mm) long.

Fallon Naval Air Station, Nevada, USA, 1980 and 1981.   In these projects, FRC was placed over a bituminous bond breaker and was found to perform extremely well. A grid roller was used to push the fibers that were near the surface, thus creating a smooth (mortar) top surface.

Taoyuan Air Base, Taiwan, 1984.   In this project, a 6 in. (150 mm) thick bonded overlay was placed using 2 in. (50 mm) long crimped steel fiber. All the joints in the old pavement were matched in the overlay. Performance data are not available for this project.

Frankfurt Airport, Germany, 1983 [17.15].   In this application, deformed (HAREX) fibers were used at a dosage of 101 lb/yd^3 (60 kg/m^3). The pavements were reported to be in good condition.

17.2.2   Highway and street pavements

This section deals with FRC used for roadways. The applications include residential streets, roads in rural and urban areas, and highways with high traffic density.

Cedar Rapids, Iowa, USA, 1972 [17.16].   Two residential streets were overlaid using FRC over badly cracked and spalled 7 in. (175 mm) thick reinforced-concrete pavement. The surface preparation involved only broom cleaning and wetting. The thickness of the slab varied from 2.5 to 4 in. (62 to 100 mm). One in. (25 mm) long steel fibers were used at a dosage of 175, 200, or 250 lb/yd^3 (104, 120, or 150 kg/m^3). The slabs developed minor cracks within a few months. Other than these cracks, the slabs were reported to be providing good service. One street was repaved because of complaints about abrasions suffered by children.

Fayetteville, Arkansas, USA, 1971 [17.17].   A 2.5 in. (62 mm) thick overlay was placed in different locations using 0.75 in. (19 mm) long fibers. The fiber dosage was 100 lb/yd^3 (60 kg/m^3). Two of the five sections developed some failure zones after 10 years of service.

Detroit, Michigan, USA, 1971 [17.18].   In this project, a minimum 3 in. (75 mm) thick FRC overlay was cast over an existing reinforced-concrete pavement. The slab was cast without any special surface preparations. Two fiber dosages of 120 and 200 lb/yd^3 (71 and 120 kg/m^3) were tried. The resurfaced lanes were opened to traffic in two days. In addition, the curing temperatures were considerably lower than the optimum.

The lanes were subjected to heavy urban traffic. The higher fiber–content FRC performed well, whereas the one with the lower fiber content developed serious problems. However, the problems were attributed to poor construction practices resulting in thicknesses as low as 1.25 in. (31 mm) instead of 3 in. (75 mm). Some sections were also subjected to incomplete curing.

Greene County, Iowa, USA, 1973 [17.19].   Thirty-three 400×20 ft (120×6.0 m) sections were resurfaced using 2 to 3 in. (50 to 75 mm) thick FRC sections. In addition, four and five sections were overlayed, respectively, with 3 to 4 in. (75 to 100 mm) and 4 to 5 in. (100 to 125 mm) thick sections. The variables evaluated were fiber sizes, cement and fiber content, and bonding between old and new surfaces. The fiber contents and lengths were 60, 100, and 160 lb/yd^3 (36, 60, and 95 kg/m^3) and 1 and 2.5 in. (25 and 62 mm). Fully bonded, partially bonded, and unbonded conditions were studied.

The debonding techniques were found to decrease the formation of transverse cracks considerably. Unbonded sections developed fewer than 2 cracks, whereas partially or fully bonded sections developed from 8 to 15 cracks. The 2 in. (50 mm) thick section was found to develop more problems in terms of cracking and curling than the thicker sections. Sections with higher fiber content were also found to perform better than the slabs with lower fiber content.

Calgary, Alberta, Canada, 1973, 1976, 1977 [17.20, 17.21].   Three test sections were constructed in 1973, 1976, and 1977 using steel fiber–reinforced concrete. The first section, which was 3 to 7 in. (75 to 175 mm) thick, was constructed over a granular subbase. Two fiber contents of 67 and 133 lb/yd^3 (40 and 79 kg/m^3) were compared with plain concrete. These fiber contents worked out to be 0.5% and 1.0% in volume. The fibers were 0.75 in. (19 mm) long and 0.01 in. (0.25 mm) in diameter.

Fiber-reinforced concrete containing 0.5% fiber was judged to have a life four times that of plain concrete. Concrete with 1% fiber was found to be four times better than the concrete with 0.5% fiber.

In the second project, 12 steel fiber–reinforced, 3 conventionally reinforced, and 3 unreinforced sections were cast. Three fiber types, namely, 1 in. (25 mm) long slit-sheet, 1.25 in. (31 mm) long melt-extract, and 0.75 in. (19 mm) long wavy-cut fibers were used at fiber dosages of 75 and 125 lb/yd^3 (44 and 74 kg/m^3). The comparative evaluation indicated that greater joint spacings are possible with fiber-reinforced concrete. In addition, a higher flexural strength was found to enhance the performance of fiber-reinforced concrete.

The third project consisted of a 2 in. (50 mm) thick overlay on 11 in. (275 mm) thick lean concrete. Brass-coated wire and two types of slit sheets were used for reinforcement. The fiber content was 125 lb/yd^3 (74 kg/m^3). Some sections were cast without causing a cold joint, and some sections were bonded using epoxy. After seven years of service, 9 of the 33 slabs were deemed to have developed significant damage. In most cases, the damage was caused by only a single transverse crack.

Motorway, M10, United Kingdom [17.22].   Two overlays, 2.4 and 3.2 in. (60 and 80 mm) thick, were constructed over a 15 year old concrete slab using 1.3%, 2.2%, and 2.7% (by weight) steel fiber. The fibers were 0.02 in. (0.5 mm) in diameter and 1.5 in. (38 mm) long. Expansion joints were placed at 120 ft (36 m) intervals to match the existing joints. Transverse joints were spaced at 40 ft (12 m). No tie or dowel bars were used for the joints.

Hairline and reflective cracks were developed on both fully and partially bonded overlays. Longitudinal cracks were found to be more pronounced in partially bonded sections. Thinner sections were found to provide better overall performance.

17.2.3   Industrial floors and pavements used by heavy trucks

Industrial floors can also be considered as pavements subjected to heavy traffic. In some instances, the presence of chemicals and higher amounts of abrasion can result in faster deterioration.

Niles, Michigan, USA, 1968, 1970 [17.23].   A 4 in. (100 mm) thick fiber-reinforced concrete was placed on one lane of a roadway leading to an industrial plant and a 7.5 in. (180 mm) thick plain concrete was placed on the other. The plain concrete section developed transverse cracks, but the FRC was crack-free.

Thirteen test sections constructed in 1970 involved the use of various slab thicknesses and fiber contents. A minimum thickness of 4 in. (100 mm) and a (straight steel) fiber content of 200 lb/yd^3 (120 kg/m^3) were needed for crack-free surfaces. Thinner sections developed cracks to various extents. The inspection was done after a period of two years.

Kashima Works, Japan [17.24].   A 6 in. (150 mm) thick fiber-reinforced section was used to carry forklifts with a gross weight of 47 tonnes (52 tons). One inch (25 mm) long fibers were used at a dosage of 222 lb/yd^3 (131 kg/m^3). An 8 in. (200 mm) thick conventional plain concrete section was used for comparative evaluation.

After one year, one large and several minor cracks developed in the plain concrete, and the FRC had only one hairline crack. It was concluded that FRC provides much longer pavement life and that larger spacings could be used for expansion joints when fibers are used in concrete.

Midlothian, Texas, USA [17.25].   Concrete reinforced with steel bars and glass fiber was evaluated for use in the road leading to a quarry of a cement plant. The plain concrete slab was 8 in. (200 mm) thick. The reinforcement consisted of 0.375 in. (9 mm) diameter bars spaced at 24 in. (600 mm). The glass fiber–reinforced sections were 4 and 6 in. (100 and 150 mm) thick. The volume fraction of 1 in. (25 mm) long fibers ranged from 0.97% to 1.34%. It was concluded that glass fibers do not provide effective reinforcement.

Burnassum Project, Holland.   A large area of 2.7×10^6 ft^2 (250,000 m^2) was paved with 7 in. (175 mm) thick steel fiber–reinforced concrete. A relatively low fiber content of 50 lb/yd^3 (30 kg/m^3) was used because the fibers were bent at the ends to provide better efficiency. The slab was found to perform well and hence replaced 7.9 in. (198 mm) thick reinforced-concrete sections.

Kidston Gold Mine Shop, Australia.   A 4.7 in. (120 mm) thick steel fiber–reinforced section was successfully used for a gold mine shop and maintenance building floor that carried heavy equipment. The floor replaced a wire-reinforced 6 in. (150 mm) thick concrete floor.

17.3   Bridge Deck Overlays

Fiber-reinforced concrete has also been used extesivley for bridge deck overlays. Some of these applications are briefly described in the following sections.

Winona, Minnesota, USA, 1971 [17.26].   One of the first uses of FRC for a bridge deck overlay was on a precast bridge in Winona. The 2.5 to 4.0 in. (61 to 100 mm) overlay was placed on a severely deteriorated precast deck after removing the surface scale to a depth of 1.5 in. (38 mm). The fiber content for the 0.5 in. (13 mm) long, 0.01 in. (0.25 mm) diameter fibers was 200 lb/yd^3 (120 kg/m^3). The overlay was reported to be in good condition in spite of deicing salts and studded tires.

New Cumberland, Pennsylvania, USA, 1971 [17.27].   FRC was used to replace asphalt overlays on a steel truss bridge that was 155 ft (47 m) long and 40 ft (12 m) wide. The overlay, which was 2 to 5 in. (50 to 125 mm) thick, consisted of concrete containing 200 lb/yd^3 (120 kg/m^3) straight steel fibers. The fibers were 1 in. (25 mm) long and 0.01 in. (0.25 mm) in diameter. After one year of service, a few minor cracks were observed at locations where the overlay thickness varied abruptly. The overlay was still in service in 1984 in spite of the heavy traffic involving about 13,700 vehicles per day.

Cedar Rapids, Iowa, USA, 1971 [17.28].   In this project, FRC was cast on wood plank decking. The overlay was separated from the deck using two polyethylene sheets. The slab, which was 152 ft (46 m) long and 11 ft (3.3 m) wide, had no joints. The 2.5 in. (62 mm) long, 0.025 in. (0.62 mm) diameter fibers were used at a dosage of 150 lb/yd^3 (90 kg/m^3). The overlay was reported to be essentially crack-free after three years.

New York City, New York, USA, 1973 [17.29].   A relatively thick 10 to 12 in. (250 to 300 mm) FRC bonded overlay was placed over the existing deck. The deck was reinforced with a 4 × 12 in. (100 × 300 mm) wire mesh at the midheight of slab and with 150 to 200 lb/yd^3 (90 to 120 kg/m^3), 2.5 in. (62 mm) long fibers.

Pittsburgh, Pennsylvania, USA, 1973 [17.29].   In this project the FRC overlay was bonded to the old deck using an epoxy bonding agent. The 3 in. (75 mm) thick overlay contained 200 lb/yd^3 (120 kg/m^3), 1 in. (25 mm) long fibers. Some shrinkage cracks appeared after curing. These cracks did not open further. In 1984 the overlay was reported to be in service in satisfactory condition even though some crazing and spalling existed in a small area.

Jefferson, Iowa, USA, 1973 [17.30].   The 3 in. (75 mm) thick overlay containing 160 lb/yd^3 (95 kg/m^3), 1 in. (25 mm) long fibers was bonded to the old deck using cement paste. An inspection in 1983 revealed that the overlay was completely debonded but was still in good service condition.

Rome, Georgia, USA, 1974 [17.31].   A 2 in. (50 mm) thick FRC overlay was used in order to stiffen a 24 × 600 ft (7.1 × 180 m) long bridge deck. The fiber content was 175 lb/yd^3 (103 kg/m^3). The overlay did not stiffen the bridge but provided good service in spite of oscillatory deflections and vibrations.

17.4   Patching

Patching is similar to overlay except that these repairs are applied in small areas and must be compatible with the existing concrete to provide long service life. Fiber-reinforced concrete patching has been tried for cast, in-situ repair and in the form of precast slabs. In most cases, the composite provided good service.

Eleven small patches were installed along a key joint on a runway used by Boeing 747s at Chicago's O'Hare International Airport. The patch sizes varied from 2 × 3 ft to 1 × 11 ft (0.6 × 0.9 to 0.3 × 3.3 m).

They were 3 to 6 in. (75 to 150 mm) deep. The patches were sawed and cleaned with forced air. The repair, done using 1 in. long and 0.016 in. (0.40 mm) diameter fibers, was performing well even though similar patches filled with epoxy proved unsatisfactory.

Precast steel fiber–reinforced slabs were used in a number of instances for quick repairs in New York City. The slabs were 3 ft (0.9 m) square and 2 in. (50 mm) thick. The mortar matrix was reinforced with 245 lb/yd^3 (145 kg/m^3), 1 in. (25 mm) long, 0.01 in. (0.25 mm) diameter straight steel fibers. In some cases, the slabs were set on a base made of quick-set materials. Most of the time, the slabs were open to traffic in less than five hours.

17.5   Other Cast-In-Place Applications

In addition to millions of square feet of pavement and industrial floor, FRCs have been used for various other applications. The following is a sample.

An FRC containing 120 lb/yd^3 (71 kg/m^3) of 2 in. (50 mm) long, 0.02 in. (0.5 mm) diameter hooked-end fibers was used for an impact-resistant encasement of a turbine test facility for Westinghouse Electric Corporation in Philadelphia, Pennsylvania, USA [17.32]. The thickness of the encasement was reduced by one-third, taking advantage of the increased impact and fatigue resistance provided by the FRC.

The upstream facing of Bar Lake Dam near Denver, Colorado, USA, was repaired with 4 in. (100 mm) thick steel fiber–reinforced concrete in 1984 [17.33]. The approximate area covered was 500,000 ft^2 (46,000 m^2). The fibers used were 2.4 in. (60 mm) long and 0.24 in. (0.6 mm) in diameter. The fiber volume fraction was 0.6%. The fiber-reinforced concrete could be easily pumped up to 47 ft (14 m) and placed at a slope of 2.5 to 1.0.

Other representative applications of FRCs are

- Repair and new construction of hydraulic structures, including dams [17.34]

- Repair and rehabilitation of structures exposed to a marine environment [17.35]

- Slip-formed cast-in-place tunnel linings [17.36]

- Latex-modified FRC bridge deck overlays [17.37]

- Roller-compacted fiber-reinforced concrete for pavements [17.38]

17.6   The Use of FRC in Precast Form

Here again, FRCs are used to take advantage of their improved impact resistance. In some cases, fibers are also used as a replacement for continuous (bar) reinforcement.

Tetrapods (Dolosse).   Steel fibers were used in lieu of conventional reinforcing bars to improve the wave-impact resistance of the tetrapods [17.39]. More than 1500 units weighing 42 tons (38 tonnes) were manufactured using 30,000 yd^3 (22,900 m^3) of FRC between 1982 and 1985.

Mine Crib Blocks.   This is a unique application in which the ductility of FRC in compression is utilized to avoid catastrophic failure. The blocks made with fibers were found to provide much better ductility when used for roof support structures in coal mines [17.40].

Vaults and Safes.   Fibers are being used extensively for precast panels used for vaults and safes. By using fibers, the thickness can be reduced by as much as two-thirds. Fiber volume fractions for this application range from 1% to 3%.

Tilt-Up Panels.   Panels up to 24 ft (7.2 m) high have been cast without using conventional reinforcement; FRC was found to provide considerable savings in labor cost [17.41].

Precast Garages.   Steel fiber–reinforced concrete has also been used to precast complete automobile garages in Europe.

Manhole Covers.   Both steel and synthetic fiber–reinforced concrete have been successfully used for manhole covers [17.42, 17.43].

17.7   References

17.1   Packard, R. G.; and Ray, G. K. "Performance of Fiber Reinforced Concrete Pavement," *Fiber Reinforced Concrete: International Symposium,* SP-81, American Concrete Institute, Detroit, Michigan. 1984, pp. 325–339.

17.2   Gray, B. H. "Fiber Reinforced Concrete—A General Discussion of Field Problems and Applications," Technical Manuscript M-12, April 1972, Construction Engineering Research Laboratory, Champaign, Illinois.

17.3   Gray, B. H.; and Rice, J. L. "Fibrous Concrete for Pavement Applications," Preliminary Report M-13, April 1972, Construction Engineering Research Laboratory, Champaign, Illinois.

17.4   Gray, B. H.; and Rice, J. L. "Pavement Performance Applications," Proceedings, Conference M-28, Fibrous Concrete—Construction Material for the Seventies, Construction Engineering Research Laboratory, Champaign, Illinois, December 1972, pp. 147–157.

17.5   Parker, F., Jr. "Steel Fibrous Concrete for Airport Pavement Applications," Technical Report S-74-12, U.S. Army Engineer Waterways Experiment Station, Vicksburg, Mississippi, November 1974.

17.6 Denson, R. H. "Report of Trip to Reno, Nevada to Observe Fiber-Reinforced Concrete Paving Project," *Memorandum for Record,* U.S. Army Engineer Waterways Experiment Station, Structures Laboratory, Vicksburg, Mississippi, May 1975, 2 pp.

17.7 "Fibrous Concrete Cuts Airport Overlay to 6 Inches," *Engineering News-Record,* Vol. 196, No. 24, 10 June 1976, p. 21.

17.8 "Fibrous Concrete Airport Pad Draws Worldwide Inquiries," *Modern Concrete, Industry News,* Vol. 40, No. 12, April 1977, p. 9–10.

17.9 "Fibrous Concrete Overlay at Las Vegas Airport," Newsletter, *American Concrete Paving Association,* Oakbrook, Illinois, Vol. 12, No. 6, June 1976, p. 2.

17.10 "Fibrous Concrete—Pavement of Tomorrow," Newsletter, *American Concrete Paving Association,* Oakbrook, Illinois, Vol. 8, No. 10, October. 1972, pp. 1–6.

17.11 Lankard, D. R.; and Walker, A. J. "Pavement Applications for Steel Fibrous Concrete," *Transportation Engineering Journal,* American Society of Civil Engineers, Vol. 101, No. TE1, February 1975, pp. 137–153.

17.12 Luke, C. E. "Driveway, Road and Airport Slabs," Proceedings, Conference M-28, Fibrous Concrete—Construction Material for the Seventies, Construction Engineering Research Laboratory, Champaign, Illinois, December 1972, pp. 199–208; also *Highway Focus,* Vol. 4, No. 5, October 1971, pp. 65–70.

17.13 Parker, F., Jr. "Construction of Fibrous Concrete Overlay: Tampa International Airport," Proceedings, Conference M-28, Fibrous Concrete—Construction Material for the Seventies, Construction Engineering Research Laboratory, Champaign, Illinois, December 1972, pp. 177–197.

17.14 Parker, F., Jr. "Construction of Fibrous Concrete Overlay Test Slabs, Tampa International Airport, Florida," Report No. FAA-RD-72-119, Federal Aviation Administration, U.S. Department of Transportation, Washington, DC, October 1972.

17.15 Grondziel, M. "Restoration of Concrete Floors with Steel-Fiber Concrete for Aircraft at Frankfurt Airport, West Germany," *Fiber Reinforced Cements and Concretes: Recent Developments,* Elsevier, 1989, pp. 610–619.

17.16 Yrjanson, W. A.; and Halm, H. J. "Field Applications of Fibrous Concrete Pavements," American Concrete Paving Association, Oak Brook, Illinois, November 1973.

17.17 Hanna, Amir, *"Steel Fiber Resurfacing Concrete Properties and Resurfacing Applications,"* Portland Cement Association, Skokie, Illinois. Research and Development Bulletin, RDO49P, 1977.

17.18 Arnold, C. J.; and Brown, M. G. "Experimental Steel Fiber-Reinforced Concrete Overlay," Research Report No. R-852, April 1973, Michigan State Highway Commission, Lansing, Michigan.

17.19 Knutson, M. J. "Green County, Iowa, Concrete Overlay Research Project," *Roadways and Airport Pavements,* SP-51 American Concrete Institute, Detroit, Michigan, 1975, pp. 175–195.

17.20 Johnston, C. D. "Steel Fiber Reinforced Concrete Pavement—Construction and Interim Performance Report," *Roadways and Airport Pavements,* SP-51 American Concrete Institute, Detroit, Michigan, 1975, pp. 161–173.

17.21 Johnston, C. D. "Steel Fiber Reinforced Concrete Pavement Trials," *Concrete International: Design & Construction,* Vol. 6, No. 12, December 1984, pp. 39–43.

17.22 Gregory, J. M.; Galloway, J. W.; and Raithby, K. D. "Full Scale Trials of a Wire-Fibre-Reinforced Concrete Overlay on a Motorway," Proceedings, RILEM Symposium, Fibre Reinforced Cement and Concrete, Construction Press, Lancaster, England, September 1975, pp. 383–394.

17.23 Luke, C. E.; Waterhouse, B. L.; and Wooldbridge, J. F. "Steel Fiber-Reinforced Concrete Optimization and Applications," *Fiber Reinforced Concrete,* SP-44 American Concrete Institute, Detroit, Michigan, 1974, pp. 393–413.

17.24 Nishioka, K.; Kakima, N.; Yamakawa, S.; and Shirakawa, K. "Effective Applications of Steel Fibre-Reinforced Concrete," Proceedings, RILEM Symposium, Fibre Reinforced Cement and Concrete, Construction Press, Lancaster, England, September 1975, pp. 425–433.

17.25 Buckley, E. L. "Accelerated Trials of Glass Fiber Reinforced Rigid Pavements," Construction Research Center, The University of Texas at Arlington, Texas, April 1974.

17.26 Sather, W. R.; and Wilson, J. R. "A New Dimension in Bridge Deck Construction," *Concrete Construction,* July 1973, pp. 321–324.

17.27 Gramling, W. L.; and Nichols, T. H. "Steel Fiber Reinforced Concrete," Report 71-3, Pennsylvania Department of Transportation, December 1972; also Special Report 148, Transportation Research Board, Washington, DC, 1974, pp. 160–165.

17.28 Galinat, M. A. "An Alternative Bridge Deck Renewal System," Paper No. IBC-84-21, Presented at the International Bridge Conference, Pittsburgh, Pennsylvania, June 1984, Mitchell Fibercon, Inc., Pittsburgh, Pennsylvania.

17.29 Lankard, D. R.; and Walker, A. J. "Bridge Deck and Pavement Overlays with Steel Fibrous Concrete," *Fiber Reinforced Concrete,* SP-44, American Concrete Institute, Detroit, Michigan, 1974, pp. 375–392.

17.30 "Concrete Means Better Pavement," Newsletter, *American Concrete Paving Association,* Oakbrook, Illinois, Vol. 9, No. 10, 1973, pp. 1–6.

17.31 Hoff, G. C. "Use of Steel Fiber Reinforced Concrete in Bridge Decks and Pavements," *Steel Fiber Concrete,* Elsevier, 1986, pp. 67–108.

17.32 Tatnall, P. C. "Steel Fibrous Concrete Pumped for Burst Protection," *Concrete International—Design and Construction,* Vol. 6, No. 12, 1984, pp. 48-51.

17.33 Rettberg, W. A. "Steel Fiber Reinforced Concrete Makes Older Dam Safer, More Reliable," *Hydro-Review,* Spring, 1986, pp. 18–22.

17.34 ICOLD Bulletin 40, *Fiber Reinforced Concrete,* International Commission on Large Dams, 1989, Paris, 23 pp.

17.35 Schupack, M. "Durability of SFRC Exposed to Severe Environments," *Steel Fiber Concrete,* Elsevier, 1986, pp. 479–496.

17.36 Jury, W.A. "In-Site Concrete Linings—Integrating the Package," *Tunnels and Tunneling,* July 1982, pp. 27–33.

17.37 "Bridge Deck Overlay Combines Steel Fiber and Latex," *Civil Engineering,* ASCE, March 1983, p. 12.

17.38 Nanni, A.; and Johari, A. "RCC Pavement Reinforced with Steel Fibers," *Concrete International—Design and Construction,* Vol. 11, No. 3, 1989, pp. 64–69.

17.39 *Engineer Update,* Vol. 8, No. 10, October 1984, 3 pp.

17.40 Mason, R. H. "Concrete Crib Block Bolster Longwall Roof Support," *Coal Mining and Processing,* October 1982, pp. 58–62.

17.41 "Stack-Cast Sandwich Panels," *Concrete International: Design and Construction,* Vol. 6, No. 12, pp. 57–61.

17.42 Rajagopalan, K. "Fiber Reinforced Concrete Access Hole Covers," *Fiber Reinforced Concrete Properties and Applications,* ACI SP-105, American Concrete Institute, Detroit, Michigan, 1987, pp. 391–401.

17.43 Ratra, O. P. "Plastic Fiber Reinforced Concrete (PFRC) Composite Manhole Cover Technology," *Fiber Reinforced Cements and Composites: Recent Applications,* Elsevier, 1989, pp. 630–639.

Appendix

Factors for Conversion to SI

Unit	SI Equivalent
Inch (in.)	25.4 Millimeter (mm)
Foot (ft)	0.3048 Meter (m)
Yard (yd)	0.9144 Meter (m)
Square inch (in.^{2})	6.451 Square centimeter (cm^2)
Square foot (ft^2)	0.0929 Square meter (m^2)
Square yard (yd^2)	0.8361 Square meter (m^2)
Cubic inch (in.^{3})	16.4 Cubic centimeter (cm^3)
Cubic foot (ft^3)	0.02832 Cubic meter (m^3)
Cubic yard (yd^3)	0.7646 Cubic meter (m^3)
Ounce	29.57 Cubic centimeter (cm^3)
Gallon	0.003785 Cubic meter (m^3)
Pound-force (lb)	4.448 Newton (N)
Kip-force (k)	4448 Newton (N)
Pound-force/square inch (psi)	6.895 Kilopascal (kPa)
Pound-force/square foot (psf)	47.88 Pascal (Pa)
Kip-force/square inch (ksi)	6.895 Megapascal (MPa)
Inch-pound-force (in-lb)	0.113 Newton-meter (Nm)
Foot-pound-force (ft-lb)	1.356 Newton-meter (Nm)
Ounce-mass (avoirdupois)	28.34 Gram (g)
Pound-mass (avoirdupois)	0.436 Kilogram (kg)
Degree Fahrenheit (°F)	(°F-32)/1.8 Degree Celsius (°C)

Illustration Source Notes

The following list provides the sources from which some of the tables and figures were taken for this book. In most instances, the figures were modified. In almost all cases, alternate units (either SI or inch - lb) were added. When the information was taken from more than one source, both sources are given.

Figs. 1.1, 1.2, 1.3: Shah, S. P. "Fiber Reinforced Concrete," *Concrete International,* Vol. 12, No. 3, 1990, pp. 81–82.

Figs. 2-4, 2-5: Wang, Y.; Backer, S.; and Li, V. C.; "An Experimental Study of Synthetic Fiber Reinforced Cementitious Composites," *Journal of Material Science,* Vol. 22, 1987, pp. 4281–91.

Figs. 2-6, 2-7, 2-8, 2-9, 2-10, 2-11: Gopalaratnam, V.; and Shah, S. P.; "Failure Mechanisms and Fracture of Fiber Reinforced Concrete", *ACI-SP105,* pp. 1–25.

Figs. 3-5, 3-6, 3-7: Shah, S. P.; Stroeven, P.; Dalhuison, D.; and Van Steleelenburg, P. "Complete Stress-Strain Curves for Steel Fiber–Reinforced Concrete in Uniaxial Tension and Compression," Proceedings, RILEM Symposium on Testing and Test Methods of Fiber Cement Composites, Construction Press, U.K. 1978, pp. 399–408.

Baggot, R., "Polypropylene Fiber Reinforcement of Lightweight Cementitious Matrices," *International Journal of Cement Composites and Lightweight Concrete,* Vol. 5, No. 2, 1983, pp. 105–114.

Fig. 3-8: Shah, S. P.; and Rangan, V. B. "Fiber Reinforced Concrete Properties," *Journal of American Concrete Institute,* Vol. 68, 1971, No. 2, pp. 126–135.

Figs. 3-9, 3-10, 3-11, 3-12: Gopalaratnam, V. S.; and Shah, S. P. "Softening Response of Plain Concrete in Direct Tension," *Journal of American Concrete Institute,* Vol. 82, No. 3, 1985, pp. 310–323.

Figs. 3-14, 3-15: Krenchel, H.; and Stang, H. "Stable Microcracking in Cementitious Materials," Brittle Matrix Composites, Proceedings, Second International Symposium on Brittle Matrix Composites (BMC2), Cedzyna, Poland, Sept. 1988, pp. 20–33.

Mobasher, B. "Reinforcing Mechanism of Fibers in Cement Based Composites" Ph.D. Dissertation, Northwestern University, Evanston, Illinois, June 1990.

Figs. 3-16, 3-17, 3-18, 3-19, 3-22, 3-23, 3-24: Stang, H.; Mobasher, B.; and Shah, S. P. "Quantitative Damage Characterization in Polypropylene Fiber Reinforced Concrete," *Cement and Concrete Research,* Vol. No. 20, 1990, pp. 540–558.

Mobasher, B.; Stang, H.; and Shah, S. P. "Microcracking in Fiber Reinforced Concrete," *Cement and Concrete Research,* Vol. 20, 1990, pp. 665–676.

Figs. 3-20, 3-21: Mobasher, B.; Castro-Montero, A.; and Shah, S. P. "A

Study of Fracture in Fiber Reinforced Cement-Based Composite Using Laser Holographic Interferometry." *Experimental Mechanics,* Vol. 30, 1990, pp. 286–294.

Fig. 3-32: Li, Z.; Mobasher, B.; and Shah, S. P. "Characterization of Interfacial Properties of Fiber Reinforced Cementitious Composites," *Journal of the American Ceramic Society,* Vol. 74, No. 9, 1991, pp. 2156–64.

Figs. 4-3, 4-4, 4-6, 4-7, 4-8, 4-9, 4-10: Gopalaratnam, V. S.; Shah, S. P.; Batson, S. P.; Criswell, M.; Ramakrishnan, V.; and Wecharatna, H. "Fracture Toughness of Fiber Reinforced Concrete," *ACI Materials Journal,* Vol. 88, No. 4, 1991, pp. 339–353. (Also Report, Task Group on CMRC NSF Research/ACI Committee 544.)

Table 5.4: Soroushian, P.; and Marikunte, S. "Reinforcement of Cement-Based Materials with Cellulose Fibers," *Thin-Section Fiber Reinforcement Concrete and Ferrocement,* SP-24, American Concrete Institute, Detroit, 1990, pp. 99–124.

Table 6.3: Ramakrishnan, V.; and Coyle, W. V. "Steel Fiber Reinforced Superplasticized Concretes for Rehabilitation of Bridge Decks and Highway Pavements," No. DOT/RSPA/DMA-50/84-2, 1983, 408 pp.

Figs. 7-1, 7-2: ACI Committee 544, "Measurements of Properties of Fiber Reinforced Concrete," American Concrete Institute, Detroit, Michigan, ACI544-2R-89, 1988, 11 pp.

Tables 7.1, 7.2, 7.3, 7.4 and Fig. 7-11: Balaguru, P.; and Ramakrishnan, V. "Comparison of Slump Cone and V-B Tests as Measures of Workability for Fiber Reinforced and Plain Concrete," *Cement, Concrete and Aggregates,* Vol. 9, No. 1, Summer 1987, pp. 3–11.

Ramakrishnan, V.; and Coyle, W. V. "Steel Fiber Reinforced Superplasticized Concretes for Rehabilitation of Bridge Decks and Highway Pavements," Final Report to U.S. Department of Transportation, DOT/RSPA/DMA-50/84-2, 1983, 408 pp.

Figs. 7-3, 7-4, 7-5, 7-6: Balaguru, P.; and Ramakrishnan, V. "Properties of Fiber Reinforced Concrete: Workability, Behavior under Long-Term Loading, and Air-Void Characteristics," *ACI Materials Journal,* Vol. 85, May-June 1988, pp. 189–196.

Tables 7.7, 7.8: Vondran, G. L.; Nagabhushanam, M.; and Ramakrishnan, V. "Fatigue Strength of Polypropylene Fiber Reinforced Concretes," *Fiber Reinforced Cements and Concretes: Recent Developments,* Elsevier, New York, 1989, pp. 533–543.

Figs. 8-3: Balaguru, P. "Fiber-Reinforced Rapid-Setting Concrete," *Concrete International,* American Concrete Institute, Detroit, Michigan, Vol. 14, No. 2, 1992, pp. 64–67.

Table 8.1, 8.2 and Figs 8-20, 8-21, 8-22, 8-23: Balaguru, P.; and Ramakrishnan, V. "Freeze-Thaw Durability of Fiber Reinforced Concrete," *ACI Journal,* Vol. 83, 1986, pp. 374–382.

Fig. 9-2: Batson, G.; Ball, C.; Bailey, L.; Landers, E.; and Hooks, J. "Flexural Fatigue Strength of Steel Fiber Reinforced Concrete Beams," *ACI Journal,* Vol. 69, No. 11, 1972, pp. 673–677.

Shah, S. P.; and Skarendahl, A. (eds.) *Steel Fiber Concrete,* U.S.–Sweden Joint Seminar (NSF–STU), Swedish Cement and Concrete Research Institute, Stockholm, 1985, 520 pp.

Figs. 9-3, 9-4, 9-5, 9-6: Wu, G. Y.; Shivaraj, S. K.; and Ramakrishnan, V. "Flexural Fatigue Strength, Endurance Limit, and Impact Strength of Fiber Reinforced Refractory Concretes," *Fiber Reinforced Cements and Concretes: Recent Developments,* Elsevier, New York, 1989, pp. 261–273.

Table 9.1: Ramakrishnan, V.; Oberling, G.; and Tatnall, P. "Flexural Fatigue Strength of Steel Fiber Reinforced Concrete," *Fiber Reinforced Concrete Properties and Applications,* SP-105, American Concrete Institute, Detroit, Michigan, 1987, pp. 225–245.

Table 9.2: Ramakrishnan, V.; Gollapudi, S.; and Zellers, R. "Performance Characteristics and Fatigue Strength of Polypropylene Fiber Reinforced Concrete," *Fiber Reinforced Concrete Properties and Applications,* SP-105, American Concrete Institute, Detroit, Michigan, 1987, pp. 159–177.

Table 9.3 and Fig. 9-7: Kormeling, H. A.; Reinhardt, H. W.; and Shah, S. P. "Static and Fatigue Properties of Concrete Beams Reinforced with Continuous Bars and with Fibers," *ACI Journal,* Vol. 77, No. 1, 1980, pp. 36–43.

Figs. 9-8, 9-9, 9-10, 9-11: ACI Committee 544; "Measurement of Properties of Fiber Reinforcement Concrete," American Concrete Institute, Detroit, Michigan, 1989, 11 pp.

Fig. 9-12: Ramakrishnan, V.; Brandshaug, T.; Coyle, W. V.; and Shrader, E. K. "A Comparative Evaluation of Concrete Reinforced with Straight Steel Fibers and Fibers with Deformed Ends Glued Together into Bundles," *ACI Journal,* Vol. 77, No. 3, 1980, pp. 135–143.

Fig. 9-13: Ramakrishnan, V.; Coyle, W. V.; Kulandaisamy, V.; and Schrader, E. K. "Performance Characteristics of Fiber Reinforced Concrete with Low Fiber Contents," *ACI Journal,* Vol. 78, No. 5, 1981, pp. 388–394.

Figs. 9-20, 9-21, 9-22: Namaan, A. E.; and Gopalaratnam, V. S. "Impact Properties of Steel Fiber Reinforced Concrete in Bending," *International Journal of Cement Composites and Lightweight Concrete,* Vol. 5, No. 4, 1983, pp. 225–233.

Figs. 9-23, 9-24 and Tables 9.4, 9.5, 9.6: Gopalaratnam, V. S.; and Shah, S. P. "Properties of Fiber Reinforced Concrete Subjected to Impact Loading," *ACI Journal,* Vol. 83, No. 1, 1986, pp. 117–126.

Table 9.7, 9.8: Bentur, A.; Mindess, S.; and Skalny, J. "Reinforcement of Normal and High Strength Concretes with Fibrillated Polypropylene Fibers," *Fiber Reinforced Cements and Concretes: Recent Developments,* Elsevier, New York, 1989, pp. 229–239.

Figs. 10-1, 10-2, 10-7, 10-8 and Table 10.1: Balaguru, P.; and Ramakrishnan, V. "Properties of Fiber Reinforced Concrete: Workability, Behavior under Long-Term Loading, and Air-Void Characteristics," *ACI Materials Journal,* Vol. 85, No. 3, 1988, pp. 189–196.

Figs. 10-3, 10-4, 10-5, 10-6 and Table 10.2: Chern, J. C.; and Young, C. H. "Compressive Creep and Shrinkage of Steel Fiber Reinforced Concrete," *The International Journal of Cement Composite and Lightweight Concrete,* Vol. 11, No. 4, 1989, pp. 205–214.

Fig. 10.9: Swamy, R. N. "Steel Fiber Concrete for Bridge Deck and Building Floor Applications," *Steel Fiber Concrete,* Elsevier, New York, 1986, pp. 443–478.

Table 10.3 and Figs. 10-10, 10-11: Mangat, P. S.; and Grusamy, K. "Permissible Crack Widths in Steel Fiber Reinforced Marine Concrete," *Materials* *and Structures,* RILEM, Vol. 20, 1987, pp. 338–347.

Fig. 10-12: Hannant, D. J. "Additional Data on Fiber Corrosion in Cracked Beams and Theoretical Treatment of the Effect of Fiber Corrosion on Beam Load Capacity," RILEM Symposium on Fiber Reinforced Cement and Concrete, 1975, Vol. II, pp. 533–538.

Figs. 10-13, 10-14, 10-15 and Table 10.4: Hannant, D. J. "Ten Year Flexural Durability Tests on Cement Sheets Reinforced with Fibrillated Polypropylene Networks," *Fiber Reinforced Cements and Concretes— Recent Developments,* Elsevier, 1989, pp. 572–563.

Fig. 11-2: Dahl, P. A. "Plastic Shrinkage and Cracking Tendency of Mortar and Concrete Containing Fiber Mesh," Report No. STF65A85039, SINTEF Div. FCB, Trondheim, Norway, September 1985.

Fig. 11-8: Grzybowski, M. "Determination of Crack Arresting Properties of Fiber Reinforced Cementitious Composites," Publication TRITA-BRO-8908, ISSN 1100-648X, Royal Institute of Technology, Stockholm, Sweden, June 1989, 190 pp.

Fig. 11-9: Paillere, A. M.; Buil, M.; and Serrano, J. J. "Effect of Fiber Addition on the Autogeneous Shrinkage of Silica Fume Concrete," *ACI Materials Journal,* Vol. 86, No. 2, March-April 1989, pp. 139–144.

Fig. 11-10: Pan, K. W., et al. "A Study on Restrained Shrinkage Cracking of Fly Ash, Cementitious Materials," Private communication, 1987.

Fig. 11-11: Pihlajawaara, S. E.; and Pihlman, E. "Results of Long-Term Deformation Tests of Glass Fiber Reinforced Concrete," NORDFORSK-FRC Project 1974–76, Delrapport,

Technical Research Center of Finland, Otaniemi, 1978.

Figs. 11-12, 11-13, 11-14, 11-15, 11-16, 11-17, 11-18, 11-19, 11-20, 11-21, 11-22, 11-23, 11-24, 11-26, Table 11.1: Grzybowski, M.; and Shah, S. P. "Shrinkage Cracking of Fiber Reinforced Concrete," *ACI Materials Journal,* Vol. 87, 1990, pp. 138–148.

Table 11.2: Grzybowski, M.; and Shah, S. P. "A Model to Predict Cracking in Fiber Reinforced Concrete Due to Restrained Shrinkage," *Magazine of Concrete Research* (London), Vol. 41, No. 148, 1989, pp. 125–135.

Fig. 11-25: Gopalaratnam, V. S.; and Shah, S. P. "Softening Response of Plain Concrete in Direct Tension," *ACI Journal,* Vol. 82, No. 3, 1985, pp. 310–323.

Fig. 12-1: ACI Committee 506. "Specification for Materials, Proportioning and Application of Shotcrete," ACI 506.2-90, *American Concrete Institute,* Detroit, 1990, 8 pp.

Ramakrishnan, V. "Steel Fiber Reinforced Shotcrete, a State-of-the-Art Report," *Steel Fiber Concrete,* S. P. Shah and A. Skarendahl (eds.), Elsevier, New York, 1986, pp. 7–24. (Also published by Swedish Cement and Concrete Research Institute, Stockholm, 1985.)

Tables 12.1, 12.2: Morgan, D. R.; and McAskill, N. "Rocky Mountain Tunnels Lined with Steel Fiber Reinforced Shotcrete," *Concrete International: Design and Construction,* Vol. 6, No. 12, 1984, pp. 33–38.

Figs. 12-2, 12-3, 12-4: Sandell, N. O.; Dir, M.; and Westerdahl, B. "System BESAB for High Strength Steel Fiber Reinforced Shotcrete," in *Steel Fiber Concrete,* S. P. Shah and A. Skarendahl (eds.), Elsevier, New York, 1986, pp. 25–39.

Tables 12.3, 12.4, 12.5, 12.6, 12.10, 12.12 and Figures 12-5, 12-6, 12-7: Ramakrishnan, V.; Coyle, W. V.; Dhal, L. F.; and Schrader, E. K. "A Comparative Evaluation of Fiber Shotcrete," *Concrete International: Design and Construction,* Vol. 3, No. 1, 1981, pp. 56–69.

Tables 12.7, 12.8, 12.9, 12.11, 12.13: Morgan, D. R. "Steel Fiber Shotcrete—A Laboratory Study," *Concrete International: Design and Construction,* Vol. 3, No. 1, 1981, pp. 70–74.

Figs. 13-2, 13-4, 13-8 through 13-13 and Tables 13.3 through 13.8 : Shah, S. P.; Daniel, J. I.; and Ludiraja, D. "Toughness of Glass Fibers Reinforced Concrete Panels Subjected to Accelerated Aging," *PCI Journal,* September-October 1987, pp. 82–99.

Fig. 13-5: Litherland, K. L., Oakley, D. R.; and Proctor, B. A. "The Use of Accelerated Aging Procedures to Predict the Long-Term Strength of GFRC Composites," *Journal of Cement and Concrete Research,* Vol. 11, 1981, pp. 455–466.

Daniel, J. I. "Glass Fiber Reinforced Concrete," Fiber Reinforced Concrete, Report Nos. 2493D and 2614D, Construction Technology Laboratories, Inc., Skokie, Illinois, 1988, pp. 5.1–5.30.

Fig. 13-6: Proctor, B. A. "Past Development and Future Prospect for GRC Materials," Proceedings of the International Congress on Glass Fiber Reinforced Cement, Paris, No. 1981.

Daniel, J. I. "Glass Fiber Reinforced Concrete," Fiber Reinforced Concrete, Report Nos. 2493D and 2614D, Construction Technology Laboratories, Inc., Skokie, Illinois, 1988, pp. 5.1–5.30.

Figs. 13-14, 13-15, 13-16: "Properties of GRC: Ten-Year Results," *BRE Information Paper,* IP36/79, Building Research Station, U.K., Department of Environment, Garston, Hertford, November 1979, 4 pp.

Daniel, J. I. "Glass Fiber Reinforced Concrete," Fiber Reinforced Concrete, Report Nos. 2493D and 2614D, Construction Technology Laboratories, Inc., Skokie, Illinois, 1988, pp. 5.1–5.30.

Figs. 13-17, 13-18, 13-19: Hayashi, M.; Sato, S., and Fujii, H., "Some Ways to Improve Durability of GRC," Proceedings of Durability of Glass Fiber Reinforced Concrete Symposium, PCI, Chicago, Illinois, 1986, pp. 270–284.

Daniel, J. I. "Glass Fiber Reinforced Concrete," Fiber Reinforced Concrete, Report Nos. 2493D and 2614D, Construction Technology Laboratories, Inc., Skokie, Illinois, 1988, pp. 5.1–5.30.

Figs. 13-20, 13-21, 13-22, 13-23, 13-24, 13-25: Akihama, S., Suenga, T., Tanaka, M., and Hayashi, M. "Properties of GFRC with Low Alkaline Cement," *ACI SP105,* 1987, pp. 189–210.

Figs. 13-26 through 13-32: PCI (Prestressed Concrete Institute), Recommended Practice for Glass Fiber Reinforced Concrete Panels, PCI, Chicago, Illinois, October 1987, 87 pp.

Daniel, J. I. "Glass Fiber Reinforced Concrete," Fiber Reinforced Concrete, Report Nos. 2493D and 2614D, Construction Technology Laboratories, Inc., Skokie, Illinois, 1988, pp. 5.1–5.30.

Table 14.1 and Figs. 14-1, 14-2, 14-3: Naaman, A. E.; Shah, S. P.; and Throne, J. L. "Some Developments in Polypropylene Fibers for Concrete," *Fiber Reinforced International Symposium,* SP-81, American Concrete Institute, Detroit, Michigan 1984, pp. 375–396.

Figs. 14-4, 14-5, 14-6, 14-7, 14-8: Dave, N. J.; and Ellis, D. G. "Polypropylene Fiber Reinforced Cement," *International Journal of Cement Composites,* Vol. 1, No. 1, May 1979, pp. 19–28.

Table 14.2, 14.3, 14.4 and Figs. 14-9, 14-10: Keer, J. G.; and Thorne, A. "Performance of Polypropylene-Reinforced Cement Sheeting Element," Fiber Reinforced Concrete-International Symposium, SP-81, American Concrete Institute, Detroit, Michigan 1984, pp. 213–231.

Figs. 14-11, 14-12, 14-13, 14-14: Gardiner, T.; and Currie, B. "Flexural Behavior of Composite Cement Sheets Using Woven Polypropylene Mesh Fabrics," *International Journal of Cement Composites and Lightweight Concrete,* Vol. 5, 1983, pp. 193–197.

Fig. 14-15 and Table 14.5: Swamy, R. N.; and Hussin, M. W. "Woven Polypropylene Fabrics—An Alternative to Asbestos for Thin Sheet Applications," *Fiber Reinforced Cements and Concrete Recent Development,* Elsevier 1989, pp. 90–100.

Figs. 14-16, 14-17, 14-18, 14-19, 14-20, 14-21, 14-22, 14-23, 14-24, 14-25: Akihama, S.; Suenaga, T.; and Banno, T. "Mechanical Properties of Carbon Reinforced Cement Composites," *International Journal of Cement Composites and Lightweight Concrete,* Vol. 8, No. 1, 1986, pp. 21–33.

Fig. 14-26: Soroushian, P., Bayasi, Z.; and Nagi, M. "Effects of Curing Procedures on Mechanical Properties of Carbon Fiber Reinforced Cement," *Fiber Reinforced Cements and Concretes: Recent Developments,* Elsevier 1959, pp. 167–178.

Figs. 14-27, 14-28, 14-29*, 14-30* and Table 14.6: Soroushian, P. and Marikante, S. "Reinforcement of Cement Based Materials with Cellulose Fibers," *Thin-Section Fiber Reinforced Concrete and Ferrocement,*

SP-124, American Concrete Institute, Detroit, Michigan, 1991, pp. 99–124. Coutts, R. S. P. "Air-Cured Wood Pulp, Fiber-Cement Mortars," *Composites,* Vol. 18, No. 4, 1987, pp. 325–328.

Figs. 14-31, 14-32, 14-33, 14-34 and Table 14.7: Vinson, K. D.; and Daniel J. I. "Advances in the Development of Specialty Cellulose Fibers Specifically Designed for the Reinforcement of Cement Matrices," *Thin-Section Fiber Reinforced Concrete and Ferrocement,* SP-124, American Concrete Institute, Detroit, Michigan, 1991, pp. 1–18.

Table 14.8: Keer, J. G. "Performance of Non-Asbestos Fiber Cement Sheeting," *Thin-Section Fiber Reinforced Concrete and Ferrocement,* SP-124, American Concrete Institute, Detroit, Michigan, 1991, pp. 19–38.

Fig. 14-35: Cook, D. J.; Paina, R. P.; and Weerasingle, H. L. S. D. "Coir Fiber Reinforced Cement as a Low Cost Roofing Material," *Building and Environment,* Vol. 13, 1983, pp. 193–198.

Fig. 14-36: Gram, H. E. "Durability Studies of Natural Organic Fibers in Concrete, Mortar and Cement," Third International Symposium on Developments in Fiber Reinforced Cement and Concrete, RILEM Symposium, FRC 86, Vol. 2, July 1986.

Figs. 15-1, 15-2, 15-3, 15-8, 15-20a, 15-24, 15-27: Lankard, D. R. "Preparation, Properties and Applications of Cement-Based Composites Containing 5% to 20% Steel Fiber," *Steel Fiber Concrete,* Elsevier, 1986, pp. 199–217. (Also published by Swedish Cement and Concrete Research Institute, Stockholm, 1985).

Tables 15.1, 15.2, 15.9, 15.10, 15.11: Homrich, J.; and Naaman, A. E. "Stress-Strain Properties of SIFCON in compression," *Fiber Reinforcement*

Concrete Properties and Applications. SP-105, American Concrete Institute, Detroit, Michigan, 1987, pp. 283–304.

Figs. 15-4, 15-5: Mandragon, R. "SIFCON in compression," *Fiber Reinforced Concrete Properties and Applications.* SP-105, American Concrete Institute, Detroit, Michigan, 1987, pp. 269–282.

Figs. 15-12, 15-13, 15-21, 15-22, 15-23: Balaguru, P.; and Kendzulak, J. "Mechanical Properties of Slurry Infiltrated Fiber Concrete (SIFCON)," *Fiber Reinforced Properties and Applications.* SP-105, American Concrete Institute, Detroit, Michigan, 1987, pp. 247–268.

Figs. 15-14, 15-15: Reinhardt, H. N.; and Fritz, C. "Optimization of SIFCON Mix," *Fiber Reinforced Cements and Concretes—Recent Developments,* Elsevier, 1989, pp. 11–20.

Figs. 15-16, 15-17: Naaman, A. E.; and Homrich, J. R. "Tensile Stress-Strain Properties of SIFCON," *ACI Materials Journal,* Vol. 86, No. 3, 1989, pp. 244–251.

Figs. 15-18, 15-19 and Tables 15.4, 15.5, 15.6: Balaguru, P.; and Kendzulak, J. "Flexural Behavior of Slurry Infiltrated Fiber Concrete (SIFCON) Made Using Condensed Silica Fume," *Fly Ash, Silica Fume, Slag, and Natural Pozzalans in Concrete,* SP-60, American Concrete Institute, Detroit, Michigan, 1986, pp. 1215–1230.

Figs. 15-20b, 15-25: Marchand, K. "Tests and Analysis of the Localized Response of SIFCON and CRC Subjected to Blast and Fragment Loading," Presented at SIFCON 1989, held in NMERI, Albuquerque, New Mexico, October 1989.

Fig. 15-26: Naman, T.; Wight, J.; and Abdou, H. "SIFCON Connections for Seismic Resistant Frames," *Concrete International,* Vol. 9, 1987, pp. 34–39.

Figs. 16-1, 16-2, 16-3, 16-4, 16-5: Craig, R. "Flexural Behavior and Design of Reinforced Fiber Concrete Members," *Fiber Reinforced Concrete Properties and Applications,* American Concrete Institute, Detroit, Michigan, 1987, pp. 517–564.

Fig. 16-6: Jindal, R. L. "Shear and Moment Capacities of Steel Fiber Reinforced Concrete Beams," *Fiber Reinforced Concrete, International Symposium* SP-81, American Concrete Institute, Detroit, Michigan, 1984, pp. 1–16.

Figs. 16-7, 16-8: Batson, G. "Use of Steel Fibers for Shear Reinforcement and Ductility," *Steel Fiber Concrete,* Elsevier, 1985, pp. 377–419.

Figs. 16-9, 16-10 and Table 16.1: Mansur, M. A.; and Paramasivam, P. "Steel Fiber Reinforced Concrete Beams in Pure Torsion," *International Journal of Cement Composites,* Vol. 4, No. 1, February 1982, pp. 39–45.

Figs. 16-11, 16-12, 16-13, 16-14, 16-15: Craig, J. R.; Dunya, S.; Riaz, J.; and Shirazi, H. "Torsional Behavior of Reinforced Fibrous Concrete Beams," *Fiber Reinforced Concrete, International Symposium* SP-81, American Concrete Institute, Detroit, Michigan, 1984, pp. 17–49.

Fig. 16-16: Kaushik, S. K.; and Sasturkar, P. J. "Simply Supported Steel Fiber Reinforced Concrete Beams under Combined Torsion, Bending and Shear," *Fiber Reinforced Cements and Concretes: Recent Developments,* Elsevier, 1989, pp. 687–698.

Fig. 16-17 and Table 16.2: Al-Ausi, M. A.; Abdul-Wahab, H. M. S.; and Khidair, R. M. "Effect of Fibers on the Strength of Reinforced Concrete Beams under Combined Loading,"

Fiber Reinforced Cements and Concretes: Recent Developments, Elsevier, 1989, pp. 664–675.

Tables 16.3, 16.4: Roberts, T. M.; and Ho, N. L.; "Shear Failure of Deep Fiber Reinforced Concrete Beams," *International Journal of Cement Composites,* Vol. 4, No. 3, August 1982, pp. 145–52.

Fig. 16-18: Fatthui, N. I.; and Hughes, B. P. "Ductility of Reinforced Concrete Corbels Containing Either Steel Fibers or Stirrups," *ACI Materials Journal,* Vol. 86, No. 6, November-December 1989, pp. 644–651.

Fig. 16-19: Fatthui, N. I.; and Hughes, B. P. "Reinforced Steel Fiber Concrete Corbels with Various Shear Span-to-Depth Ratios," *ACI Structural Journal,* Vol. 86, No. 6, November-December 1989, pp. 590–596.

Fig. 16-20: Sood, V.; and Gupta, S. "Behavior of Steel Fibrous Concrete Beam-Column Connections," *Fiber Reinforced Concrete Properties and Applications,* American Concrete Institute, Detroit, Michigan, 1987, pp. 437–474.

Fig. 16-21: Jindal, R.; and Sharma, V. "Behavior of Steel Fiber Reinforced Concrete Knee-Type Beam-Column Connections," *Fiber Reinforced Concrete Properties and Applications,* American Concrete Institute, Detroit, Michigan, 1987, pp. 475–492.

Fig. 16-22: Craig, J. R.; Mahader, S.; Patel, C. C.; Viteri, M.; and Kertesz, C. "Behavior of Joints Using Reinforced Fibrous Concrete," *Fiber Reinforced Concrete, International Symposium* SP-81, American Concrete Institute, Detroit, Michigan, 1984, pp. 125–167.

Figs. 16-23, 16-24, 16-25, 16-26: Yashior, H.; Tanaka, Y.; Ro, Y.; and Hirose, K. "Study of Shear Failure of Steel Fiber Reinforced Concrete Short Columns in Consideration of Arrangement of Ties," *Fiber Reinforced Cements and Concretes: Recent Developments,* Elsevier, 1989, pp. 489–498.

Figs. 16-27, 16-28: Balaguru, P.; and Ezeldin, A. "Behavior of Partially Prestressed Beams Made with High Strength Fiber Reinforced Concrete," *Fiber Reinforced Concrete Properties and Applications,* American Concrete Institute, Detroit, Michigan, 1987, pp. 419–434.

Figs. 16-29, 16-30: Ezeldin, A. S.; and Balaguru, P. "Stress-Strain Behavior of Normal and High Strength Fiber Reinforced Concrete under Compression," *ASCF Materials Engineering Journal,* 1992, in press.

Figs. 16-32, 16-33: Ezeldin, A; and Balaguru, P. "Bond Performance of Reinforcing Bars Embedded in Fiber Reinforced Concrete and Subjected to Monotonic and Cyclic Loads," ASCE Materials Engineering Congress, Denver, Colorado, August 1990, pp. 145–154.

Table 17.1: Packard, R. G.; and Ray, G. K. "Performance of Fiber Reinforced Concrete Pavement," *ACI SP-81,* 1984, pp. 325–339.

Index

ABOUT THE AUTHORS

Perumalsamy N. Balaguru is professor of civil engineering at Rutgers University, New Brunswick, New Jersey. He has done extensive research in the areas of reinforced and prestressed concrete, construction management, and the uses of new materials for construction. He is the chairman of ACI Committee 549, Ferrocement and Other Thin Sheet Products and a member of Committees 544, Fiber-Reinforced Concrete; 214, Evaluation of Strength Tests; 215, Fatigue; and, 440 FRP Bar and Tendon Reinforcement. He has written more than 115 publications in the area of concrete, new materials, and construction management, and has been a visiting professor at Northwestern University.

Surendra P. Shah is professor of civil engineering and director of the National Science Foundation's Center for Advanced Cement Based Materials at Northwestern University, Evanston, Illinois. He is past chairman of the American Concrete Institute's Committee on Fiber-Reinforced Concrete, and has written more than 300 technical publications and edited more than 10 publications on concrete technology, fiber-reinforced concrete, concrete structures, and fracture of quasi-brittle materials. Professor Shah has received many awards for his contributions, including the RILEM gold medal, the ACI Anderson award, and the ASTM Thomson award. He is a Fellow of ACI, NATO senior visiting scientist to France, and has been awarded the U.S. distinguished visiting scientist Von Humboldt award in Germany.